THERMAL METAMATERIALS

Controlling the Flow of Heat

Other Titles by Liujun Xu

Transformation Thermotics and Extended Theories: Inside and Outside Metamaterials

THERMAL METAMATERIALS

Controlling the Flow of Heat

Ying Li
Zhejiang University, China

Liujun Xu
Graduate School of China Academy of Engineering Physics, China

Cheng-Wei Qiu
National University of Singapore, Singapore

World Scientific

NEW JERSEY • LONDON • SINGAPORE • BEIJING • SHANGHAI • HONG KONG • TAIPEI • CHENNAI • TOKYO

Published by

World Scientific Publishing Co. Pte. Ltd.
5 Toh Tuck Link, Singapore 596224
USA office: 27 Warren Street, Suite 401-402, Hackensack, NJ 07601
UK office: 57 Shelton Street, Covent Garden, London WC2H 9HE

Library of Congress Cataloging-in-Publication Data
Names: Li, Ying (Professor of electrical and computer engineering), author. |
Xu, Liujun (Professor of engineering), author. | Qiu, Cheng-Wei, author.
Title: Thermal metamaterials : controlling the flow of heat / Ying Li, Liujun Xu, Cheng-Wei Qiu.
Description: New Jersey : World Scientific, 2025. | Includes bibliographical references and index.
Identifiers: LCCN 2024042806 | ISBN 9789811294402 (hardcover) |
ISBN 9789811294419 (ebook for institutions) | ISBN 9789811294426 (ebook for individuals)
Subjects: LCSH: Metamaterials--Thermal properties.
Classification: LCC TA418.52 .L5 2025 | DDC 620.1/15--dc23/eng/20241214
LC record available at https://lccn.loc.gov/2024042806

British Library Cataloguing-in-Publication Data
A catalogue record for this book is available from the British Library.

For any available supplementary material, please visit
https://www.worldscientific.com/worldscibooks/10.1142/13874#t=suppl

Desk Editor: Shaun Tan Yi Jie

Typeset by Stallion Press
Email: enquiries@stallionpress.com

Preface

The title of the book was inspired by a famous text in photonics — *Photonic Crystals: Molding the Flow of Light* by Joannopoulos, Johnson, Winn, and Meade. We use the title as a homage to this classic work, which guided us in studying artificial structures. We would also like to express our best regards to Prof. Jiping Huang and Prof. Baowen Li, who pioneered thermal metamaterials and phononics. We thank their enlightening discussions with us.

The book contains three parts dedicated to thermal metamaterials that control the steady-state, time-harmonic, and transient heat transfer. This arrangement highlights the unique conceptual and theoretical frameworks for each kind of temporal evolution of the thermal field (constant, oscillatory, or decaying). The perspective is thus different from those of existing texts. While the other works mainly focus on the functionalities of thermal metamaterials, this book is more intended to introduce the basic ideas behind them.

The content of the book is mostly about conductive and convective heat transfer, which similar diffusive governing equations can describe. Metamaterials manipulating radiative heat transfer, either far-field or near-field, are regretfully not covered due to the limited space. We are by no means unaware of the exciting progress in the field of thermal photonics, like thermophotovoltaics and radiative cooling.

We are particularly grateful to our postdoctoral researchers and graduate students for their contributions to the writing of the major part of the book. Chapter 3 was mainly written by Qiang-Kai-Lai Huang. Chapter 4 was mainly written by Ran Ju. Chapter 5 was mainly written by Dr. Pei-Chao Cao with input from Haoran Yan. Chapter 6 was mainly written by Dong Wang with the help of Dr. Pei-Chao Cao. Chapter 7 was mainly written by Minghong Qi. We also sincerely thank editor Shaun Y.J. Tan for his patience and great help in handling many publication matters.

Finally, we acknowledge the financial support by the National Natural Science Foundation of China under Grants No. 92163123 and No. 52250191, and the Zhejiang Provincial Natural Science Foundation of China under Grant No. LZ24A050002.

About the Authors

Professor Ying Li is Assistant Professor in the College of Information Science and Electronic Engineering at Zhejiang University, China. He is a Member of the Chinese Physical Society and a Senior Member of the Chinese Institute of Electronics. He has published more than 60 peer-reviewed papers in *Science*, *Nature Materials*, *Physical Review Letters*, and other high-impact journals. Many of them are highly cited and/or cover articles, and featured on news websites such as phys.org, *Science Daily* and *Asian Scientist.*

Dr. Liujun Xu is Research Assistant Professor at the Graduate School of China Academy of Engineering Physics. He has published over 70 peer-reviewed papers in journals like *Reviews of Modern Physics*, *Nature Reviews Physics*, *Physical Review Letters*, and *PNAS*. Many are selected as the Featured Article, Editor's Pick, Editor's Suggestion, Best Paper, Highlight, and Inside Front Cover.

Professor Cheng-Wei Qiu is Provost's Chair Professor in the College of Design and Engineering at the National University of Singapore (NUS). He is an elected Fellow of the American Physical Society, Optica, SPIE, The Electromagnetics Academy, Chinese Optical Society, and the ASEAN Academy of Engineering and Technology. He has published more than 600 peer-reviewed papers, many in top journals. He has been named a Highly Cited Researcher for five consecutive years (2019–2023) by Web of Science, with a h-index of 113 as of January 2024. His work was selected as "Top 10 Breakthroughs 2020" by *Physics World* and as "Optics in 2021" by *Optica*. In 2022, he published two papers on brainwave-based metasurfaces, which was read more than 10 million times.

Professor Qiu has received many prestigious awards, including the URSI Young Scientist Award in 2008, NUS Young Investigator Award in 2010, MIT TR35@Singapore Award in 2012, NUS Faculty Young Research Award in 2012, Young Scientist Award by Singapore National Academy of Science in 2013, Young Investigator Award by Optical Society of America in 2017, NUS Young Engineering Research Award in 2018, SPIE Rising Researcher Award in 2018, NUS Engineering Researcher Award in 2019, World Scientific Medal by Institute of Physics Singapore in 2021, Achievement in Asia Award (Robert T. Poe Prize) by International Organization of Chinese Physicists and Astronomers in 2023, and Singapore's highest science accolade — the President's Science Award — in 2023.

Contents

Part II Time-Harmonic Heat Transfer 85

Chapter 1

Introduction

1.1. Heat Transfer and Its Control

Heat is a product of almost all physical processes because of imperfect energy conversion efficiency. It follows that heat transfer is a ubiquitous phenomenon around us. This thermal energy transportation can be favorable or undesirable in different situations. For example, enhancing heat transfer is essential for cooling and heating food. On the other hand, suppressing heat transfer between a building and its environment helps maintain a comfortable temperature. Controlling heat transfer for such purposes has been a central issue for human beings since the Stone Age.

The efficiency of heat transfer strongly depends on materials. For example, thermal conductivity κ is a key material property that determines the efficiency of heat conduction. According to *Fourier's law*, the heat flux $\boldsymbol{q}$ through a material equals $\boldsymbol{q} = -\kappa\nabla T$, where ∇T is the temperature gradient. Enhancing heat conduction relies on a good thermal conductor with large thermal conductivity, while suppressing heat conduction needs a good thermal insulator with small thermal conductivity. The thermal conductivity of natural bulk materials at room temperature ranges from ~0.1 W/m · K for wood and plastics to ~100 W/m·K for metals [1]. That is why a pan usually has a metal body and a wood or plastic handle (Fig. 1.1a).

Today, heat is more involved in human activities. Better control over heat transfer is crucial for energy utilization and information processing. On the one hand, roughly half of global energy input is dissipated as heat without being converted to useful work [2], which is a huge waste and a cause of global warming. On the

(a) (b)

Fig. 1.1. Controlling heat transfer efficiency with (a) materials and (b) structures.

other hand, the power consumption and heat generation of electronic devices is growing rapidly, leading to near-junction heat flux that can be as large as $\sim$1000 W/cm^2, comparable to the heat flux at the sun's surface [3]. The thermal throttling issue is thus a major challenge in developing next-generation chips. To meet such demands, people have developed novel structures such as heat sinks and heat pipes (Fig. 1.1b) for fast heat dissipation. However, the efficiency of heat transfer is not the only concern. We also expect more accurate, systematic, and flexible control over all aspects of heat transfer, especially its spatiotemporal distribution. One promising way towards this goal is to use metamaterials.

1.2. Metamaterials

The term metamaterial was coined by Rodger M. Walser in 1999, where meta means "beyond" or "after." As its name suggests, metamaterial refers to something superior to natural material in certain aspects. For example, the term was originally used to describe left-handed or negative-index materials, first proposed by Victor Veselago in 1968 [4]. These materials have both negative permittivity and permeability, so they have an unusual response to electromagnetic (EM) waves. How to realize such kinds of materials? They must be made of artificial structures. For example, the first experimental demonstration of negative-index refraction [5] uses an array of split-ring resonators, like those shown in Fig. 1.2. Just like natural crystalline materials, such kind of metamaterial is a periodic

Fig. 1.2. Structure of a metamaterial (photo provided by Prof. Zuojia Wang).

lattice of unit structures dubbed meta-atoms, which play the role of atoms in crystals. In early studies, metamaterials were often required to have three features: (1) properties rarely found in nature, (2) periodic structures, and (3) sub-wavelength-scale unit cells. These three features are thus considered as a definition of metamaterials.

The metamaterial concept has broadened significantly during the past decades, so its definition should be updated. First, metamaterials are not only used to surpass natural materials but also mimic them in cases where the desired natural materials are not easily available or processed. Second, like how there are both crystals and noncrystals in nature, metamaterials can also have nonperiodic patterns. A typical example is graded metamaterial, whose meta-atoms have spatially varying structures [6]. They are more like devices with particular functions, such as cloaking. Finally, loosening the sub-wavelength requirement and allowing larger unit cells is helpful. For EM waves, such structures are like a hybridization of traditional metamaterials and photonic crystals [7], where both meta-atoms and their collective behaviors make sense. The nonlocality or spatial dispersion of these metamaterials can perform better than traditional metamaterials in controlling reflection and refraction [8]. Another issue about the sub-wavelength requirement is that some metamaterials do not target waves. For example, some mechanical metamaterials are designed to possess unusual responses to static forces [9]. In many cases, heat conduction and convection are not regarded as waves, so the wavelength definition for thermal metamaterials is often unclear. We will come back to this topic soon.

It thus appears difficult, if not impossible, to give a conclusive definition to this fast-evolving term. We can only loosely state that metamaterials are artificial structures with some desirable properties. In our opinion, whether a particular structure can be treated as a metamaterial often depends on its source of inspiration. If its design principle, property, or functionality is inspired by previous metamaterials, it is likely to share some of their features and can be called a metamaterial. For the same reason, we are not going to define thermal metamaterials either. The following section will briefly introduce the history and some basic properties of thermal metamaterials. After being acquainted with the concept, readers are expected to form their judgments, given the detailed context.

Before we discuss thermal metamaterials, it is helpful to illustrate the functionality of metamaterials with a typical example. In 2006, Sir Pendry *et al.* [10] and Leonhardt [11] independently proposed two theoretical frameworks for the fascinating possibility of invisible cloaks. As shown in Fig. 1.3a, an EM cloak is a closed structure with an interior hole. It has two effects. One is that the incident EM wave cannot penetrate its interior, so any object put inside remains untouched by the exterior EM field. The other effect is that the cloak itself is perfectly transparent to the incident wave, such that the exterior EM field is undisturbed as if the cloak were absent. These two effects make the cloak with anything hidden inside invisible to outside observers. The idea was experimentally demonstrated in the same year [12] at microwave frequencies. The measured magnetic field distribution of an EM cloak [13] is shown in Fig. 1.3c. One can see that the pattern outside a bare object (left) is distorted, while that outside the cloak (right) is similar to a plane wave in free space.

It should be noted that there are still some traces in the exterior field of the cloak [12], indicating its existence, mainly because of two issues. First, an ideal EM cloak has extreme requirements on the material parameters: its electric permittivity $\boldsymbol{\varepsilon}(\boldsymbol{r})$ and magnetic permeability $\boldsymbol{\mu}(\boldsymbol{r})$ should be anisotropic, nonuniform, and even

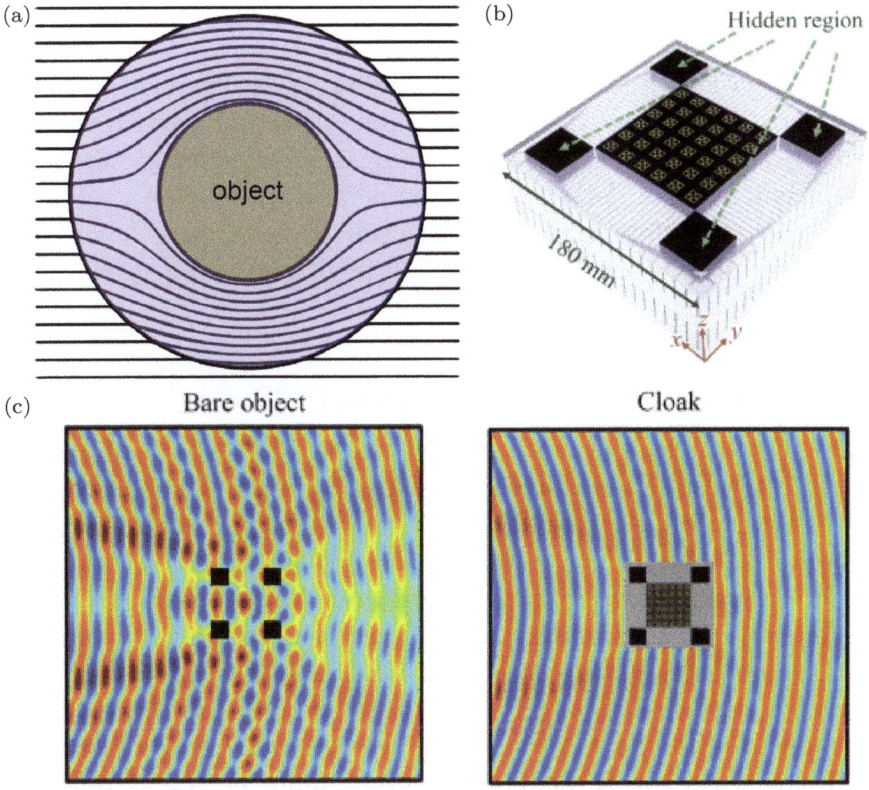

Fig. 1.3. EM cloaking. (a) Schematic of an EM cloak. (b) A prototype of cloaking at microwave frequencies [13]. (c) The magnetic field distribution around a bare object (left) and the cloak (right) [13].

approaching infinity at some positions. The prototype [12] has to adopt a reduced version of the parameters for fabrication, which leads to unwanted scattering. Second, the compositions have an intrinsic loss, which leads to unwanted absorption. Another severe problem is the strong temporal dispersion of structures, leading to a narrow bandwidth. We will see later that thermal cloaking is comparably more conveniently realizable with better performance.

Except for cloaking, EM metamaterials have many other potential applications in imaging, sensing, and communication [14]. The field has also been well extended to infrared, visible, and other wavelengths. Two other major types of metamaterials are acoustic [15] and mechanical [16] metamaterials targeting sound and other mechanical fields, respectively. Despite their different physical natures, these fields follow similar wave dynamics, so many concepts like wavelength, resonance, reflection, and interference are universally applicable. However, thermal metamaterials targeting heat conduction differ since the temperature field obeys diffusive dynamics. We thus face the difficulties of developing a suitable theoretical framework and the possibilities of uncovering novel effects and mechanisms.

1.3. Thermal Metamaterial: Thermal Cloaking

The study on thermal metamaterials was initiated by Prof. Jiping Huang's team at Fudan University from a theoretical perspective. In 2008, Fan *et al.* [17] proved that the macroscopic heat conduction equation at a steady state (the Laplace equation) is form invariant under coordinate transformations, which is the basis of Sir Pendry's theory for EM cloaking [10]. Therefore, they proposed the concept of thermal cloaking. Like the effects of EM cloaking, a thermal cloak prevents heat flux from entering its interior without introducing any perturbation to its exterior temperature field. As a result, any thermal measurement outside the cloak cannot detect its presence or any object hidden inside, such that the object is made thermally invisible. A schematic of this effect is shown in Fig. 1.4a, adapted from the paper that Chen *et al.* [18] published a few months later.

Both Refs. [17] and [18] derive an anisotropic and nonuniform thermal conductivity tensor distribution $\boldsymbol{k}(\boldsymbol{r})$ without suggesting a potential method to realize it. In 2010, Li *et al.* [19] proposed a composite structure with elliptical inclusions (Fig. 1.4b). Based on the Bruggeman effective medium theory, the anisotropic thermal conductivity at each local spot can be approximated by the structure, thanks to the anisotropic shape of the inclusions. However, preparing and aligning these different inclusions is challenging, so the method was only numerically verified.

The first experimental demonstration of thermal cloaking was performed by Narayana and Sato [20] in 2012. Their method is not directly based on the transformation theory but on the effective

(a)

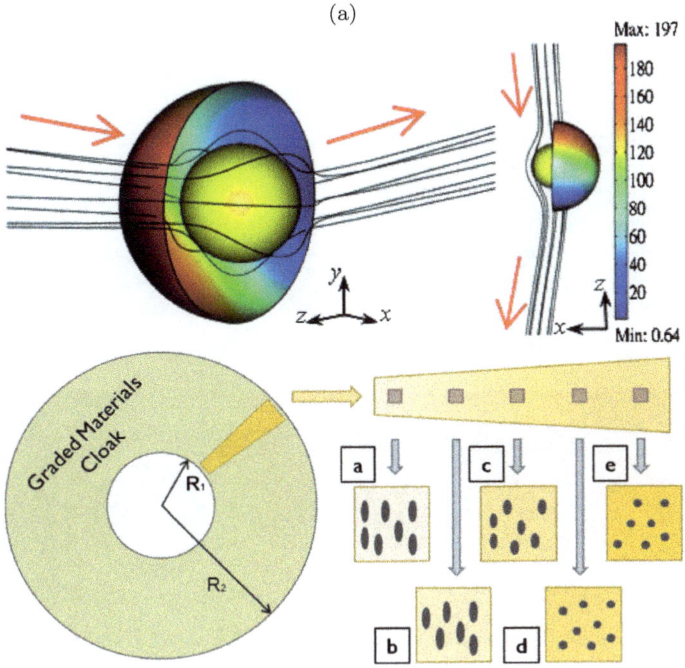

(b)

Fig. 1.4. (a) Effect of thermal cloaking [18]. (b) Proposed structure of a thermal cloak with elliptical inclusions [19].

medium theory. As Ref. [19] indicates, a composite structure can be approximated as a material with certain local effective thermal conductivity. An extension of this idea is to consider the global effective thermal conductivity of the cloak itself, which should equal the thermal conductivity of the background for thermal invisibility. Heat flux should not penetrate the central hole of the cloak. They designed and fabricated a prototype [20] that satisfies both requirements using a layered structure of thermal conductor (Material A) and thermal insulator (Material B), as shown in Fig. 1.5a. The measured temperature distribution around the cloak is shown in Fig. 1.5b. The temperature field outside the cloak is similar to that on a uniform background, which is a linear distribution. Notably, such invisibility for electrical conduction has been discussed earlier. In 2003, Greenleaf *et al.* showed that objects with the same shape but different

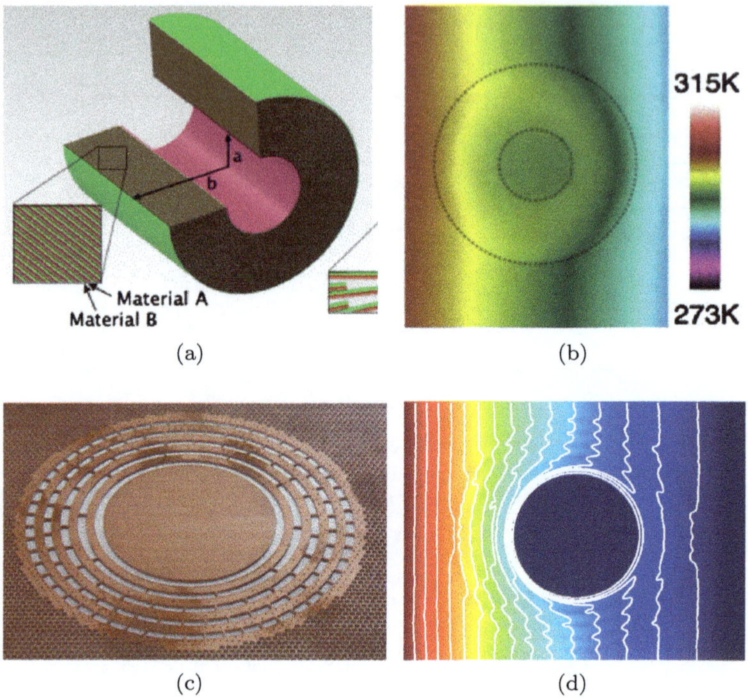

Fig. 1.5. (a) Multilayered thermal cloak [19]. (b) Measured temperature distribution around (a) at a steady state [19]. (c) Transformation-based thermal cloak [23]. (d) Measured temperature distribution around (c) at a transient state [23].

conductivity distributions can exhibit the same effective conductivity [21] and counter-detect electric impedance tomography [22].

Also, the thermal transformation theory, or transformation thermotics, was further developed in 2012. Guenneau *et al.* [23] proved the form invariance of the governing equation for transient heat conduction. It is found that to realize thermal cloaking for a time-varying temperature field, the product of density ρ and specific heat capacity C of the cloak should also be nonuniform $\rho C(\boldsymbol{r})$, in addition to the thermal conductivity $\boldsymbol{\kappa}(\boldsymbol{r})$. In 2013, Schittny *et al.* [24] fabricated a prototype that almost strictly follows the transformation theory (Fig. 1.5c), except for some homogenization on the density and specific heat capacity. The device performs excellent transient and steady-state thermal cloaking, as shown in Fig. 1.5d.

1.3.1. *Comparing Thermal and Electromagnetic Cloaking*

The above results show that, as we mentioned before, thermal cloaking has a better effect than EM cloaking in experiments. This fact is due to several reasons.

First, the requirements of the transformation theory are different. For an EM cloak, we must realize two nonuniform distributions $\boldsymbol{\varepsilon}(\boldsymbol{r})$ and $\boldsymbol{\mu}(\boldsymbol{r})$, where both are anisotropic tensors. For a thermal cloak, the distributions are $\boldsymbol{\kappa}(\boldsymbol{r})$ and $\rho C(\boldsymbol{r})$, where the former is an anisotropic tensor, and the latter is a scalar. Two more material parameters must be considered for EM cloaking in 3D, assuming the tensors are diagonalizable.

Second, the requirements of the composite materials are different. For an EM cloak, normal nonmagnetic materials can only have near-unity relative permeabilities, so complicated sub-wavelength structures must be used to achieve the desired permeability. For a thermal cloak, the required thermal conductors and insulators can be easily found in metals and nonmetals. The requirements for density and heat capacity are also moderate. In addition, the intrinsic losses of many common materials significantly contribute to the imaginary parts of $\boldsymbol{\varepsilon}(\boldsymbol{r})$ and $\boldsymbol{\mu}(\boldsymbol{r})$, making them deviate from the ideal values. On the other hand, the imaginary part of the $\boldsymbol{\kappa}(\boldsymbol{r})$ or $\rho C(\boldsymbol{r})$ represents some wave-like response. For macroscopic heat conduction in common materials, the thermal wave effects [25] can safely be ignored.

Finally, the requirements on the field outside the cloak are different. For an EM cloak, the disturbance to the outside field is exhibited as an interference between the incident and scattered waves. Suppressing such disturbance requires a very small scattering coefficient. For a thermal cloak, the disturbance to the outside field can also be considered interference. Still, the scattered field exponentially decays from the clock and varies slowly in the angular direction (see Chapter 4). Therefore, the requirement on the scattering coefficient is comparably not so stringent. In 2014, Han *et al.* [26] and Xu *et al.* [27] made use of this fact to independently propose a much-simplified design of thermal cloaking with just two layers of thermal insulator and thermal conductor. 2D and 3D prototypes were fabricated, showing good performances at a steady state. These bilayer thermal cloaks cancel out the scattered field to the first order in the angular coordinate, so the method is called scattering cancellation.

The above points also apply to wave fields such as acoustic or elastic cloaking, reflecting some fundamental differences between metamaterials for waves and metamaterials for diffusion.

1.3.2. *Potential Applications*

After seeing its performance, people may ask how a thermal cloak can be used in real applications. Indeed, the early studies were mainly driven by interest in the possibility rather than by some specific needs. The first thought may be that thermal cloaking can be used against thermal detection in analogy to EM cloaking. In particular, it is expected to work against techniques like infrared thermography and thermal nondestructive testing [28]. Another potential usage of a thermal cloak is to provide thermal protection by maintaining its interior region at a uniform temperature without heat flux. It also protects its exterior since no disturbance is introduced to the environment.

A natural extension of the idea is to modify the first function while keeping the second. In Ref. [20], two other devices were thus designed and fabricated. One is the thermal concentrator, which gathers heat flux into its interior region to enhance its temperature gradient. It can collect thermal energy for storage or conversion (e.g., with a thermoelectric generator). The other device is the thermal rotator, which changes the heat flux direction through its interior. It is expected to work with other functional devices for heat guidance, collection, and dissipation. Both devices have no environmental disturbance, so they can be easily integrated with the thermal cloak in a single platform for more complicated demands on heat flux patterns or temperature distributions without worrying about their crosstalk. In 2015, Dede *et al.* [29] implemented this idea to design a printed circuit board (PCB) with customized thermal management effects (Fig. 1.6).

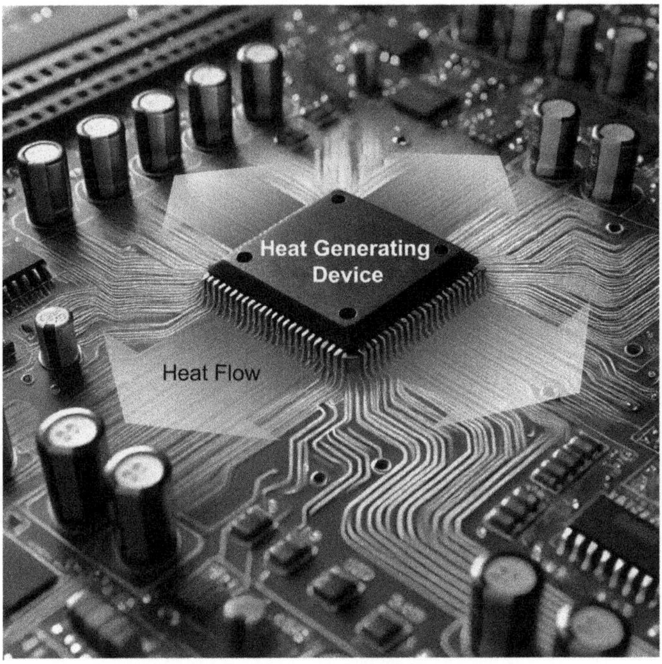

Fig. 1.6. Schematic of a PCB with customized thermal managing effects for devices on it.

In thermal management, one does not only focus on the heat flux for shielding or dissipation. Maintaining a suitable temperature for a region is also important. Although no temperature gradient is inside a thermal cloak, the initial and boundary conditions determine the temperature value. In 2016, Shen *et al.* [30] found that it is possible to stabilize the interior temperature by replacing the common composite materials with nonlinear materials, whose thermal conductivities strongly depend on the temperature. As a result, the interior temperature can be maintained close to the desired value despite a changing ambient temperature gradient (Fig. 1.7a). Since the effect utilizes the heat sources and sinks in the environment without consuming additional energy, the device is called an energy-free thermostat. In 2023, Zhou *et al.* [31] further developed an adaptive multi-temperature control system, which can be used to store goods that require different thermal conditions (Fig. 1.7b).

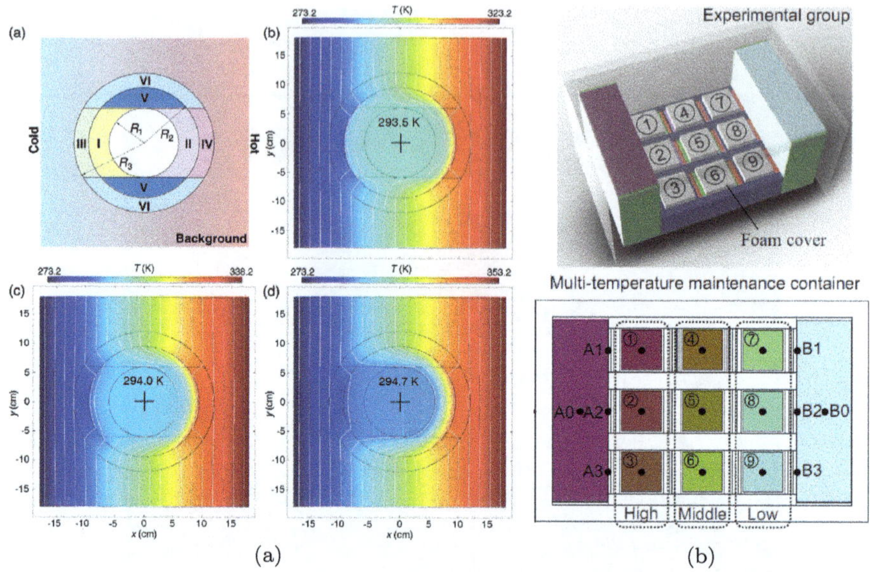

Fig. 1.7. (a) Energy-free thermostat [30]. (b) Multi-temperature control system [31].

The above examples show that thermal metamaterials are useful to control the spatial distribution of heat transfer, which will be the theme of Part I (Chapters 2 and 3). More recent studies bring two other new possibilities.

1.4. Thermal Metamaterial: Asymmetric Heat Transfer

Heat transfer in common materials is often symmetric in two senses. One is that the thermal conductivities of most common materials are isotropic, so heat transfer is equally efficient in all directions. The other symmetry is that heat transfer is equally efficient in the forward and backward directions, before and after swapping the positions of the hot and cold sides. The second symmetry is closely related to the reciprocity of the Green function for the heat conduction equation. Therefore, in the following, we will refer to the first symmetry as isotropy and the second as reciprocity.

The microscopic reversibility and weak temperature dependence of common materials determine that the heat transfer in them is reciprocal. In 2002, Terraneo *et al.* [32] studied heat conduction on nonlinear 1D lattices and showed that thermal rectification is possible in such models. In 2004, Li *et al.* [33] proposed a more realistic model with two coupled nonlinear 1D lattices, which can act as a thermal diode that allows heat flow in one direction but blocks it in the opposite direction. In 2006, Chang *et al.* [34] realized a solid-state thermal rectifier with nanotubes, but the rectification ratio is small. Like its electronic counterpart, the thermal diode inspires the possibility of using heat as an information carrier for communication and computation. Following designs of thermal memory, thermal transistors, thermal logic gates, and other functional devices were thus proposed, forming the field of phononics [35].

Phononics focuses on microscopic structures whose realizations are challenging. Therefore, the above theoretical designs have not been fully implemented in experiments. Different design principles can be used for macroscopic thermal metamaterials, and the fabrication is much easier. In 2015, Li *et al.* [36] proposed and realized a macroscopic thermal diode based on transformation thermotics. The underlying mechanism of combining different nonlinear components is similar to the phononic designs, but the nonlinearity is effectively achieved by the mechanical deformation of shape-memory alloys (Fig. 1.8). The deformation-triggered modulation of the effective thermal conductivity is strong enough to promote the rectification ratio significantly.

Instead of comparing the heat fluxes in opposite directions at a steady state, one can also compare the total amounts of heat transferred in a period for unsteady states. The time-harmonic case is particularly interesting, where the temperature field oscillates in time at a certain frequency. This effect can be induced by periodically modulating either the temperatures of the heat baths or the material properties. In 2008, Li *et al.* [37] proposed the concept of thermal rachet, which conveys a finite amount of heat with a non-biased alternating bath temperature. The rachet is a thermal diode made of a nonlinear lattice coupled with a Harmonic lattice. In 2009,

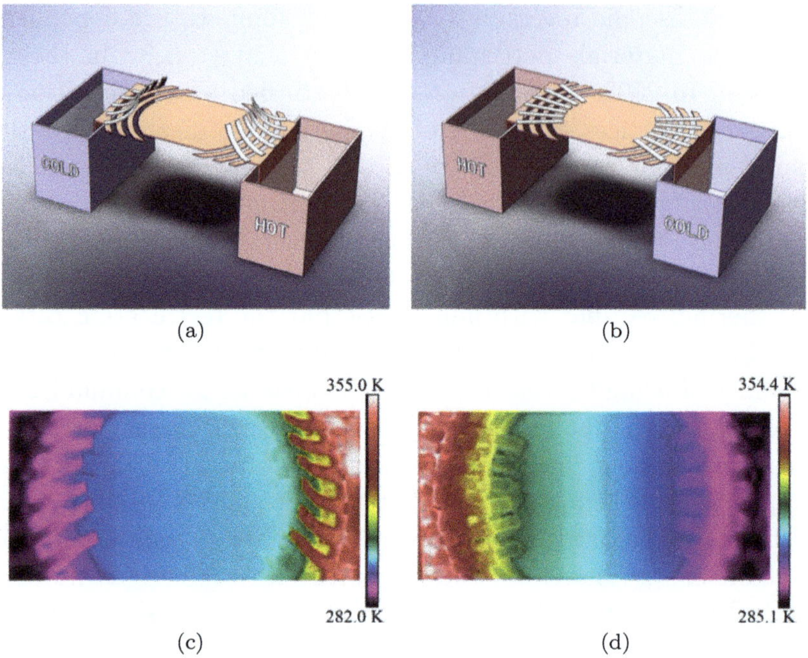

Fig. 1.8. Macroscopic thermal diode [36]. (a, b) Configurations. (c, d) The measured temperature distributions. (a, c) Backward heat transfer. (b, d) Forward heat transfer.

Li *et al.* [38] proposed a similar heat shuttling effect that can be realized in a uniform nonlinear lattice. In 2010, Ren *et al.* [39] found that such a heat-pumping effect has a geometric contribution from the Berry phase. At the macro scale, heat shuttling based on phase-change materials was numerically studied by Ordonez-Miranda *et al.* [40] in 2022.

The experimental verifications of the above theoretical schemes are difficult. In 2018, Torrent *et al.* [41] considered different routes for dynamically modulating the density and thermal conductivity of a macroscopic thermal material. The material was predicted to work as a thermal diode even with constant bath temperatures. In 2022, Li *et al.* [42] fabricated a thermal metamaterial with rotating components to realize the modulation but did not observe nonreciprocal heat transfer for constant bath temperatures. The negative

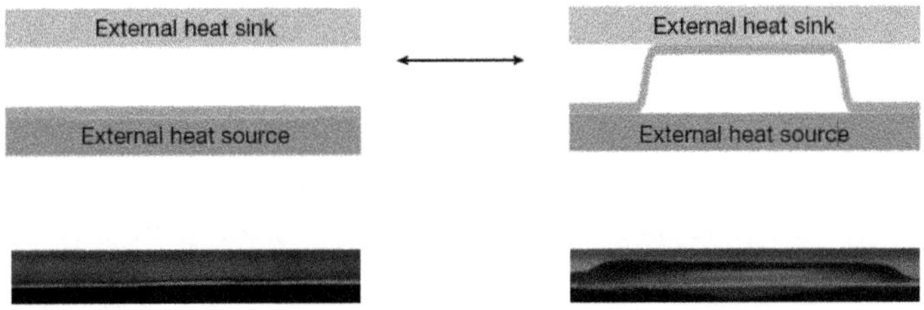

Fig. 1.9. Cycle of the electrocaloric refrigerator [45].

result was attributed to the convective heat introduced for density modulation. On the other hand, the asymmetric heat pumping effect can be realized for varying bath temperatures. In 2022, Wang *et al.* [43] performed experiments on a system periodically made in contact and isolated from the heat baths, which were maintained at the same varying temperature. A finite amount of heat was pumped through the system.

The nonreciprocal heat transfer also has practical applications in addition to thermal computing. For example, it can be used in refrigeration systems [44]. In 2024, Han *et al.* [45] made a refrigerator prototype with self-oscillating electrocaloric material, which alternatively contacted the heat sink and source (Fig. 1.9). The device showed a high energy efficiency with a small footprint.

In 2021, Li *et al.* [46] showed that the steady-state and dynamic (in terms of the bath temperatures) thermal nonreciprocity are closely related. They can be studied within the same thermal scattering framework, where steady state is a limiting case as the working frequency approaches zero. Therefore, studying the time-harmonic response of thermal metamaterials is important and will be the theme of Part II (Chapters 4 and 5).

1.5. Thermal Metamaterial: Spatiotemporal Evolution

In previous discussions, the evolution of the temperature field in time is not the central concern. Although a transformation theory has been developed for transient heat conduction, most existing experiments focus on steady-state performance. It is possible to detect a thermal cloak through its transient response [47]. In reality, many processes are not in a steady or time-harmonic state. For example, power electronic components often undergo pulsed or irregular loads, leading to complex transient behaviors of the temperature field [48].

A fundamental form of transient heat transfer is the free evolution in the absence of any heat source. In such cases, a certain temperature profile or thermal mode decays exponentially with a corresponding decay rate. A heat transfer system can have many thermal modes with different decay rates. Studying them gives the actual temperature field a linear combination, similar to the mode and spectrum analysis for EM metamaterials and photonic crystals. Indeed, by simply multiplying an imaginary unit by the temporal partial derivative, the diffusion equation becomes the Schrödinger equation, so the diffusion and wave dynamics share some common features.

In 2019, inspired by the above correspondence, Li *et al.* [49] proposed the effective Hamiltonian of a heat transfer system to analyze and engineer its spatio-temporal evolution. The system is not purely diffusive but has convection to enrich the dynamics. It comprises two coupled counter-rotating rings (Fig. 1.10a) with an anti-parity-time (APT) symmetric effective Hamiltonian. Under

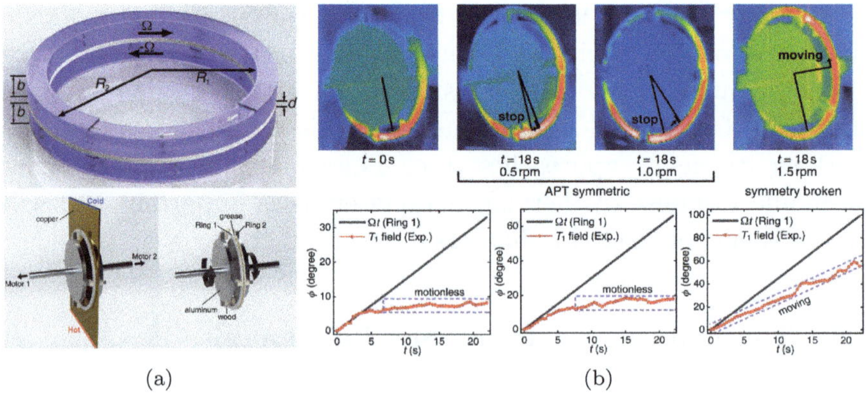

(a) (b)

Fig. 1.10. APT symmetric heat transfer [48]. (a) The model and experimental setup. (b) Measured temperature profiles (upper) and their rotation angles (lower).

the protection of APT symmetry, the temperature profiles do not follow the rotation of the rings but remain motionless (Fig. 1.10b). Rotating temperature profiles can be observed after a symmetry-breaking phase transition as the rotation speed of the rings exceeds a critical value at the exceptional point.

The rotation of the thermal field shows that the phase can be a new degree of freedom in heat transfer. In 2024, Cao *et al.* [50] designed and realized a heat transfer system that is parity-time (PT) symmetric. The thermal fields on it demonstrate a stable phase lag that could potentially be used in communication and detection. The PT- and APT-symmetric Hamiltonians are two special cases of non-Hermitian Hamiltonians, which have drawn great interest recently. In 2021, Xu *et al.* [51] demonstrated a typical effect in non-Hermitian physics, the mode switching induced by exceptional point encircling. The proposed heat transfer system presents the mode switching as a robust π-phase flipping after each encircling.

The effective Hamiltonian is a useful tool for studying and designing the transient behaviors of thermal metamaterials with spatial periodicity. Just like for electronic and other wave lattices, the band structure of thermal lattices can also reveal many fundamental properties, particularly the topological effects. In 2021, Yoshida and Hatsugai [52] made a theoretical prediction on the topological edge state in a 1D diffusion lattice, which mimics the Su-Schrieffer-Heeger model for quantum systems [53]. In 2022, Hu *et al.* [54] and Qi *et al.* [55] independently observed topologically protected thermal edge states with different boundary conditions. The thermal edge state is roughly a hot spot well localized at the boundary of a 1D lattice, whose temperature decays exponentially in time (Fig. 1.11). The spatial-temporal behaviors are potentially favorable for thermal management techniques that require a small influence on the neighboring region and fast heat dissipation. Recently, 2D thermal lattices were also designed and fabricated by Wu *et al.* [56] and Liu *et al.* [57] to host corner states.

Another interesting phenomenon, the non-Hermitian skin effect, is protected by the winding topology of the complex eigenvalues. In 2021, Cao *et al.* [58] proposed that a quasi-periodic 1D thermal lattice

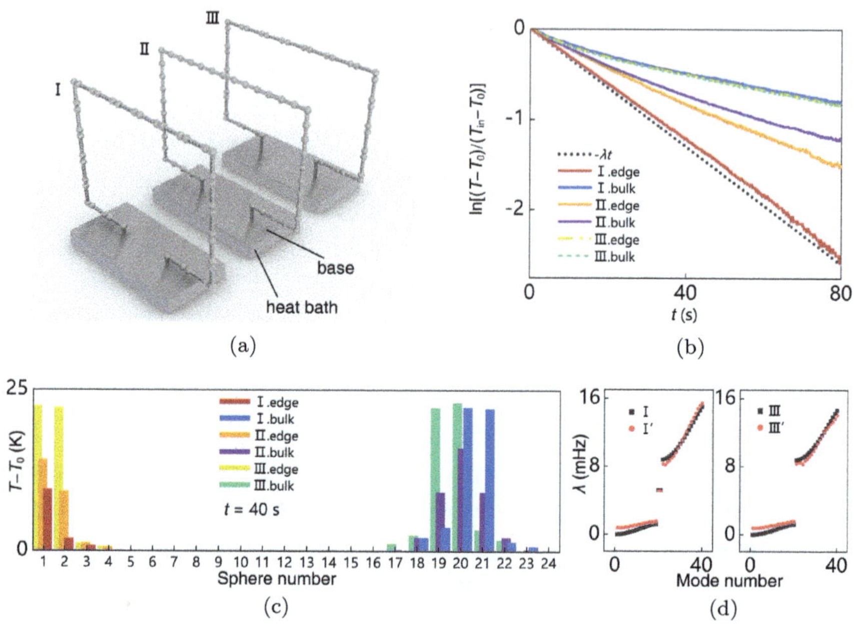

Fig. 1.11. Topological thermal lattice [55]. (a) The samples. (b) Decay of the hot spot temperatures in time. (c) Temperature distributions at $t = 40\,\text{s}$. (d) The decay-rate spectrums.

with gradient density can be mapped to an effective Hamiltonian model with nonreciprocal coupling. The non-Hermitian skin effect was numerically demonstrated as the directional concentration of the temperature gradient. The diffusive non-Hermitian skin effect or topological heat funneling effect was experimentally verified in 2024 by Liu *et al.* [59]. The effect could potentially be used for thermal energy collection and conversion.

The above thermal lattices are purely conductive systems. In 2022, Xu *et al.* [60] showed that a 1D lattice with convection induced by fluid flow possesses rich topological properties. Thermal edge states were also observed in the lattice, which can be modified by the fluid velocity. In 2023, Xu *et al.* [61] further designed a 2D lattice with rotating meta-atoms to construct thermal quadrupole moment, experimentally observing the higher-order topological corner states.

The studies on transient heat transfer reveal many novel mechanisms and interesting possibilities. They show the link between the distinct diffusion and wave dynamics and uncover the symmetry and topological properties of heat transfer. The fundamental theories and recent progress on the theme will be covered in Part III (Chapters 6 and 7).

1.6. Structure and Scope of this Book

1.6.1. *Structure*

Thermal metamaterials are certainly not limited to the above three types but are the most representative. Indeed, most existing thermal metamaterials can be categorized based on the time dependence of the targeted heat transfer processes. The works in different types use different design principles and methods. That is why the following chapters are organized based on the time dependency of the targeted heat transfer process. There is a major difference between this book and other existing monographs on similar themes, including three books by Huang *et al.* [62–64] and one book on thermal cloaking by Yeung and Yang [65]. The detailed arrangement is as follows.

Chapter 1: Introduction

The present chapter gives an overview of the history and the concept of thermal metamaterials.

Part I: Steady-state heat transfer

This part introduces the basic methods behind thermal cloaks and other related thermal metamaterials. Their common target is to control the heat transfer at a steady state, thereby achieving the desired temperature or heat flux distributions in space. The heat transfer can be conduction, convection, or those involving multi-physical processes. Chapter 2 is on the transformation theory for designing the functionalities of these thermal metamaterials. Chapter 3 discusses the effective medium theory for realizing certain thermal material properties.

Part II: Time-harmonic heat transfer

This part introduces the basic framework for designing and analyzing thermal diodes and other related thermal metamaterials. As mentioned, a thermal diode is often supposed to work for steady-state heat transfer. For example, the macroscopic thermal diodes based on transformation theory are covered in Part I: Chapter 2. However, it can be shown that the mechanism of breaking thermal reciprocity can be better understood for the time-harmonic case, and the steady-state behaviors can be obtained by taking a zero-frequency limit. A general thermal scattering theory is introduced in Chapter 4 to set a useful framework. The time-modulated heat transfer is discussed in Chapter 5, emphasizing nonreciprocity and heat shuttling.

Part III: Transient heat transfer

This part introduces the method of effective Hamiltonian and some unique transient heat transfer behaviors predicted by it. The target of the related thermal metamaterials is to control the entire spatio-temporal evolution of the thermal fields. In Chapter 6, effective Hamiltonians with different symmetries are introduced to reveal the symmetry-protected or symmetry-breaking phenomena. In Chapter 7, effective Hamiltonians of thermal lattices are introduced to explore the topologically protected heat transfer effects.

Chapter 8: Outlook

This chapter gives an outlook on this field. We emphasize the open questions on multi-physical, topological, disorder, and nonlinear effects, with potential possibilities to explore and utilize them.

1.6.2. *Scope*

This book is intended as an introduction and tutorial to thermal metamaterials, so the major concern is presenting the frameworks for their designs rather than comprehensively reviewing the field. We only introduce a few works for each topic to derive the theory and demonstrate the effects. Many interesting and important studies are thus not mentioned or discussed in detail. Fortunately, several high-quality review articles have been published recently [66–71]. The review by Li *et al.* [66] covers a broad range, including thermal metamaterials targeting both macroscopic and microscopic heat transfer in the forms of conduction and radiation, as well as multiphysical processes. Yang *et al.* [67] focus on manipulating macroscopic heat transfer in conduction, convection, and radiation. Fan *et al.* [68] give a more detailed and up-to-date account of similar topics from the static and dynamic perspectives. Zhu *et al.* [69] highlight the role of machine learning in the design and optimization of thermal metamaterials. Zhang *et al.* [70] focus on the diffusive nature of thermal conduction and its hybridization with convection or radiation. Metamaterials used to manipulate particle diffusion are also discussed. Yang *et al.* [71] give a more in-depth and comprehensive review of the same theme. Readers who would like to get a more complete view of the field of thermal or diffusion metamaterials may refer to these works.

It is worth noting that two important fields, nanophononics and thermal photonics, will not be discussed in this book. The artificial structures designed in the two fields are also used to control heat transfer in phonon transport and far-field/near-field thermal radiation. Therefore, they can also be regarded as thermal metamaterials. We choose not to include them because the governing equations for their targeted processes are distinct from those that will be discussed. The central governing equation of this book is the diffusion equation at steady state or transient state. Convection and other multiphysical effects modify the equation by adding more heat source/sink terms or tuning its material properties. On the other hand, nanophononics and thermal photonics are based on the Boltzmann and Maxwell equations, respectively. In addition, categorizing the processes into steady-state, time-harmonic, and transient heat transfers is inappropriate for these two fields. It is thus more suitable to discuss them elsewhere for the sake of clarity and unity of this book.

1.7. Remarks and Conclusion

1.7.1. *Characteristic Length*

The introduction to EM metamaterials mentioned that they are commonly expected to have sub-wavelength meta-atoms. Many would thus be curious about the length scales of thermal metamaterials. First of all, what is the wavelength for a thermal conduction process?

To illustrate the issue, let us consider a 1D conductive system with uniform material properties. Suppose that its left end ($x = 0$) is connected to a heat sink with the oscillating temperature of $T(0, t) = T_0 + \mathrm{Re}[A \exp(-i\omega t)]$. The temperature distribution on the system is solved as

$$T(x,t) = T_0 + \mathrm{Re}\{A \exp[i(kx - \omega t)]\}, \tag{1.1}$$

where T_0 is a constant temperature, A is the oscillating amplitude, k is the wavenumber, ω is the frequency, and i is the imaginary unit. Equation (1.1) appears to be a right-going plane wave. However, the wavenumber k is a complex number,

$$k = (1+i)\sqrt{\omega/2D}, \tag{1.2}$$

where $D = \kappa/\rho c$ is the diffusivity of the system. Therefore, the thermal wave amplitude decays quickly. The corresponding wavelength is then $\lambda = \mathrm{Re}(k) = \sqrt{2D/\omega}$. For example, the diffusivity of aluminum is around $97\,\mathrm{mm}^2/\mathrm{s}$. Under a frequency of 0.1 Hz, the thermal wavelength is calculated to be 17 mm. A unit cell of size ~1 mm would be justified as sub-wavelength.

A real-world transient process usually consists of multiple frequencies. It is more convenient to consider the purely imaginary frequency $\omega = -ig$, where the decay rate g is a positive real number. The same dispersion relation Eq. (1.2) gives a real wavenumber,

$$k = \sqrt{\gamma/D}. \tag{1.3}$$

Typically, without an oscillating source, the transient process can be decomposed into multiple modes with different wavenumber and decay rates. The modes with large k or small wavelengths will decay quickly and become unobservable. The precision and temporal resolution of the measuring device thus determine the characteristic wavelength. For example, consider the free decay of a hot spot whose temperature is 50 K higher than its surroundings. An infrared thermal camera with ~0.1 K precision and ~10 Hz refresh rate is used. If we want to observe the evolution for at least 10 frames, the maximum measurable decay rate is $g = 6.2$ rad/s. The corresponding minimum wavelength in aluminum is then $l = 4$ mm. The minimum wavelength is not a constant, which will gradually increase with time as the short-wavelength components decay until the system reaches a steady state.

At a steady state, the frequency or decay rate approaches zero, so the wavelength approaches infinity. That is partially why a bilayer thermal cloak with no unit-cell structure also works. For other steady-state thermal metamaterials, it does not mean that there is no constraint on their length scales. After all, the requirement of subwavelength meta-atoms is imposed to avoid spatial dispersion. The k-independent effective material parameter is not always a necessity for thermal metamaterials. On the other hand, if we are interested in the temperature field of the thermal metamaterial, the unit cell should not be too much larger than the spatial resolution of the measuring device. Otherwise, the measured field will be too coarse and dependent on the size and shape of the meta-atoms. For an infrared thermal camera with 640×480 pixels, the length of the unit cell should ideally be no larger than 10 pixels, or $\sim 1/50$ the length of the entire system. However, using bigger unit cells to fabricate prototypes conveniently is common practice.

1.7.2. *Miniaturizing Thermal Metamaterials*

After discussing the length scales, it is a good place to briefly review recent efforts to design and fabricate microscale thermal metamaterials. As mentioned before, the nanoscale heat transfer can be controlled by nanophononic metamaterials. In 2014, Davis and Hussein [72] proposed a nano-pillar array that hosts local resonances to reduce the phononic thermal conductivity. In the same year, Han *et al.* [73] suggested a similar possibility by engineering an array of atomic defects. In the following, we mainly focus on works inspired by the macroscopic thermal metamaterials and targeted at their miniaturization.

In 2016, Ye and Cao [74] used molecular dynamics to design and study a nanoscale thermal cloak. It was proposed to use partial chemical functionalization to reduce the local thermal conductivity of graphene, thereby creating the required thermal protection on a 10 nm-diameter region. In the same year, Loke *et al.* [75] proposed a CMOS-integrable, thermal-guiding structure on silicon, which can perform not only thermal shielding (on a 40 nm-diameter region) but also some sought-after thermal Boolean logic operations. The device is proposed to be made of silica and graphene layers and fabricated by etching and deposition. The numerical study was performed with the finite-element method, where the non-Fourier effect was not considered. In 2019, Choe *et al.* [76] experimentally demonstrated the first microscale thermal cloak with a 12 mm outer diameter. The thermal cloak is made of a monolithic silicon membrane, whose thermal conductivity is locally, continuously, and reversibly tuned by focused ion irradiation (Fig. 1.12). Under the irradiation of the He^+ ion beam, single-crystal silicon can be transformed into an amorphous state, suppressing thermal conductivity from $65\,W/m \cdot K$ down to $2\,W/m \cdot K$. The change can be reversed by

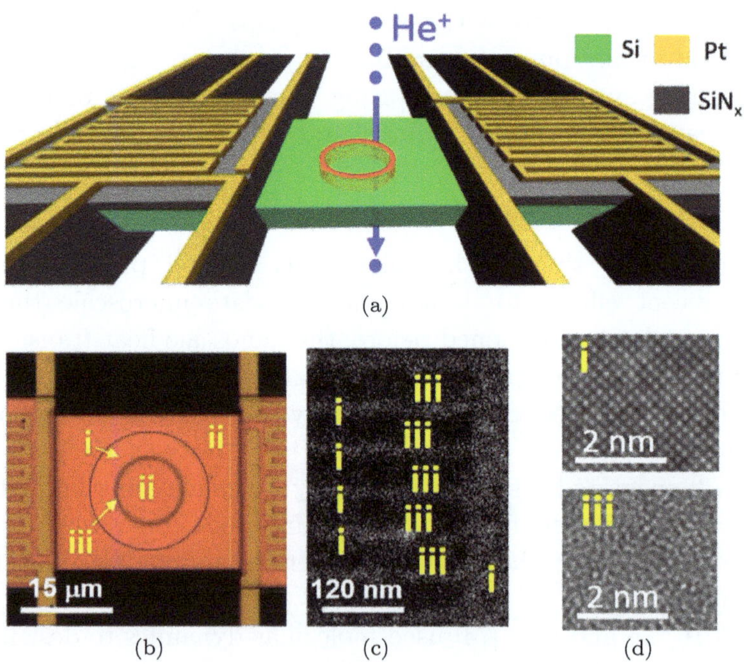

Fig. 1.12. Ion-written microscale thermal cloak [76]. (a) Schematic of the setup. (b) The sample. (c) Ion-written patterns. (d) Atomic structures of different regions.

thermal annealing the sample in the N_2 environment. In the same year, Liu *et al.* [77] used molecular dynamics to study a similar structure of smaller size. In 2021, Xiao *et al.* [78] numerically studied the thermal conductivities of different nanoporous structures and designed thermal camouflaging devices of ~20 mm in size. At this length scale, the Fourier law is supposed to be valid such that the finite-element method was used. The role of non-Fourier heat transfer in nanoscale thermal metamaterials was studied by Li *et al.* [80] in 2024. The heat transfer through a 2D thermal lattice was calculated using the finite-element method and extended-plane-wave-expansion method for comparison.

1.7.3. *Conclusion*

We have shown growing demands on controlling all aspects of heat transfer, where finding appropriate materials is the key issue. The emergence of EM metamaterials shows the possibility to mimic and surpass natural materials with artificial structures. Thermal metamaterials were thus designed and fabricated. They show intriguing new capabilities to control the spatial distribution, symmetry, and temporal evolution of thermal fields. These studies inspire potential applications in thermal management, thermal energy utilization, and thermal computation. They also reveal new physical mechanisms in diffusive processes that help us understand the fundamental phenomenon from different perspectives. The following chapters will introduce the basic ideas behind these thermal metamaterials.

References

1. Hahn, D. W. and Özisik, M. N. (2012). *Heat Conduction.* John Wiley & Sons.
2. Firth, A., Zhang, B., and Yang, A. (2019). Quantification of global waste heat and its environmental effects. *Appl. Energy*, 235, pp. 1314–1334.
3. Joshi, S. N., Zhou, F., Liu, Y. D., Lohan, J., Ukegawa, H., Lee, J., and Dede, E. M. (2023). A review of select patented technologies for cooling of high heat flux power semiconductor devices. *IEEE Trans. Power Electron.*, 38, pp. 6790–6794.

4. Veselago, V. G. (1968). The electrodynamics of substances with simultaneously negative values of ε and μ. *Sov. Phys. Usp.*, 10, pp. 509–514.
5. Shelby, R. A., Smith, D. R., and Schultz, S. (2001). Experimental verification of a negative index of refraction. *Science*, 292, pp. 77–79.
6. Smith, D. R., Mock, J. J., Starr, A. F., and Schurig, D. (2005). Gradient index metamaterials. *Phys. Rev. E*, 71, pp. 036609.
7. Joannopoulos, J. D., Johnson, S. G., Winn, J. N., and Meade, R. D. (2008). *Photonic Crystals: Molding the Flow of Light (Second Edition)*. Princeton University Press.
8. Asadchy, V. S., Albooyeh, M., Tcvetkova, S. N., Díaz-Rubio, A., Ra'di, Y., and Tretyakov, S. A. (2016). Perfect control of reflection and refraction using spatially dispersive metasurfaces. *Phys. Rev. B*, 94, pp. 075142.
9. Lee, J.-H., Singer, J. P., and Thomas, E. L. (2012). Micro-nanostructured mechanical metamaterials. *Adv. Mater.*, 24, pp. 4782–4810.
10. Pendry, J. B., Schurig, D., and Smith, D. R. (2006). Controlling electromagnetic fields. *Science*, 312, pp. 1780–1782.
11. Leonhardt, U. (2006). Optical conformal mapping. *Science*, 312, pp. 1777–1780.
12. Schurig, D., Mock, J. J., Justice, B. J., Cummer, S. A., Pendry, J. B., Starr, A. F., and Smith, D. R. (2006). Metamaterial electromagnetic cloak at microwave frequencies. *Science*, 314, pp. 977–980.
13. Zheng, B., Lu, H., Qian, C., Ye, D., Luo, Y., and Chen, H. (2023). Revealing the transformation invariance of full-parameter omnidirectional invisibility cloaks. *Electromag. Sci.*, 1, pp. 0020092.
14. Cui, T. J., Smith, D. R., and Liu, R. (2010). *Metamaterials: Theory, Design, and Applications*. Springer.
15. Cummer, S. A., Christensen, J., and Alù, A. (2016). Controlling sound with acoustic metamaterials. *Nat. Rev. Mater.*, 1, pp. 16001.
16. Yu, X., Zhou, J., Liang, H., Jiang, Z., and Wu, L. (2018). Mechanical metamaterials associated with stiffness, rigidity and compressibility: A brief review. *Prog. Mater. Sci.*, 94, pp. 114–173.
17. Fan, C. Z., Gao, Y., and Huang, J. P. (2008). Shaped graded materials with an apparent negative thermal conductivity. *Appl. Phys. Lett.*, 92, pp. 251907.
18. Chen, T., Weng, C.-N., and Chen, J.-S. (2008). Cloak for curvilinearly anisotropic media in conduction. *Appl. Phys. Lett.*, 93, pp. 114103.
19. Li, J. Y., Gao, Y., and Huang, J. P. (2010). A bifunctional cloak using transformation media. *J. Appl. Phys.*, 108, pp. 074504.

20. Narayana, S. and Sato, Y. (2012). Heat flux manipulation with engineered thermal materials. *Phys. Rev. Lett.*, 108, pp. 214303.
21. Greenleaf, A., Lassas, M., and Uhlmann, G. (2003). On nonuniqueness for Calderón's inverse problem. *Math. Res. Lett.*, 10, pp. 685–693.
22. Greenleaf, A., Lassas, M., and Uhlmann, G. (2003). Anisotropic conductivities that cannot be detected by EIT. *Physiol. Meas.*, 24, pp. 413–419.
23. Guenneau, S., Amra, C., and Veynante, D. (2012). Transformation thermodynamics: cloaking and concentrating heat flux. *Opt. Express*, 20, pp. 8207.
24. Schittny, R., Kadic, M., Guenneau, S., and Wegener, M. (2013). Experiments on transformation thermodynamics: molding the flow of heat. *Phys. Rev. Lett.*, 110, pp. 195901.
25. Joseph, D. D. and Preziosi, L. (1989). Heat waves. *Rev. Mod. Phys.*, 61, pp. 41–73.
26. Han, T., Bai, X., Gao, D., Thong, J. T. L., Li, B., and Qiu, C.-W. (2014). Experimental demonstration of a bilayer thermal cloak. *Phys. Rev. Lett.*, 112, pp. 054302.
27. Xu, H., Shi, X., Gao, F., Sun, H., and Zhang, B. (2014). Ultrathin three-dimensional thermal cloak. *Phys. Rev. Lett.*, 112, pp. 054301.
28. Vavilov, V. and Burleigh, D. (2020). *Infrared Thermography and Thermal Nondestructive Testing*. Springer.
29. Dede, E. M., Schmalenberg, P., Nomura, T., and Ishigaki, M. (2015). Design of anisotropic thermal conductivity in multilayer printed circuit boards. *IEEE Trans. Compon. Packag. Manuf. Technol.*, 5, pp. 1763–1774.
30. Shen, X., Li, Y., Jiang, C., and Huang, J. (2016). Temperature trapping: energy-free maintenance of constant temperatures as ambient temperature gradients change. *Phys. Rev. Lett.*, 117, pp. 055501.
31. Zhou, X., Xu, X., and Huang, J. (2023). Adaptive multi-temperature control for transport and storage containers enabled by phase-change materials. *Nat. Commun.*, 14, pp. 5449.
32. Terraneo, M., Peyrard, M., and Casati, G. (2002). Controlling the energy flow in nonlinear lattices: a model for a thermal rectifier. *Phys. Rev. Lett.*, 88, pp. 094302.
33. Li, B., Wang, L., and Casati, G. (2004). Thermal diode: rectification of heat flux. *Phys. Rev. Lett.*, 93, pp. 184301.
34. Chang, C. W., Okawa, D., Majumdar, A., and Zettl, A. (2006). Solid-state thermal rectifier. *Science*, 314, pp. 1121–1124.
35. Li, N., Ren, J., Wang, L., Zhang, G., Hänggi, P., and Li, B. (2012). Colloquium: Phononics: Manipulating heat flow with electronic analogs and beyond. *Rev. Mod. Phys.*, 84, pp. 1045–1066.

36. Li, Y., Shen, X., Wu, Z., Huang, J., Chen, Y., Ni, Y., and Huang, J. (2015). Temperature-dependent transformation thermotics: from switchable thermal cloaks to macroscopic thermal diodes. *Phys. Rev. Lett.*, 115, pp. 195503.
37. Li, N., Hänggi, P., and Li, B. (2008). Ratcheting heat flux against a thermal bias. *EPL*, 84, pp. 40009.
38. Li, N., Zhan, F., Hänggi, P., and Li, B. (2009). Shuttling heat across one-dimensional homogenous nonlinear lattices with a Brownian heat motor. *Phys. Rev. E*, 80, pp. 011125.
39. Ren, J., Hänggi, P., and Li, B. (2010). Berry-phase-induced heat pumping and its impact on the fluctuation theorem. *Phys. Rev. Lett.*, 104, pp. 170601.
40. Ordonez-Miranda, J., Anufriev, R., Nomura, M., and Volz, S. (2022). Net heat current at zero mean temperature gradient. *Phys. Rev. B*, 106, pp. L100102.
41. Torrent, D., Poncelet, O., and Batsale, J.-C. (2018). Nonreciprocal thermal material by spatiotemporal modulation. *Phys. Rev. Lett.*, 120, pp. 125501.
42. Li, J., Li, Y., Cao, P.-C., Qi, M., Zheng, X., Peng, Y.-G., Li, B., Zhu, X.-F., Alù, A., Chen, H., and Qiu, C.-W. (2022). Reciprocity of thermal diffusion in time-modulated systems. *Nat. Commun.*, 13, pp. 167.
43. Wang, Z., Chen, J., and Ren, J. (2022). Geometric heat pump and no-go restrictions of nonreciprocity in modulated thermal diffusion. *Phys. Rev. E*, 106, pp. L032102.
44. Wehmeyer, G., Yabuki, T., Monachon, C., Wu, J., and Dames, C. (2017). Thermal diodes, regulators, and switches: Physical mechanisms and potential applications. *Appl. Phys. Rev.*, 4, pp. 041304.
45. Han, D., Zhang, Y., Huang, C., Zheng, S., Wu, D., Li, Q., Du, F., Duan, H., Chen, W., Shi, J., Chen, J., Liu, G., Chen, X., and Qian, X. (2024). Self-oscillating polymeric refrigerator with high energy efficiency. *Nature*, 629, pp. 1041–1046.
46. Li, Y., Li, J., Qi, M., Qiu, C.-W., and Chen, H. (2021). Diffusive nonreciprocity and thermal diode. *Phys. Rev. B*, 103, pp. 014307.
47. Sklan, S. R., Bai, X., Li, B., and Zhang, X. (2016). Detecting thermal cloaks via transient effects. *Sci. Rep.*, 6, pp. 32915.
48. Bagnall, K. R. and Wang, E. N. (2018). Theory of thermal time constants in GaN high-electron-mobility transistors. *IEEE Tran. Comp. Pack. Man.*, 8, pp. 606–620.
49. Li, Y., Peng, Y.-G., Han, L., Miri, M.-A., Li, W., Xiao, M., Zhu, X.-F., Zhao, J., Alù, A., Fan, S., and Qiu, C.-W. (2019). Anti-parity-time symmetry in diffusive systems. *Science*, 364, pp. 170–173.

50. Cao, P.-C., Ju, R., Wang, D., Qi, M., Liu, Y.-K., Peng, Y.-G., Chen, H., Zhu, X.-F., and Li, Y. (2024). Observation of parity-time symmetry in diffusive systems. *Sci. Adv.*, 10, pp. eadn1746.
51. Xu, G., Li, Y., Li, W., Fan, S., and Qiu, C.-W. (2021). Configurable phase transitions in a topological thermal material. *Phys. Rev. Lett.*, 127, pp. 105901.
52. Yoshida, T. and Hatsugai, Y. (2021). Bulk-edge correspondence of classical diffusion phenomena. *Sci. Rep.*, 11, pp. 888.
53. Su, W. P., Schrieffer, J. R., and Heeger, A. J. (1979). Solitons in polyacetylene. *Phys. Rev. Lett.*, 42, pp. 1698–1701.
54. Hu, H., Han, S., Yang, Y., Liu, D., Xue, H., Liu, G.-G., Cheng, Z., Wang, Q. J., Zhang, S., Zhang, B., and Luo, Y. (2022). Observation of topological edge states in thermal diffusion. *Adv. Mater.*, 34, pp. 2202257.
55. Qi, M., Wang, D., Cao, P.-C., Zhu, X.-F., Qiu, C.-W., Chen, H., and Li, Y. (2022). Geometric phase and localized heat diffusion. *Adv. Mater.*, 34, pp. 2202241.
56. Wu, H., Hu, H., Wang, X., Xu, Z., Zhang, B., Wang, Q. J., Zheng, Y., Zhang, J., Cui, T. J., and Luo, Y. (2023). Higher-order topological states in thermal diffusion. *Adv. Mater.*, 35, pp. 2210825.
57. Liu, Z., Cao, P.-C., Xu, L., Xu, G., Li, Y., and Huang, J. (2024). Higher-order topological in-bulk corner state in pure diffusion systems. *Phys. Rev. Lett.*, 132, pp. 176302.
58. Cao, P.-C., Li, Y., Peng, Y.-G., Qi, M., Huang, W.-X., Li, P.-Q., and Zhu, X.-F. (2021). Diffusive skin effect and topological heat funneling. *Commun. Phys.*, 4, pp. 230.
59. Liu, Y.-K., Cao, P.-C., Qi, M., Huang, Q.-K.-L., Gao, F., Peng, Y.-G., Li, Y., and Zhu, X.-F. (2024). Observation of non-Hermitian skin effect in thermal diffusion. *Sci. Bull.*, 69, pp. 1228–1236.
60. Xu, G., Yang, Y., Zhou, X., Chen, H., Alù, A., and Qiu, C.-W. (2022). Diffusive topological transport in spatiotemporal thermal lattices. *Nat. Phys.*, 18, pp. 450–456.
61. Xu, G., Zhou, X., Yang, S., Wu, J., and Qiu, C.-W. (2023). Observation of bulk quadrupole in topological heat transport. *Nat. Commun.*, 14, pp. 3252.
62. Huang, J.-P. (2020). *Theoretical Thermotics: Transformation Thermotics and Extended Theories for Thermal Metamaterials.* Springer.
63. Xu, L.-J., and Huang, J.-P. (2023). *Transformation Thermotics and Extended Theories.* Springer.
64. Yang, F.-B. and Huang, J.-P. (2024). *Diffusionics: Diffusion Process Controlled by Diffusion Metamaterials.* Springer.

65. Yeung, W.-S. and Yang, R.-J. (2021). *Introduction to Thermal Cloaking: Theory and Analysis in Conduction and Convection.* Springer.
66. Li, Y., Li, W., Han, T., Zheng, X., Li, J., Li, B., Fan, S., and Qiu, C.-W. (2021). Transforming heat transfer with thermal metamaterials and devices. *Nat. Rev. Mater.*, 6, pp. 488–507.
67. Yang, S., Wang, J., Dai, G., Yang, F., and Huang, J. (2021). Controlling macroscopic heat transfer with thermal metamaterials: Theory, experiment and application. *Phys. Rep.*, 908, pp. 1–65.
68. Fan, C., Wu, C.-L., Wang, Y., Wang, B., and Wang, J. (2024). Thermal metamaterials: From static to dynamic heat manipulation. *Phys. Rep.*, 1077, pp. 1–111.
69. Zhu, C., Bamidele, E.A., Shen, X., Zhu, G., and Li, B. (2024). Machine learning aided design and optimization of thermal metamaterials. *Chem. Rev.*, 124, pp. 4258–4331.
70. Zhang, Z., Xu, L., Qu, T., Lei, M., Lin, Z.-K., Ouyang, X., Jiang, J.-H., and Huang, J. (2023). Diffusion metamaterials. *Nat. Rev. Phys.*, 5, pp. 218–235.
71. Yang, F., Zhang, Z., Xu, L., Liu, Z., Jin, P., Zhuang, P., Lei, M., Liu, J., Jiang, J.-H., Ouyang, X., Marchesoni, F., and Huang, J. (2024). Controlling mass and energy diffusion with metamaterials. *Rev. Mod. Phys.*, 96, pp. 015002.
72. Davis, B. L. and Hussein, M. I. (2014). Nanophononic metamaterial: thermal conductivity reduction by local resonance. *Phys. Rev. Lett.*, 112, pp. 055505.
73. Han, H., Potyomina, L. G., Darinskii, A. A., Volz, S., and Kosevich, Y. A. (2014). Phonon interference and thermal conductance reduction in atomic-scale metamaterials. *Phys. Rev. B*, 89, pp. 180301.
74. Ye, Z.-Q. and Cao, B.-Y. (2016). Nanoscale thermal cloaking in graphene via chemical functionalization. *Phys. Chem. Chem. Phys.*, 18, pp. 32952–32961.
75. Loke, D., Skelton, J. M., Chong, T.-C., and Elliott, S. R. (2016). Design of a nanoscale, CMOS-integrable, thermal-guiding structure for boolean-logic and neuromorphic computation. *ACS Appl. Mater. Interfaces*, 8, pp. 34530–34536.
76. Choe, H. S., Prabhakar, R., Wehmeyer, G., Allen, F. I., Lee, W., Jin, L., Li, Y., Yang, P., Qiu, C.-W., Dames, C., Scott, M., Minor, A., Bahk, J.-H., and Wu, J. (2019). Ion write microthermotics: programing thermal metamaterials at the microscale. *Nano Lett.*, 19, pp. 3830–3837.
77. Liu, Y., Cheng, Y., Hu, R., and Luo, X. (2019). Nanoscale thermal cloaking by in-situ annealing silicon membrane. *Phys. Lett. A*, 383, pp. 2296–2301.

78. Xiao, Y., Chen, Q., and Hao, Q. (2021). Inverse thermal design of nanoporous thin films for thermal cloaking. *Mater. Today Phys.*, 21, pp. 100477.
79. Zhang, J., Zhang, H., Li, Y., and Huang, Z. (2024). Mechanism analysis and optimization of dynamic functional response of nanoscale thermal cloak. *Appl. Thermal Eng.*, 236, pp. 121555.
80. Li, Z.-Y., Mellmann, M., Wang, Y., Ma, T.-X., Yan, D., Golub, M. V., Hosseini, S. M., Liu, D., Wei, P., and Zhang, C. (2024). Non-Fourier heat conduction in 2D thermal metamaterials. *Mater. Today Commun.*, 38, pp. 107828.

Part I
Steady-State Heat Transfer

Chapter 2

Transformation Thermotics

2.1. Introduction

Unlike wave propagation, thermal conduction is a diffusion process described by the Laplace equation. Whether transformation optics [1, 2], proposed in 2006, applies to diffusion systems was a big enigma. Professor Jiping Huang's group first overcame this enormous challenge in 2008 by establishing the theory of transformation thermotics and the concept of thermal cloaking [3], facilitating a main branch of metamaterials, i.e., diffusion metamaterials [4–7]. This chapter introduces transformation thermotics and its development.

In the first work in 2008 [3], researchers considered steady and passive heat conduction in an isotropic background and proposed the transformation principle for thermal conductivity (i.e., transformation thermotics). The same year, another group discussed an anisotropic substrate [8]. Due to the similar equations for describing thermal and electric currents, a multiphysics transformation theory was established to manipulate thermal and electric fields simultaneously [9]. At that time, transformation thermotics remained at a theoretical level. The first experimental demonstration of transformation thermotics appeared in 2012, realizing thermal cloaking, concentrating, and rotating [10]. Yet transformation thermotics has still been in a steady state. In 2012, researchers discussed transient heat conduction containing thermal sources [11], which was experimentally verified in the following year [12]. These studies all assumed linear thermal conductivity that is independent of temperature. Since many materials are nonlinear, i.e., dependent on temperature, temperature-dependent transformation thermotics was developed to make functionalities switchable [13, 14]. Besides designing parameters in space, the temporal dimension was also considered to deal with dynamic systems [15–17]. These approaches generally produce anisotropic thermal conductivity. To reduce material anisotropy to isotropy, researchers developed diffusive pseudo-conformal mapping and realized perfect interface matching [18].

Besides thermal conduction, heat convection is another crucial mode induced by mass transfer. These two modes are often considered together because convection usually accompanies conduction. In 2015, researchers proposed a transformation theory to manipulate thermal convection by directly transforming convective velocity [19]. Since the direct control of fluids is not easy, a transformation theory based on porous media was proposed to control convective velocity by the permeability of porous media [20], which was also improved to deal with transient heat convection [21]. Besides porous media, researchers discussed creeping flows directly and proposed a corresponding transformation theory [22, 23]. In addition to conduction and convection, thermal radiation is also a primary way of heat transfer. Since thermal radiation is electromagnetic waves, transformation optics can principally manipulate it [1, 2]. Yet it is not easy to implement in practice due to the complex radiation process. Therefore, researchers considered a simplified radiation model based on the Rosseland diffusion approximation and proposed the principle of transformation multithermotics, i.e., conduction and radiation [24]. Finally, a universal model containing conduction, convection, and radiation was established [25].

Heat transfer can also be coupled with other physical fields. One typical example is thermoelectric transport, where the Seebeck effect makes a temperature gradient generate an electric current. To manipulate thermoelectric transport flexibly, researchers proposed a transformation theory to design the cloaking functionality [26]. Other applications were also developed, such as concentrating [27, 28]. These studies considered linear parameters irrelevant to temperature. Then, temperature-dependent transformation multiphysics was proposed by considering nonlinear parameters, making multiphysics metamaterials ambient-adaptive [29]. In addition to electric fields, thermal fields can also interact with elastic fields, i.e., thermoelastic fields. A transformation theory was established to design thermoelastic cloaking [30]. These findings indicate significant potential in applying the coupling effect between heat transfer and other physical domains.

In what follows, we introduce transformation thermotics from the three perspectives mentioned earlier, i.e., single heat transfer mode (heat conduction), multiple heat transfer modes (heat convection), heat transfer coupled with other physical fields (thermoelectric transport), and a new design method (conformality-assisted tracing). Then, we detail several typical coordinate transformations for realizing cloaking, concentrating, rotating, camouflaging, and rectifying. We finally conclude this chapter and give an outlook on transformation thermotics.

2.2. Transforming Heat Conduction

Transformation thermotics designs material parameters to manipulate heat flux at will. See Fig. 2.1. For example, we can tailor material parameters in real space to guide heat flux to bypass the central region.

We start with the transient heat conduction equation containing thermal sources,

$$\rho C \frac{\partial T}{\partial t} + \nabla \cdot (-\kappa \nabla T) = Q, \tag{2.1}$$

where ρ is mass density, C is heat capacity, κ is thermal conductivity, Q is thermal power density, T is temperature, and t is time. The component form in the Cartesian coordinate system x^i is

$$\rho C \partial_t T + \partial_{x^i}(-\kappa \partial_{x^j} T) = Q. \tag{2.2}$$

We suppose a space transformation for the desired heat flux manipulation. Form invariance suggests that the governing equation in the curvilinear coordinate system u^i should be

$$\rho C \partial_t T + \partial_{u^i}(-\kappa \partial_{u^j} T) = Q, \tag{2.3}$$

which directly replaces x^i and x^j with u^i and u^j. Therefore, identical coordinates no longer represent the same positions due to the

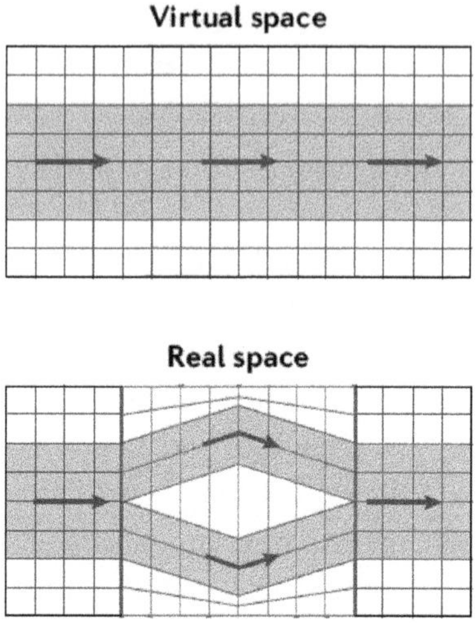

Fig. 2.1. Schematic of transformation thermotics. Adapted from Ref. [6].

transformed basis vectors. We further rewrite the equation from the curvilinear coordinate system to the Cartesian coordinate system,

$$\rho C \partial_t T + \frac{1}{\sqrt{g}} \partial_{x^i}(-\kappa \sqrt{g} g^{ij} \partial_{x^j} T) = Q, \tag{2.4}$$

with the definitions of $J_{ij} = \partial x^i / \partial u^j$, $g^{ij} = (JJ^\dagger)_{ij}$, and $\sqrt{g} = 1/\det J$. The superscript dagger denotes transpose. Identical positions have different coordinates, which are connected by the coordinate transformation $x^i = f(u^j)$. We can further derive

$$\sqrt{g} \rho C \partial_t T + \partial_{x^i}(-\kappa \sqrt{g} g^{ij} \partial_{x^j} T) = \sqrt{g} Q. \tag{2.5}$$

Finally, the transformed parameters (with the superscript prime) in physical space are

$$\rho' C' = \frac{\rho C}{\det J}, \tag{2.6a}$$

$$\kappa' = \frac{J \kappa J^\dagger}{\det J}, \tag{2.6b}$$

$$Q' = \frac{Q}{\det J}. \tag{2.6c}$$

These three equations are the key results of transformation thermotics. Since these parameters have no particular constraints, they can depend on temperature or time.

2.3. Transforming Heat Convection

Thermal convection is more complex than heat conduction because we must consider fluid dynamics, i.e., mass and momentum conservation. We first discuss the energy conservation equation, i.e., the heat convection equation,

$$\rho C \frac{\partial T}{\partial t} + \nabla \cdot (-\kappa \nabla T + \rho C \boldsymbol{v} T) = Q, \tag{2.7}$$

where $\boldsymbol{v}$ is the convective velocity. Following the similar derivation in heat conduction, we can obtain the transformation rule for this convective term,

$$\boldsymbol{v}' = J\boldsymbol{v}. \tag{2.8}$$

Other parameters are transformed as before. Though this equation seems simple, it does not explain the origin of convective velocity. Therefore, we have no idea about how to manipulate convective velocity effectively. To solve this problem, we need to consider more specific cases that describe the origin of convective velocity. We introduce two typical examples in the following sections.

2.3.1. *Porous Media*

Porous media, composed of fluids and solids, are often considered in thermal convection [20, 21]. The Darcy law [31] can effectively describe the convective velocity in this model (Fig. 2.2),

$$\tau \frac{\partial \boldsymbol{v}}{\partial t} + \boldsymbol{v} = -\frac{\beta}{\eta} \nabla P, \tag{2.9}$$

where τ is the characteristic time of the varying convective velocity, β is the permeability of the porous media, η is the dynamic viscosity of the fluids, and P is pressure. In most cases, the relaxation process

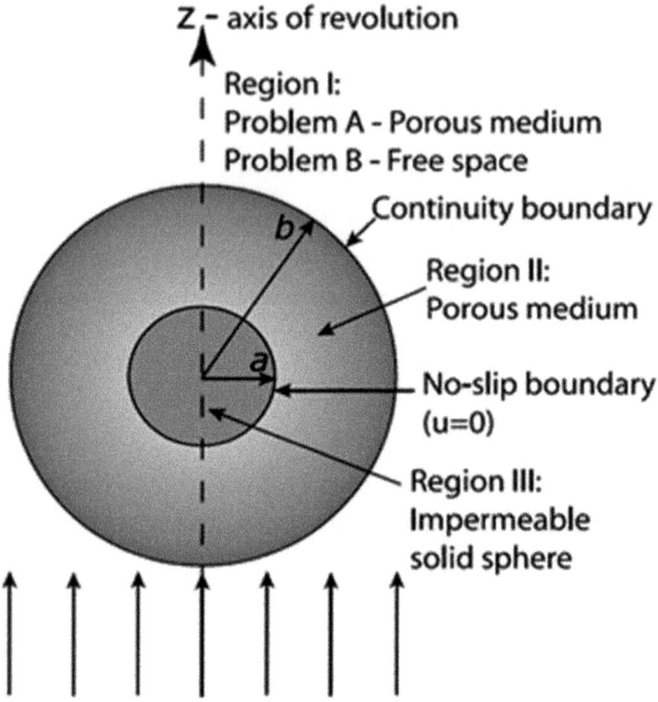

Fig. 2.2. Schematic diagram of a porous medium structure. Adapted from Ref. [31].

is speedy and can be neglected. Therefore, we can express the steady-state Darcy law as

$$\boldsymbol{v} = -\frac{\beta}{\eta}\nabla P. \tag{2.10}$$

We should also consider the mass conservation of the fluids,

$$\frac{\partial(\phi\rho_f)}{\partial t} + \nabla \cdot (\rho_f \boldsymbol{v}) = 0, \tag{2.11}$$

where ρ_f is the mass density of the fluids, and ϕ is the porosity of the porous media. Finally, we can rewrite the heat convection equation as

$$\frac{\partial\left(\rho_m C_m T\right)}{\partial t} + \nabla \cdot (\rho_f C_f \boldsymbol{v} T) = \nabla \cdot (\kappa_m \nabla T), \tag{2.12}$$

where C_f is the heat capacity of the fluids. The weighted average method calculates the product of mass density and heat capacity of the porous media $\rho_m C_m$,

$$\rho_m C_m = (1-\phi)\,\rho_s C_s + \phi \rho_f C_f, \tag{2.13}$$

where ρ_s and C_s are the mass density and heat capacity of the solids, respectively. The weighted average method also applies to the thermal conductivity of the porous media κ_m,

$$\kappa_m = (1-\phi)\,\kappa_s + \phi \kappa_f, \tag{2.14}$$

where κ_s (or κ_f) is the thermal conductivity of the solids (or fluids). Following the similar derivation in heat conduction, we can obtain the transformation principle for the mass conservation equation,

$$\phi' = \frac{\phi}{\det J}, \tag{2.15a}$$

$$\rho_f' = \rho_f, \tag{2.15b}$$

$$\beta' = \frac{J\beta J^\dagger}{\det J}, \tag{2.15c}$$

$$\eta' = \eta. \tag{2.15d}$$

We can also get the transformation rule for the heat convection equation,

$$\rho_m' C_m' = \frac{\rho_m C_m}{\det J}, \tag{2.16a}$$

$$\rho_f' C_f' = \rho_f C_f, \tag{2.16b}$$

$$\kappa_m' = \frac{J\kappa_m J^\dagger}{\det J}, \tag{2.16c}$$

$$\kappa_f' = \kappa_f. \tag{2.16d}$$

According to the weighted average method, we can derive the parameters of the solids,

$$\rho_s' C_s' = \frac{1-\phi' \det J}{(1-\phi')\det J}\rho_s C_s, \tag{2.17a}$$

$$\kappa_s' = \frac{\kappa_m' - \phi' \kappa_f'}{1-\phi'}. \tag{2.17b}$$

We derive all transformation rules for heat convection in porous media. Since we do not transform the parameters of the fluids, this model is more practical to implement because tailoring the parameters of the solids is feasible.

2.3.2. *Pure Fluids*

Porous media are not the only model. Another model was proposed to control creeping viscous potential flows [22, 23]. We discuss transient, incompressible, and creeping flows without body forces (Fig. 2.3). The equations for mass, momentum, and energy conservation are

$$\nabla \cdot \boldsymbol{v} = 0, \tag{2.18a}$$

$$\rho \frac{\partial \boldsymbol{v}}{\partial t} + \nabla P = \nabla \cdot (\eta \nabla \boldsymbol{v}), \tag{2.18b}$$

$$\rho C \frac{\partial T}{\partial t} + \rho C \boldsymbol{v} \cdot \nabla T = \nabla \cdot (\kappa \nabla T), \tag{2.18c}$$

which describes a heat convection model based on pure fluids.

The weak form of the mass conservation equation in the virtual space can be written as

$$\int_\Omega [(\nabla \cdot \boldsymbol{v})\, \varphi]\, dxdydz = 0, \tag{2.19}$$

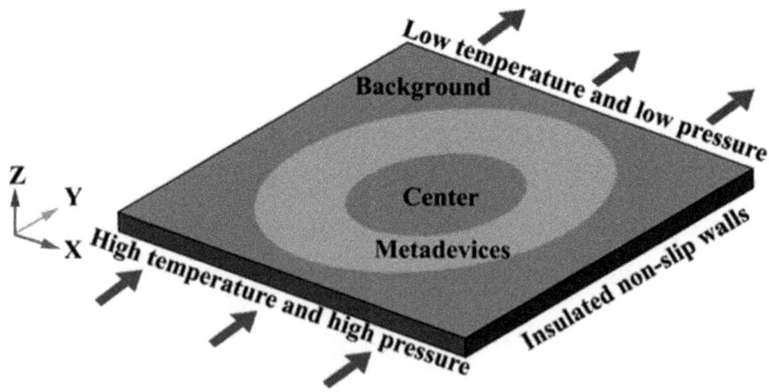

Fig. 2.3. Schematic of a pure fluid structure. Adapted from Ref. [23].

where φ and Ω are the tested function and solution domain. A coordinate transformation turns this equation into

$$\int_{\Omega'} \frac{(\det J\nabla' \cdot \boldsymbol{v}')\,\varphi}{\det J} dx'dy'dz' = 0, \tag{2.20}$$

where the prime superscripts denote the parameters after transformation. For the momentum conservation equation, we have

$$\nabla^2 \boldsymbol{v} = -\nabla \times (\nabla \times \boldsymbol{v}), \tag{2.21}$$

which further leads to

$$\nabla \cdot \nabla^2 \boldsymbol{v} = -\nabla \cdot [\nabla \times (\nabla \times \boldsymbol{v})]. \tag{2.22}$$

We can also rewrite the momentum conservation equation as

$$\frac{\rho}{\eta}\frac{\partial \nabla \cdot \boldsymbol{v}}{\partial t} + \nabla \cdot \left(\frac{1}{\eta}\nabla P\right) = \nabla \cdot \nabla^2 \boldsymbol{v}, \tag{2.23}$$

which can be simplified as

$$\nabla \cdot \left(\frac{1}{\eta}\nabla P\right) = 0, \tag{2.24}$$

whose weak form in virtual space is

$$\int_{\Omega} \left[\nabla \cdot \left(\frac{1}{\eta}\nabla P\right)\right] \varphi dxdydz = 0. \tag{2.25}$$

A coordinate transformation turns this equation into

$$\int_{\Omega'} \left[\nabla' \cdot \left(\frac{J\eta^{-1}J^{\dagger}}{\det J}\nabla' P\right)\right] \varphi\, dx'dy'dz' = 0. \tag{2.26}$$

Finally, we consider the weak form of the energy conservation equation,

$$\int_{\Omega} \left(\rho C\frac{\partial T}{\partial t} + \rho C\boldsymbol{v} \cdot \nabla T\right) \varphi\, d\Omega = \int_{\Omega} [\nabla \cdot (\kappa \nabla T)]\varphi\, d\Omega, \tag{2.27}$$

whose transformed form becomes

$$\int_{\Omega'} \frac{\left(\rho C\frac{\partial T}{\partial t} + \rho C \det J\boldsymbol{v}' \cdot \nabla' T\right)\varphi}{\det J} d\Omega' = \int_{\Omega'} \frac{[J\nabla' \cdot (\kappa J\nabla' T)]\varphi}{\det J} d\Omega'. \tag{2.28}$$

The above discussions can be concluded as

$$\eta'^{-1} = \frac{J\eta^{-1}J^{\dagger}}{\det J}, \tag{2.29a}$$

$$\rho' = \frac{\rho}{\det J}, \tag{2.29b}$$

$$\kappa' = \frac{J\kappa J^{\dagger}}{\det J}. \tag{2.29c}$$

Though this model discusses pure fluids, manipulating these parameters requires compounding other materials like micropillars.

2.4. Transforming Thermoelectric Transport

Heat transfer is often coupled with other physical processes. We discuss a typical example, i.e., thermoelectric transport, where the Seebeck effect makes a temperature gradient generate an electric current [26]. The flux conservation equations for heat and electric currents are

$$\nabla \cdot \boldsymbol{q}_1 = 0, \tag{2.30a}$$

$$\nabla \cdot \boldsymbol{q}_2 = -\nabla\mu \cdot \boldsymbol{q}_1, \tag{2.30b}$$

where $\boldsymbol{q}_1$ is the electric current density, $\boldsymbol{q}_2$ is the heat current density, and μ is the electrochemical potential. The constitutive relations can be expressed as

$$\boldsymbol{q}_1 = -\sigma\nabla\mu - \sigma S\nabla T, \tag{2.31a}$$

$$\boldsymbol{q}_2 = -\kappa\nabla T + TS^{\dagger}\boldsymbol{q}_1 = -\kappa\nabla T - TS^{\dagger}(\sigma\nabla\mu + \sigma S\nabla T), \tag{2.31b}$$

where σ is electric conductivity, and S is the Seebeck coefficient. With a similar derivation in heat conduction, we can express the transformation rules as

$$\sigma' = \frac{J\sigma J^{\dagger}}{\det J}, \tag{2.32a}$$

$$\kappa' = \frac{J\kappa J^{\dagger}}{\det J}, \tag{2.32b}$$

$$S' = (J^{\dagger})^{-1}SJ^{\dagger}. \tag{2.32c}$$

For a common non-tensorial Seebeck coefficient, we can directly derive $S' = S$. We only need to design electric and thermal conductivity. This conclusion is understandable because the uncoupled limit of the Seebeck effect, i.e., $S = 0$, is purely thermal and electric conduction. Hence, the transformation theory for thermoelectric transport must be consistent with thermal and electric conduction.

2.5. Typical Transformations and Applications

We have discussed how to control different heat transfer modes with transformation thermotics. Now, we are in a position to consider specific transformations to realize novel functionalities. See Fig. 2.4. We detail how to design cloaking, concentrating, rotating, camouflaging, and rectifying.

2.5.1. *Cloaking*

Thermal cloaking is a shell that can protect interior objects from being thermally detected (Fig. 2.4a). To this end, we transform a circle into a ring, which can be expressed in the cylindrical coordinate system as

$$r' = \frac{r_2 - r_1}{r_2} r + r_1, \tag{2.33a}$$

$$\theta' = \theta, \tag{2.33b}$$

$$z' = z, \tag{2.33c}$$

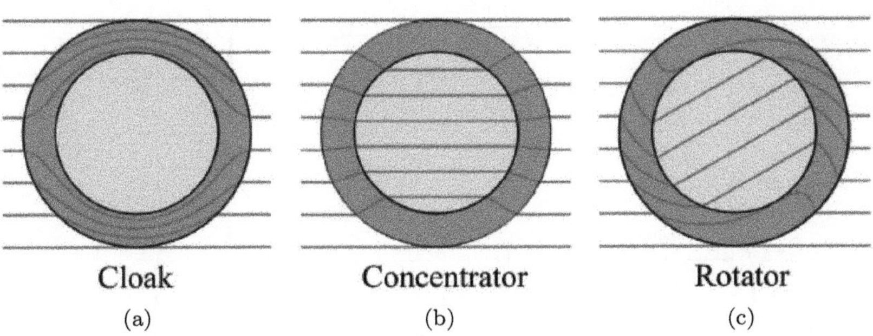

Fig. 2.4. Schematic diagram of thermal (a) cloaking, (b) concentrating, and (c) rotating. Lines denote streamlines. Adapted from Ref. [7].

or in the spherical coordinate system as

$$r' = \frac{r_2 - r_1}{r_2} r + r_1, \tag{2.34a}$$

$$\theta' = \theta, \tag{2.34b}$$

$$\varphi' = \varphi, \tag{2.34c}$$

where r_1 and r_2 are the inner and outer radii of the shell, and the prime superscript denotes the physical space. The transformation matrix in 2D can be expressed as

$$J = \begin{pmatrix} \dfrac{\partial r'}{\partial r} & \dfrac{\partial r'}{r\partial\theta} \\ \dfrac{r'\partial\theta'}{\partial r} & \dfrac{r'\partial\theta'}{r\partial\theta} \end{pmatrix}, \tag{2.35}$$

and that in 3D can be written as

$$J = \begin{pmatrix} \dfrac{\partial r'}{\partial r} & \dfrac{\partial r'}{r\partial\theta} & \dfrac{\partial r'}{r\sin\theta\partial\varphi} \\ \dfrac{r'\partial\theta'}{\partial r} & \dfrac{r'\partial\theta'}{r\partial\theta} & \dfrac{r'\partial\theta'}{r\sin\theta\partial\varphi} \\ \dfrac{r'\sin\theta'\partial\varphi'}{\partial r} & \dfrac{r'\sin\theta'\partial\varphi'}{r\partial\theta} & \dfrac{r'\sin\theta'\partial\varphi'}{r\sin\theta\partial\varphi} \end{pmatrix}. \tag{2.36}$$

Substituting the cloaking transformation into the transformation matrix, we can derive that in 2D,

$$J = \begin{pmatrix} \dfrac{r_2 - r_1}{r_2} & 0 \\ 0 & \dfrac{r_2 - r_1}{r_2}\dfrac{r'}{r' - r_1} \end{pmatrix}, \tag{2.37}$$

and that in 3D,

$$J = \begin{pmatrix} \dfrac{r_2 - r_1}{r_2} & 0 & 0 \\ 0 & \dfrac{r_2 - r_1}{r_2}\dfrac{r'}{r' - r_1} & 0 \\ 0 & 0 & \dfrac{r_2 - r_1}{r_2}\dfrac{r'}{r' - r_1} \end{pmatrix}. \tag{2.38}$$

Take heat conduction as an example. Those three crucial parameters can be expressed in 2D as

$$\rho' C' = \rho C \left(\frac{r_2}{r_2 - r_1} \right)^2 \frac{r' - r_1}{r'}, \tag{2.39a}$$

$$\kappa' = \kappa \begin{pmatrix} \dfrac{r' - r_1}{r'} & 0 \\ 0 & \dfrac{r'}{r' - r_1} \end{pmatrix}, \tag{2.39b}$$

$$Q' = Q \left(\frac{r_2}{r_2 - r_1} \right)^2 \frac{r' - r_1}{r'}, \tag{2.39c}$$

and expressed in 3D as

$$\rho' C' = \rho C \left(\frac{r_2}{r_2 - r_1} \right)^3 \left(\frac{r' - r_1}{r'} \right)^2, \tag{2.40a}$$

$$\kappa' = \kappa \begin{pmatrix} \dfrac{r_2}{r_2 - r_1} \left(\dfrac{r' - r_1}{r'} \right)^2 & 0 & 0 \\ 0 & \dfrac{r_2}{r_2 - r_1} & 0 \\ 0 & 0 & \dfrac{r_2}{r_2 - r_1} \end{pmatrix}, \tag{2.40b}$$

$$Q' = Q \left(\frac{r_2}{r_2 - r_1} \right)^3 \left(\frac{r' - r_1}{r'} \right)^2. \tag{2.40c}$$

Without loss of generality, we demonstrate simulation results in 2D (Fig. 2.5). On the one hand, the temperature gradient in the central region is zero. On the other hand, the temperature distribution in the background is not distorted over time. These features confirm excellent cloaking performance.

2.5.2. *Concentrating*

Unlike thermal cloaking, which makes the central temperature gradient zero, thermal concentrating aims to generate a more significant temperature gradient in the center (Fig. 2.4b). For this purpose, we need two transformations: first, compressing a circle with a radius of r_m into a smaller one with a radius of r_1; second, stretching a ring with inner and outer radii of r_m and r_2 into a bigger one with

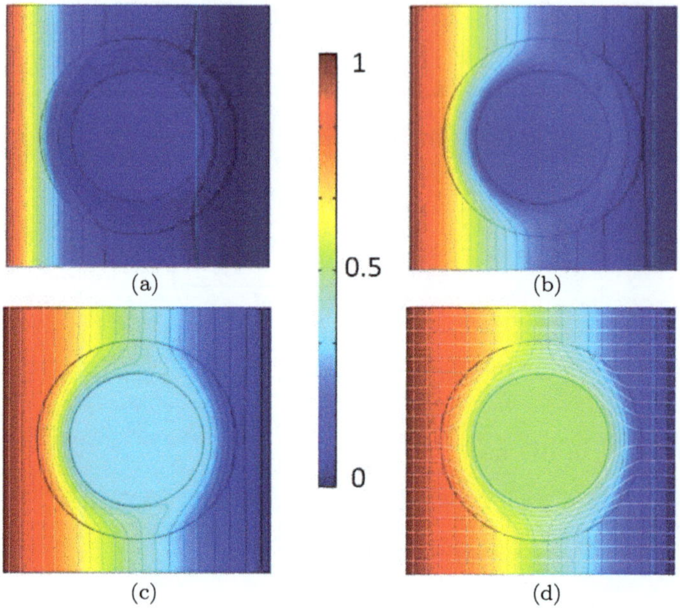

Fig. 2.5. Transient thermal cloaking. (a)–(d) Temperature profiles at 0.001, 0.005, 0.02, and 0.05 s, respectively. Adapted from Ref. [11].

inner and outer radii of r_1 and r_2. These transformations can be described as

$$r' = \frac{r_1}{r_m} r, \quad (r < r_m), \tag{2.41a}$$

$$r' = \frac{r_2 - r_1}{r_2 - r_m} r + \frac{r_1 - r_m}{r_2 - r_m} r_2, \quad (r_m < r < r_2), \tag{2.41b}$$

$$\theta' = \theta, \tag{2.41c}$$

$$z' = z, \tag{2.41d}$$

where we discuss a 2D case. The transformation matrix can be calculated by

$$J = \begin{pmatrix} \frac{r_1}{r_m} & 0 \\ 0 & \frac{r_1}{r_m} \end{pmatrix}, \quad (r' < r_1), \tag{2.42a}$$

$$J = \begin{pmatrix} \dfrac{r_2 - r_1}{r_2 - r_m} & 0 \\ 0 & \dfrac{r_2 - r_1}{r_2 - r_m} \dfrac{r'}{r' - \dfrac{r_1 - r_m}{r_2 - r_m} r_2} \end{pmatrix}, \quad (r_1 < r' < r_2). \tag{2.42b}$$

We consider transient and passive heat conduction. The corresponding parameters can be written as

$$\rho' C' = \rho C \left(\frac{r_m}{r_1} \right)^2, \quad (r' < r_1), \tag{2.43a}$$

$$\kappa' = \kappa, \quad (r' < r_1), \tag{2.43b}$$

$$\rho' C' = \rho C \left(\frac{r_2 - r_m}{r_2 - r_1} \right)^2 \frac{r' - \frac{r_1 - r_m}{r_2 - r_m} r_2}{r'}, \quad (r_1 < r' < r_2), \tag{2.43c}$$

$$\kappa' = \kappa \begin{pmatrix} \dfrac{r' - \frac{r_1 - r_m}{r_2 - r_m} r_2}{r'} & 0 \\ 0 & \dfrac{r'}{r' - \frac{r_1 - r_m}{r_2 - r_m} r_2} \end{pmatrix}, \quad (r_1 < r' < r_2). \tag{2.43d}$$

We also present simulation results in 2D (Fig. 2.6). The temperature gradient in the central region is more significant than in the background. Meanwhile, the temperature profile in the matrix is not disturbed over time. Thus, an excellent concentration effect is achieved.

2.5.3. *Rotating*

Thermal cloaking and concentrating change the magnitude of the temperature gradient. Thermal rotating can manipulate the direction of the temperature gradient (Fig. 2.4c). For this purpose, we rotate the circle with a radius of r_1 by θ_0 rad and the shell with inner and outer radii of r_1 and r_2 linearly, which can be written as

$$r' = r, \tag{2.44a}$$

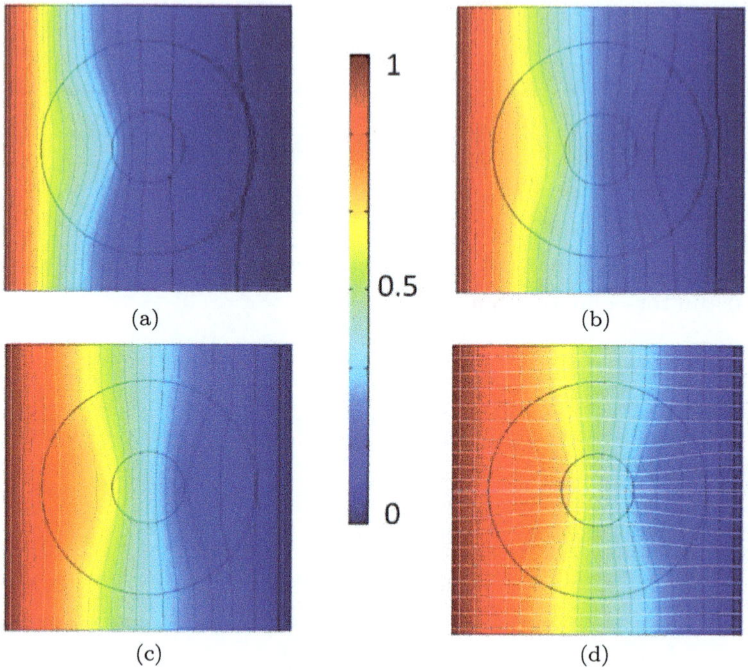

Fig. 2.6. Transient thermal concentrating. (a)–(d) Temperature profiles at 0.002, 0.005, 0.01, and 0.02 s, respectively. Adapted from Ref. [11].

$$\theta' = \theta + \theta_0, \quad (r < r_1), \tag{2.44b}$$

$$\theta' = \theta + \frac{\theta_0(r - r_2)}{r_1 - r_2}, \quad (r_1 < r < r_2), \tag{2.44c}$$

$$z' = z. \tag{2.44d}$$

The transformation matrix can be calculated by

$$J = \begin{pmatrix} 1 & 0 \\ 0 & 1 \end{pmatrix}, \quad (r' < r_1), \tag{2.45a}$$

$$J = \begin{pmatrix} 1 & 0 \\ \dfrac{r'\theta_0}{r_1 - r_2} & 1 \end{pmatrix}, \quad (r_1 < r' < r_2). \tag{2.45b}$$

The corresponding parameters are

$$\rho' C' = \rho C, \quad (r' < r_1), \tag{2.46a}$$

$$\kappa' = \kappa, \quad (r' < r_1), \tag{2.46b}$$

$$\rho' C' = \rho C, \quad (r_1 < r' < r_2), \tag{2.46c}$$

$$\kappa' = \kappa \begin{pmatrix} 1 & \dfrac{r'\theta_0}{r_1 - r_2} \\ \dfrac{r'\theta_0}{r_1 - r_2} & 1 + \left(\dfrac{r'\theta_0}{r_1 - r_2}\right)^2 \end{pmatrix}, \quad (r_1 < r' < r_2). \tag{2.46d}$$

We also exhibit simulation results in 2D (Fig. 2.7). The heat flux direction in the center rotates 90°. The temperature evolution in the substrate is undistorted. Hence, the heat flux direction can be flexibly controlled without influencing the background.

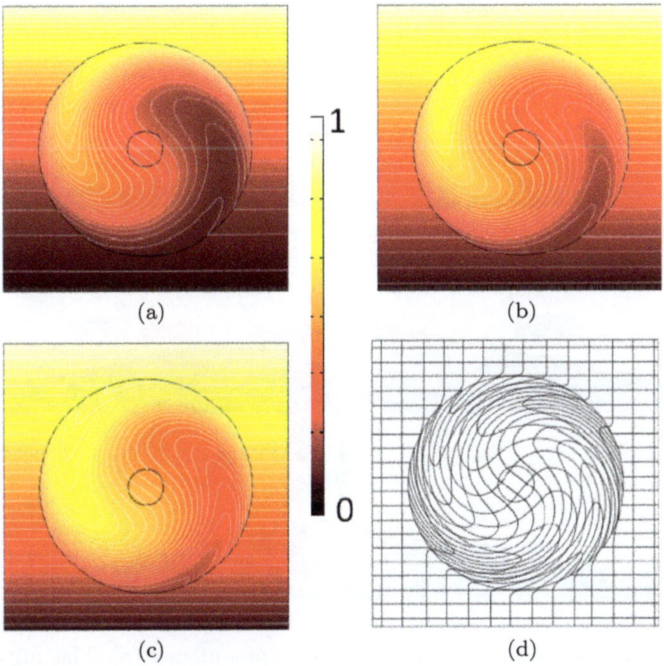

Fig. 2.7. Transient thermal rotating. (a)–(c) Temperature profiles at 0.005, 0.01, and 0.1 s, respectively. (d) Mesh of streamlines and isotherms in the steady-state limit. Adapted from Ref. [32].

2.5.4. *Camouflaging*

Most functionalities need only a one-step transformation. We further introduce a three-step transformation to achieve thermal camouflage [33]. To this end, we start with an initial space (Fig. 2.8a) and then create a new space (Fig. 2.8b). The transformation aims to stretch a small region with height δ (from x_0 to x_1) at the background

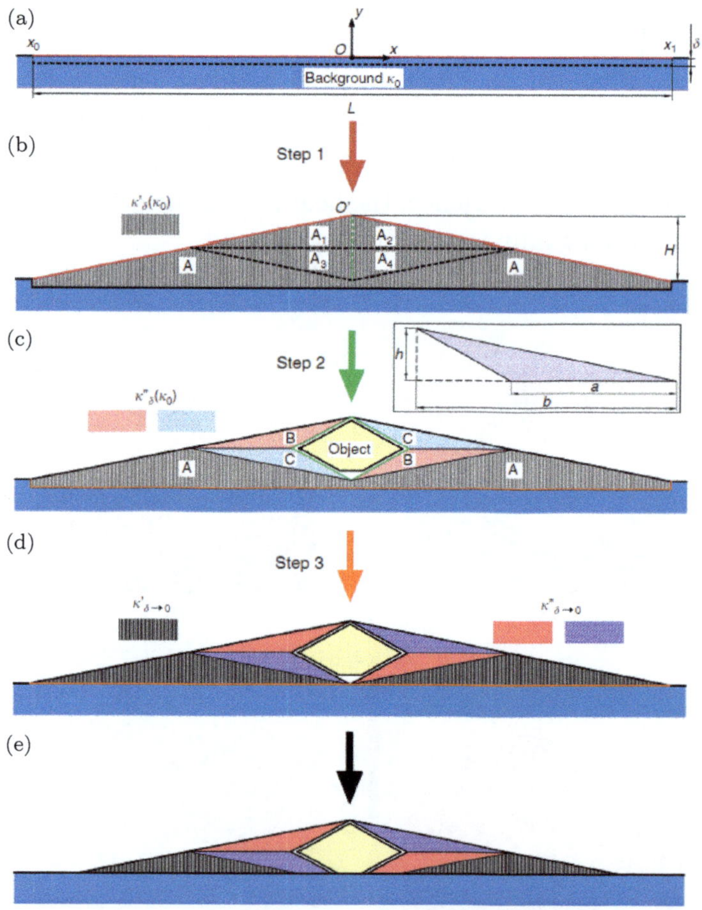

Fig. 2.8. Three steps for thermal radiative camouflage. (a) The initial space. (b) First step for space creation. (c) Second step for unidirectional thermal cloaking (shaded with pink and cerulean lines). (d) Third step for space elimination. (e) Truncate the bottom to allow contact between the background and the object. Adapted from Ref. [33].

surface to construct a wedge (shaded with grey lines), which can be expressed as

$$x' = x, \tag{2.47a}$$

$$y' = \frac{(\delta + y)(L - 2|x|)H}{\delta L} + y, \tag{2.47b}$$

where L and H are the length and height of the wedge. As long as the value of δ is small enough, Eqs. (2.47a) and (2.47b) can produce extremely anisotropic thermal conductivity with a near-zero component along the horizontal direction and a near-infinite component along the vertical direction, thus being independent of the background thermal conductivity.

The second step is creating unidirectional cloaking (Fig. 2.8c). We separate the wedge into six regions, as the dashed lines in Fig. 2.8b present. Region A is undeformed, and the other four areas are transformed based on the standard unidirectional cloaking.

The third step is removing the infinitesimal space by taking the value of δ near zero (Fig. 2.8d). In this way, the final thermal conductivity is unrelated to the thermal conductivity of the background, making this device applicable to any environment.

We may require a final step if an object is directly on the background surface and cannot be moved. In this case, we cut off the bottom of the device to make the object and the background contactable (Fig. 2.8e).

Eventually, we can construct a metasurface for radiative thermal camouflage with these steps. Of course, camouflaging [34] is a widespread concept, including other location camouflaging [35] and path-dependent functionalities [36].

2.5.5. *Rectifying*

The above functionalities remain unchanged if the positions of hot and cold sources exchange. Thermal rectifying aims to change heat flux, which is often challenging due to the symmetric feature of heat conduction. Note that thermal cloaking has zero central heat flux. Thermal rectifying might be realized if we make thermal cloaking switchable (Fig. 2.9).

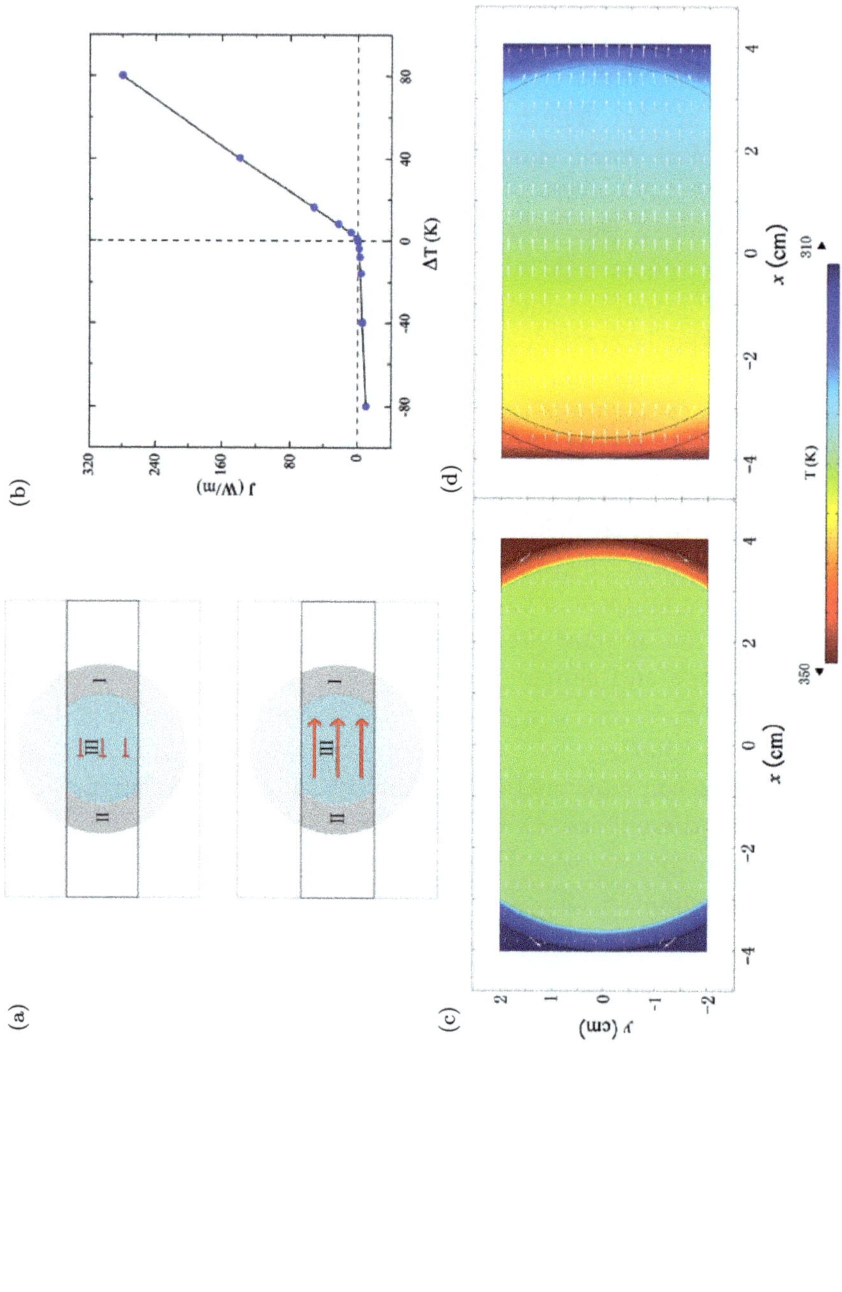

Fig. 2.9. Thermal rectifying. (a) Schematic diagram. (b) Heat flux feature. (c) Insulating case. (d) Conducting case. Adapted from Ref. [13].

For this purpose, we should consider nonlinear thermal materials whose thermal conductivity depends on temperature. Previous work has confirmed that transforming temperature-dependent thermal conductivity equals applying a temperature-dependent transformation [13]. Therefore, we design a 2D and temperature-dependent coordinate transformation,

$$r' = \frac{r_2 - r_1(T)}{r_2} r + r_1(T), \tag{2.48a}$$

$$\theta' = \theta, \tag{2.48b}$$

$$z' = z, \tag{2.48c}$$

with the definitions of $r_1(T) = r_1[1 - (1 + e^{\gamma(T-T_c)})^{-1}]$ for type-A cloak and $r_1(T) = r_1(1+e^{\gamma(T-T_c)})^{-1}$ for type-B cloak. T_c is the critical temperature at which cloaking is switched on or off. γ is a scaling coefficient. Since we consider steady and passive heat conduction, the transformed thermal conductivity can be expressed as

$$\kappa' = \kappa \begin{pmatrix} \dfrac{r' - r_1(T)}{r'} & 0 \\ 0 & \dfrac{r'}{r' - r_1(T)} \end{pmatrix}. \tag{2.49}$$

We combine these two types of thermal cloaks to realize a thermal diode (Fig. 2.9a). Regions I and II are the type-A and type-B cloak segments. Region III is a thermal conductor. The heat flux property of this structure is presented in Fig. 2.9b. When the temperature difference is lower than zero, i.e., low temperature on the left and high temperature on the right, the system almost does not allow heat flux (Fig. 2.9c). In contrast, the system has a large heat flux when the temperature difference is higher than zero (with the same absolute value, Fig. 2.9d). Thermal rectifying is achieved as long as the curve in Fig. 2.9b is not symmetric about the origin.

2.5.6. *Transforming Multiphysics*

The above discussions are about heat transfer control. We further introduce how to use transformation theory to manipulate multiphysics fields. If different physical processes are decoupled, such as

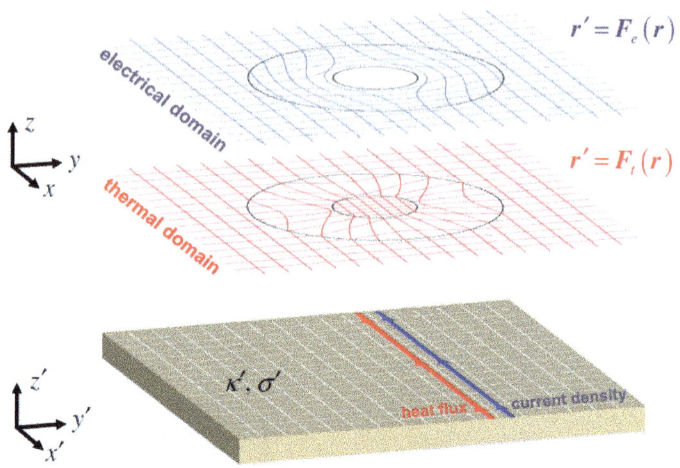

Fig. 2.10. Independent realization of electric cloaking and thermal concentrating on a single device. Adapted from Ref. [37].

electric and thermal transfer without the Seebeck effect, we only need to provide their respective parameters, and the remaining challenge is actual preparation. Since the electric and thermal conductivity of metals are proportional, the functionalities for electric and thermal control are usually identical, e.g., electric cloaking and thermal cloaking [9]. We must perform specific transformations on different physical fields if different functionalities are desired [37]. For example, if we aim to realize electric cloaking and thermal concentrating (Fig. 2.10), the following transformation works,

$$r' = F_e(r), \tag{2.50a}$$

$$r' = F_t(r), \tag{2.50b}$$

$$\theta' = \theta, \tag{2.50c}$$

$$z' = z, \tag{2.50d}$$

where $F_e(r)$ is the cloaking transformation for electric control, the same as Eq. (2.33a), and $F_t(r)$ is the concentrating transformation for thermal control, described by Eqs. (2.41a) and (2.41b). The following processes are identical to those in the *cloaking* and *concentrating* subsections.

2.5.7. *Conformality-assisted Tracing*

The recent development of the conformality-assisted tracing (CAT) method enables the precise design of material parameters for complex 3D structures, overcoming previous geometric constraints [38]. Unlike traditional inverse design techniques, CAT is a quasi-analytical method that utilizes steady-state heat conduction principles to create an orthogonal mesh. By employing a conformal mesh of orthogonal streamlines and isotherms (or isothermal surfaces), this technique quasi-analytically generates free-form metamaterials using only isotropic media, allowing for achieving the desired effects. See Fig. 2.11. By matching the boundary temperature and heat flux, the thermal conductivity of the shell layer $\kappa_{\rm SI}(x_0, y_0)$ can be determined as

$$\frac{\kappa_{\rm SI}(x_0, y_0)}{\kappa_0} = \epsilon(x_0, y_0)\frac{J_{\rm n}(y(x_0, y_0))}{J'_{\rm n}(y(x_0, y_0))}\frac{G'_t(x(x_0, y_0))}{G_t(x(x_0, y_0))}, \tag{2.51}$$

where (x_0, y_0) is coordinates, $\kappa_0\epsilon(x_0, y_0)$ is initial thermal conductivity, $J_{\rm n}$ and G_t are the boundary normal heat flux and tangential temperature gradients of pure background, $J'_{\rm n}$ and G'_t are the boundary normal heat flux and tangential temperature gradients of the shell layer, $y(x_0, y_0)$ [or $x(x_0, y_0)$] is the coordinates of the streamline (or isotherm) endpoints.

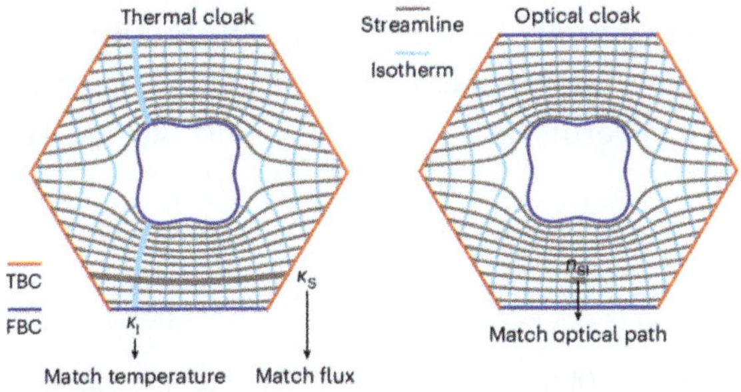

Fig. 2.11. Fundamentals of the conformality-assisted tracing method. Adapted from Ref. [38].

The CAT method has many potential applications, particularly in thermal invisibility. It enables the design of materials that achieve conformal thermal invisibility for arbitrary shapes, significantly improving upon traditional methods that struggle with complex geometries and high computational demands. Beyond thermal invisibility, the CAT method shows promise in designing optical waveguides and achieving thermoelectric transparency. By precisely controlling material parameters, it can guide light or heat in specific ways, opening up new possibilities for device design and functionality. Its versatility extends to various physical fields, such as designing materials for acoustic stealth to make objects invisible to sonar and electromagnetic waveguides to control the precise propagation of electromagnetic waves.

2.6. Conclusion

We have introduced transformation thermotics for dealing with the single heat transfer mode (heat conduction), multiple heat transfer modes (heat convection), heat transfer coupled with other physical domains (thermoelectric transport), and a new design method (conformality-assisted tracing). Transformation thermotics delicately turns the space transformation into the parameter transformation, contributing to revolutionary physical field control. Typical applications are demonstrated based on transformation thermotics, e.g., cloaking, concentrating, rotating, camouflaging, and rectifying. In this chapter, we directly apply ideal transformed parameters. The next chapter will discuss the effective medium approach to realizing these complex parameters. Transformation thermotics is expected to handle more interdisciplinary fields and design more practical functionalities.

References

1. Leonhardt, U. (2006). Optical conformal mapping. *Science*, 312, pp. 1777–1780.
2. Pendry, J. B., Schurig, D., and Smith, D. R. (2006). Controlling electromagnetic fields. *Science*, 312, pp. 1780–1782.

3. Fan, C., Gao, Y., and Huang, J. (2008). Shaped graded materials with an apparent negative thermal conductivity. *Appl. Phys. Lett.*, 92, pp. 251907.
4. Yang, F., Zhang, Z., Xu, L., Liu, Z., Jin, P., Zhuang, P., Lei, M., Liu, J., Jiang, J.-H., Ouyang, X., Marchesoni, F., and Huang, J. (2024). Controlling mass and energy diffusion with metamaterials. *Rev. Mod. Phys.*, 96, pp. 015002.
5. Zhang, Z., Xu, L., Qu, T., Lei, M., Lin, Z.-K., Ouyang, X., Jiang, J.-H., and Huang, J. (2023). Diffusion metamaterials. *Nat. Rev. Phys.*, 5, pp. 218–235.
6. Li, Y., Li, W., Han, T., Zheng, X., Li, J., Li, B., Fan, S., and Qiu, C.-W. (2021). Transforming heat transfer with thermal metamaterials and devices. *Nat. Rev. Mater.*, 6, pp. 488–507.
7. Yang, S., Wang, J., Dai, G., Yang, F., and Huang, J. (2021). Controlling macroscopic heat transfer with thermal metamaterials: Theory, experiment and application. *Phys. Rep.*, 908, pp. 1–65.
8. Chen, T., Weng, C.-N., and Chen, J.-S. (2008). Cloak for curvilinearly anisotropic media in conduction. *Appl. Phys. Lett.*, 93, pp. 114103.
9. Li, J., Gao, Y., and Huang, J. (2010). A bifunctional cloak using transformation media. *J. Appl. Phys.*, 108, pp. 074504.
10. Narayana, S. and Sato, Y. (2012). Heat flux manipulation with engineered thermal materials. *Phys. Rev. Lett.*, 108, pp. 214303.
11. Guenneau, S., Amra, C., and Veynante, D. (2012). Transformation thermodynamics: Cloaking and concentrating heat flux. *Opt. Express*, 20, pp. 8207–8218.
12. Schittny, R., Kadic, M., Guenneau, S., and Wegener, M. (2013). Experiments on transformation thermodynamics: Molding the flow of heat. *Phys. Rev. Lett.*, 110, pp. 195901.
13. Li, Y., Shen, X., Wu, Z., Huang, J., Chen, Y., Ni, Y., and Huang, J. (2015). Temperature-dependent transformation thermotics: From switchable thermal cloaks to macroscopic thermal diodes. *Phys. Rev. Lett.*, 115, pp. 195503.
14. Li, Y., Shen, X., Huang, J., and Ni, Y. (2016). Temperature-dependent transformation thermotics for unsteady states: Switchable concentrator for transient heat flow. *Phys. Lett. A*, 380, pp. 1641–1647.
15. García-Meca, C. and Barceló, C. (2016). Dynamically tunable transformation thermodynamics. *J. Opt.*, 18, pp. 044026.
16. Yang, F., Xu, L., Wang, J., and Huang, J. (2022). Transformation theory for spatiotemporal metamaterials. *Phys. Rev. Appl.*, 18, pp. 034080.
17. Lei, M., Xu, L., and Huang, J. (2023). Spatiotemporal multiphysics metamaterials with continuously adjustable functions. *Mater. Today Phys.*, 34, pp. 101057.

18. Dai, G., Yang, F., Wang, J., Xu, L., and Huang, J. (2023). Diffusive pseudo-conformal mapping: Anisotropy-free transformation thermal media with perfect interface matching. *Chaos Solitons & Fractals*, 174, pp. 113849.
19. Guenneau, S., Petiteau, D., Zerrad, M., Amra, C., and Puvirajesinghe, T. (2015). Transformed Fourier and Fick equations for the control of heat and mass diffusion. *AIP Adv.*, 5, pp. 053404.
20. Dai, G., Shang, J., and Huang, J. (2018). Theory of transformation thermal convection for creeping flow in porous media: Cloaking, concentrating, and camouflage. *Phys. Rev. E*, 97, pp. 022129.
21. Dai, G. and Huang, J. (2018). A transient regime for transforming thermal convection: Cloaking, concentrating and rotating creeping flow and heat flux. *J. Appl. Phys.*, 124, pp. 235103.
22. Wang, B., Shih, T.-M., and Huang, J. (2021). Transformation heat transfer and thermo-hydrodynamic cloaks for creeping flows: Manipulating heat fluxes and fluid flows simultaneously. *Appl. Therm. Eng.*, 190, pp. 116726.
23. Wang, H., Yao, N., Wang, B., and Wang, X. (2024). A transient transformation theory for the design of convective thermal metamaterials: Cloaks, concentrators, and rotators. *Int. J. Heat Mass Transf.*, 221, pp. 125088.
24. Xu, L., Dai, G., and Huang, J. (2020). Transformation multithermotics: Controlling radiation and conduction simultaneously. *Phys. Rev. Appl.*, 13, pp. 024063.
25. Xu, L., Yang, S., Dai, G., and Huang, J. (2020). Transformation omnithermotics: Simultaneous manipulation of three basic modes of heat transfer. *ES Energy & Environ.*, 7, pp. 65–70.
26. Stedman, T. and Woods, L. (2017). Cloaking of thermoelectric transport. *Sci. Rep.*, 7, pp. 6988.
27. Shi, W., Stedman, T., and Woods, L. (2019). Transformation optics for thermoelectric flow. *J. Phys. Energy*, 1, pp. 025002.
28. Shi, W., Stedman, T., and Woods, L. (2020). Thermoelectric transport control with metamaterial composites. *J. Appl. Phys.*, 128, pp. 025104.
29. Lei, M., Wang, J., Dai, G., Tan, P., and Huang, J. (2021). Temperature-dependent transformation multiphysics and ambient-adaptive multiphysical metamaterials. *EPL*, 135, pp. 54003.
30. Tian, Y.-Z., Wang, Y.-F., Huang, G.-Y., Laude, V., and Wang, Y.-S. (2022). Dual-function thermoelastic cloak based on coordinate transformation theory. *Int. J. Heat Mass Transf.*, 195, pp. 123128.
31. Urzhumov, Y. A. and Smith, D. R. (2011). Fluid flow control with transformation media. *Phys. Rev. Lett.*, 107, pp. 074501.

32. Guenneau, S. and Amra, C. (2013). Anisotropic conductivity rotates heat fluxes in transient regimes. *Opt. Express*, 21, pp. 6578–6583.
33. Li, Y., Bai, X., Yang, T., Luo, H., and Qiu, C.-W. (2018). Structured thermal surface for radiative camouflage. *Nat. Commun.*, 9, pp. 273.
34. Hu, R., Xi, W., Liu, Y., Tang, K., Song, J., Luo, X., Wu, J., and Qiu, C.-W. (2021). Thermal camouflaging metamaterials. *Mater. Today*, 45, pp. 120–141.
35. Hu, R., Zhou, S., Li, Y., Lei, D.-Y., Luo, X., and Qiu, C.-W. (2018). Illusion thermotics. *Adv. Mater.*, 30, pp. 1707237.
36. Su, Y., Li, Y., Yang, T., Han, T., Sun, Y., Xiong, J., Wu, L., and Qiu, C.-W. (2020). Path-dependent thermal metadevice beyond Janus functionalities. *Adv. Mater.*, 33, pp. 2003084.
37. Moccia, M., Castaldi, G., Savo, S., Sato, Y., and Galdi, V. (2014). Independent manipulation of heat and electrical current via bifunctional metamaterials. *Phys. Rev. X*, 4, pp. 021025.
38. Xu, L., Dai, G., Yang, F., Liu, J., Zhou, Y., Wang, J., Xu, G., Huang, J., and Qiu, C.-W. (2024). Free-form and multi-physical metamaterials with forward conformality-assisted tracing. *Nat. Comput. Sci.*, 4, pp. 532–541.

Chapter 3

Effective-Medium Methods

3.1. Introduction

This chapter introduces the basis of materials and structures for designing thermal metamaterials. In general, the parameters given by the transformation theory are anisotropic, inhomogeneous, and even singular, which are not met by natural materials. Effective-medium methods offer a practical approach to realizing these specific parameters.

We begin by presenting the Maxwell-Garnett method, which is applicable when one material (inclusion) is embedded within another (host). Subsequently, we introduce the Bruggeman method for analyzing cases involving two inclusions distributed randomly in space. By drawing an analogy to series and parallel models in electrical circuits, we point out that the effective-medium formulas for series and parallel materials can also be derived by setting specific values to variables within the Bruggman formula. We introduce the Rayleigh method as a solution for the effective-medium method concerning periodic composites. After that, the scattering cancellation method is introduced to determine the material parameters for scatterers to form a thermal cloak. Additionally, in the context of 2D structures, we discuss the dimensional reduction method, which allows for the modulation of thermal conductivity in higher dimensions. Finally, we demonstrate the effective-medium method of the convective structure, as convection significantly expands the meaning of effective thermal conductivity, providing a tunable heat control scheme.

This chapter introduces several methods for achieving desired material parameters (especially thermal conductivity). These methods can design various thermal metamaterials, such as thermal cloaks, thermal concentrators, thermal absorbers, and thermal transparency devices.

3.2. Maxwell-Garnett Method

The Maxwell-Garnett method can analyze the structure with inclusions randomly distributed in a host material, as shown in Fig. 3.1. This method can only accurately calculate the effective parameter when the volume fraction of the inclusions is low. We assume the inclusions are so dilute that we can ignore the interaction between two inclusive particles. In this part, we derive the Maxwell-Garnett formula in both linear and nonlinear cases.

3.2.1. *Linear Case*

The system with inclusions randomly located in the host is illustrated in Fig. 3.1, where κ_i and κ_h represent the thermal conductivity of inclusion and host, respectively. The definition of thermal conductivity is

$$\kappa_e = -\langle J_x \rangle / \langle \nabla T_x \rangle , \tag{3.1}$$

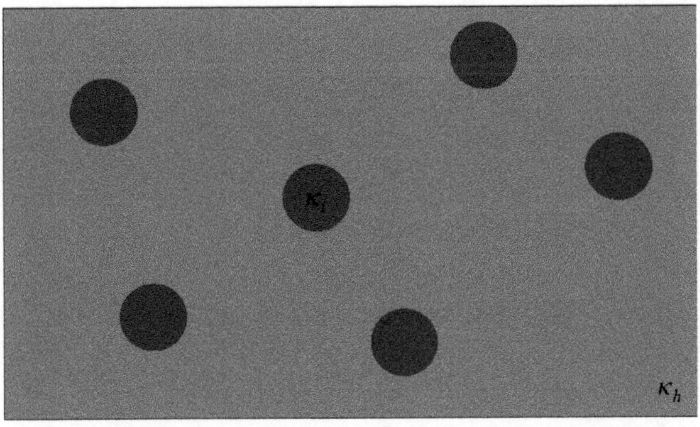

Fig. 3.1. Schematic diagram of typical composites for Maxwell-Garnett formula.

where κ_e, $< J_x >$, and $< \nabla T_x >$ denote effective conductivity, average heat flux vector, and average temperature gradient, respectively. Considering the binary composites material and Eq. (3.1), the 2D and 3D Maxwell-Garnett equations can be written as [7]

$$\begin{aligned} \frac{\kappa_e - \kappa_h}{\kappa_e + \kappa_h} &= f_i \frac{\kappa_i - \kappa_h}{\kappa_i + \kappa_h} \quad \text{two} - \text{dimension}, \\ \frac{\kappa_e - \kappa_h}{\kappa_e + 2\kappa_h} &= f_i \frac{\kappa_i - \kappa_h}{\kappa_i + 2\kappa_h} \quad \text{three} - \text{dimension}, \end{aligned} \tag{3.2}$$

where f_i is the area or volume fraction of the inclusions. Based on Eq. (3.2), we can derive effective conductivity κ_e easily.

3.2.2. *Nonlinear Case*

In the nonlinear case, thermal conductivity is temperature-dependent. By considering 2D composites where either the inclusion or host is nonlinear temperature-dependent materials, the conductivity κ_j can be written as

$$\kappa_j = \kappa_{j0} + \chi_j (T + T_{\text{rt}})^\alpha \quad (j = i, h) \tag{3.3}$$

where κ_{j0} is the linear conductivity term, χ_j represents the nonlinear coefficient, T_{rt} is the reference temperature for measuring nonlinearity, and α can be any real number according to different materials. The subscripts i and h represent inclusion and host material, respectively.

We consider the weak nonlinearity case, namely $\chi_j (T + T_{\text{rt}})^\alpha \ll \kappa_{j0}$ in Eq. (3.3). The effective conductivity can be written as

$$\kappa_e = \kappa_{e0} + \chi_e (T + T_{\text{rt}})^\alpha + O[(T + T_{\text{rt}})^{2\alpha}], \tag{3.4}$$

where κ_{e0} represents the linear solution discussed in the former part, χ_e represents the effective nonlinear coefficient, and $O[(T + T_{\text{rt}})^{2\alpha}]$ denotes the high-order nonlinear term smaller than the former two. With Eq. (3.4), the effective conductivity can be calculated.

To figure out the enhancement or reduction of nonlinearity, we define the nonlinearity enhancement ratio as

$$c = \frac{\chi_e}{\chi_j} (j = i, h), \tag{3.5}$$

which can be calculated by cutting off the series expansions of Eq. (3.4).

When $j = i$ (i.e., only the inclusion is nonlinearity: $\chi_i \neq 0$, $\chi_h = 0$), the nonlinearity enhancement ratio for the Maxwell-Garnett method is

$$c = \frac{\chi_e}{\chi_i} = \frac{4f_i}{\left(1 + \frac{\kappa_{i0}}{\kappa_{h0}} + f_i - f_i\frac{\kappa_{i0}}{\kappa_{h0}}\right)^2}. \tag{3.6}$$

When $j = h$ (i.e., only the inclusion is nonlinearity: $\chi_i = 0$, $\chi_h \neq 0$), the nonlinearity enhancement ratio for the Maxwell-Garnett method is

$$c = \frac{\chi_e}{\chi_h} = \frac{(1 - f_i^2)\left[1 + \left(\frac{\kappa_{i0}}{\kappa_{h0}}\right)^2\right] + 2(1 - f_i)^2\frac{\kappa_{i0}}{\kappa_{h0}}}{\left(1 + \frac{\kappa_{i0}}{\kappa_{h0}} + f_i - f_i\frac{\kappa_{i0}}{\kappa_{h0}}\right)^2}. \tag{3.7}$$

Secondly, we consider the case that both the inclusions and host have strong nonlinearity, i.e., $\chi_j(T + T_{\rm rt})^\alpha \gg \kappa_{j0}$. In this case, Eq. (3.3) can be rewritten as

$$\kappa_j = \kappa_{j0} + \chi_j(T + T_{\rm rt})^\alpha \approx \chi_j(T + T_{\rm rt})^\alpha \quad (j = i, h). \tag{3.8}$$

The effective conductivity is

$$\kappa_e = \chi_e(T + T_{\rm rt})^\alpha. \tag{3.9}$$

The strong nonlinearity coefficients of the Maxwell-Garnett method can be written as

$$\frac{\chi_e - \chi_h}{\chi_e + \chi_h} = f_i\frac{\chi_i - \chi_h}{\chi_i + \chi_h}. \tag{3.10}$$

Notably, Eq. (3.10) has the same form as the 2D linear case in Eq. (3.2) if we replace χ_e, χ_h, and χ_i with κ_e, κ_h, and κ_i, respectively.

3.3. Bruggeman Method

When the medium configuration of the structure is symmetric, the Bruggeman method can accurately estimate the effective material

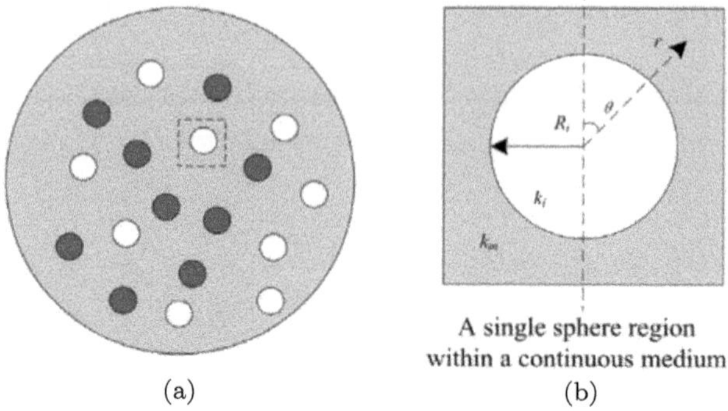

Fig. 3.2. Schematic diagram of composites. (a) Multi-inclusions material. (b) Partial magnification of (a). Adapted from Ref. [6].

parameters. The symmetry medium configuration refers to the random filling of spherical particles (with topological equivalence) throughout the entire space. Fig. 3.2a illustrates this case, where two types of inclusions irregularly fill the material space, and we derive the Bruggeman method to analyze the effective conductivity of this system. Additionally, similar to the circuit equivalent theory, the effective-medium methods for series and parallel configurations can be derived from the Bruggeman theory by simply changing the value of a variable.

3.3.1. *Linear Case*

A typical composite consisting of two kinds of particles is illustrated in Fig. 3.2a. We can assume that the inclusion is so dilute that only one sphere is included in the host background, as shown in Fig. 3.2b. κ_i and κ_m represent the thermal conductivity for the ith inclusion and host, respectively. Starting from the Laplace Equation with boundary conditions, we can derive a relation for the 3D effective-medium equation [6],

$$\sum f_i \left(\frac{\kappa_i - \kappa_e}{\kappa_i + 2\kappa_m} \right) = 0, \tag{3.11}$$

where f_i denotes the volume fraction of the ith type of inclusions, and κ denotes the thermal conductivity with subscriptions i, e, m representing the ith type of inclusions, effective medium, and host, respectively.

Considering a two-phase system, Eq. (3.11) can be simplified as

$$f_1\left(\frac{\kappa_1-\kappa_e}{\kappa_1+2\kappa_m}\right)+f_2\left(\frac{\kappa_2-\kappa_e}{\kappa_2+2\kappa_m}\right)=0. \tag{3.12}$$

It is important to note that Eq. (3.12) can take various forms depending on the value of κ_m. With $\kappa_m = \kappa_e$, Eq. (3.12) corresponds to the Bruggeman equation for binary composites, as inclusions 1 and 2 are randomly distributed. When $\kappa_m = 0$, Eq. (3.12) represents the series model, as heat flux between inclusions 1 and 2 equals zero. When $\kappa_m = \infty$, Eq. (3.12) corresponds to the parallel model, as heat preferentially transfers through inclusion 1 to inclusion 2. The expressions for these cases are listed below,

$$\begin{aligned} f_1\left(\frac{\kappa_1-\kappa_e}{\kappa_1+2\kappa_e}\right)+f_2\left(\frac{\kappa_2-\kappa_e}{\kappa_2+2\kappa_e}\right)&=0 && \text{Bruggeman Equation,}\\ \kappa_e&=\frac{1}{f_1/\kappa_1+f_2/\kappa_2} && \text{Series Model,}\\ \kappa_e&=f_1\kappa_1+f_2\kappa_2 && \text{Parallel Model.} \end{aligned} \tag{3.13}$$

Additionally, for the 2D case, the Bruggeman formula writes

$$f_1\left(\frac{\kappa_1-\kappa_e}{\kappa_1+\kappa_e}\right)+f_2\left(\frac{\kappa_2-\kappa_e}{\kappa_2+\kappa_e}\right)=0. \tag{3.14}$$

3.3.2. *Nonlinear Case*

The nonlinear case for the Bruggeman method is similar to the Maxwell-Garnett method. The nonlinear thermal conductivity is expressed in Eq. (3.3), and the effective nonlinear conductivity is shown in Eq. (3.4). We rewrite them as follows,

$$\begin{aligned} \kappa_j &= \kappa_{j0}+\chi_j(T+T_{\text{rt}})^\alpha (j=i,h),\\ \kappa_e &= \kappa_{e0}+\chi_e(T+T_{\text{rt}})^\alpha + O[(T+T_{\text{rt}})^{2\alpha}], \end{aligned} \tag{3.15}$$

where $j = i, h$ represents two distinct types of materials, rather than referring to the inclusion and the host as in Eq. (3.3). We also firstly consider the weak nonlinearity case as $\chi_j(T+T_{\text{rt}})^\alpha \ll \kappa_{j0}$. When

$j = 1$ (i.e., only the type 'i' material is nonlinearity: $\chi_i \neq 0$, $\chi_h = 0$), the nonlinearity enhancement ratio for the Bruggeman method is

$$c = \frac{\chi_e}{\chi_i} = \frac{1}{2}\left[\frac{(2f_i - 1)\left(2f_i - 2f_i\dfrac{\kappa_{h0}}{\kappa_{i0}} - 1 + \dfrac{\kappa_{h0}}{\kappa_{i0}}\right) + 2\dfrac{\kappa_{h0}}{\kappa_{i0}}}{\sqrt{\left(2f_i - 2f_i\dfrac{\kappa_{h0}}{\kappa_{i0}} - 1 + \dfrac{\kappa_{h0}}{\kappa_{i0}}\right)^2 + 4\dfrac{\kappa_{h0}}{\kappa_{i0}}}} + 2f_i - 1\right]. \tag{3.16}$$

When $j = h$ (i.e., only the type 'h' material is nonlinearity: $\chi_i = 0$, $\chi_h \neq 0$), the nonlinearity enhancement ratio for the Bruggeman method is

$$c = \frac{\chi_e}{\chi_h} = \frac{1}{2}\left[\frac{(2f_i - 1)\left(2f_i - 2f_i\dfrac{\kappa_{i0}}{\kappa_{h0}} - 1 + \dfrac{\kappa_{i0}}{\kappa_{h0}}\right) + 2\dfrac{\kappa_{i0}}{\kappa_{h0}}}{\sqrt{\left(2f_i - 2f_i\dfrac{\kappa_{i0}}{\kappa_{h0}} - 1 + \dfrac{\kappa_{i0}}{\kappa_{h0}}\right)^2 + 4\dfrac{\kappa_{i0}}{\kappa_{h0}}}} + 2f_h - 1\right]. \tag{3.17}$$

Secondly, in the case of strong nonlinearity, i.e., $\chi_j(T + T_{\mathrm{rt}})^\alpha \gg \kappa_{j0}$, the effective conductivity takes the same form as Eq. (3.9). Therefore, the effective nonlinear coefficients in the Bruggeman method for strong nonlinearity is

$$f_1\left(\frac{\chi_i - \chi_e}{\chi_i + 2\chi_e}\right) + f_2\left(\frac{\chi_h - \chi_e}{\chi_h + 2\chi_e}\right) = 0. \tag{3.18}$$

We can see that Eq. (3.18) has the same form as the Bruggeman method in Eq. (3.13) if we substitute χ_e, χ_h, and χ_i with κ_e, κ_2, and κ_1, respectively.

3.4. Rayleigh Method

The Rayleigh method is a first-principles approach to predict effective conductivity, particularly for periodic composites. In contrast to the approximate assumption of non-interaction between every two particles in the Maxwell-Garnett method, the Rayleigh method explicitly considers these interactions. Consequently, the Rayleigh method can accurately predict effective-medium parameters even in cases where the volume fraction of inclusions is high.

3.4.1. *Linear Case*

Let us begin with a system where inclusions occupy a central region within selected composite cells. We can represent this system using Laplace's equation in polar coordinates (ρ,θ). The general solutions for the temperature of the inclusion (T_i) and the host (T_h) can be expressed as follows,

$$T_i(\rho, \theta) = C_{00} + \sum_{m=1}^{\infty} C_{0m}^2 \rho^m \cos(m\theta) + C_{0m}^1 \rho^m \sin(m\theta), \quad (3.19)$$

$$T_h(\rho, \theta) = A_{00} + \sum_{m=1}^{\infty} [(A_{0m}^2 \rho^m + B_{0m}^2 \rho^{-m}) \cos(m\theta)$$
$$+ (A_{0m}^1 \rho^m + B_{0m}^1 \rho^{-m}) \sin(m\theta)], \quad (3.20)$$

where A, B, and C are constant coefficients. No ρ^{-m} term exists in Eq. (3.19) because the temperature in the position (0,0) must be finite.

Now, we use another way to describe the host temperature, which can be written as

$$T_h = \sum_{m-1}^{\infty} \rho^{-m} [B_{0m}^2 \cos(m\theta) + B_{0m}^1 \sin(m\theta)]$$
$$+ \sum_{k=1}^{\infty} \sum_{m=1}^{\infty} \rho_k^{-m} [B_{0m}^2 \cos(m\theta_k) + B_{0m}^1 \sin(m\theta_k)] - \frac{T_L - T_R}{L} x. \quad (3.21)$$

T_L and T_R represent the temperature at the left and right boundaries, and L is the distance between the two boundaries. The notation (ρ_k,θ_κ) signifies the distance from the kth cell ($k > 0$) to the 0th cell (the cell under study) in polar coordinates. The first term in Eq. (3.21) corresponds to the scattering effects attributed to the inclusion within the studied cell. The second term represents the influence from the kth cell, while the third term represents the incident temperature gradient from an external heat source. Equations (3.20) and (3.21) must equal. Hence, we can derive the

Rayleigh identity for periodic composites [4],

$$\begin{aligned} A_{00} + \sum_{m=1}^{\infty} \rho^m [A_{0m}^2 \cos(m\theta) + A_{0m}^1 \sin(m\theta)] \\ = \sum_{k=1}^{\infty} \sum_{m=1}^{\infty} \rho_k^{-m} [B_{0m}^2 \cos(m\theta_k) + B_{0m}^1 \sin(m\theta_k)] - \tfrac{T_L - T_R}{L} x. \end{aligned} \tag{3.22}$$

Based on Rayleigh identity and boundary conditions, we can derive the linear effective conductivity [1],

$$\kappa_e = \kappa_h \frac{(-\beta_1 + \beta_1 f_i + f_i^4)\kappa_h^2 - 2(\beta_1 + f_i^4)\kappa_h \kappa_i + (-\beta_1 - \beta_1 f_i + f_i^4)\kappa_i^2}{(-\beta_1 - \beta_1 f_i + f_i^4)\kappa_h^2 - 2(\beta_1 + f_i^4)\kappa_h \kappa_i + (-\beta_1 + \beta_1 f_i + f_i^4)\kappa_i^2}, \tag{3.23}$$

where $\beta_1 = 3.31248$. f_i, κ_i, and κ_h are the volume fraction, conductivity of the inclusion, and conductivity of the host, respectively.

3.4.2. *Nonlinear Case*

In the case of nonlinearity, thermal conduction varies with temperature, similar to the Maxwell-Garnett and Bruggeman methods. The effective nonlinear conductivity can be expressed as

$$\kappa_e = \kappa_{e0} + \chi_e (T + T_{\mathrm{rt}})^\alpha + O[(T + T_{\mathrm{rt}})^{2\alpha}]. \tag{3.24}$$

For the weak nonlinearity case, $\chi_j (T + T_{\mathrm{rt}})^\alpha \ll \kappa_{j0}$. When $j = i$ (i.e., only the inclusion is nonlinearity: $\chi_i \neq 0$, $\chi_h = 0$), the nonlinearity enhancement ratio for the Rayleigh method is

$$\begin{aligned} c &= \frac{\chi_e}{\chi_i} \\ &= \frac{4\beta_1 f_i \kappa_{h0}^2 [\beta_1 (\kappa_{h0} + \kappa_{i0})^2 + f_i^4 (\kappa_{h0} - \kappa_{i0})^2]}{[\beta_1 (\kappa_{h0} + \kappa_{i0})(f_i \kappa_{h0} - f_i \kappa_{i0} + \kappa_{h0} + \kappa_{i0}) - f_i^4 (\kappa_{h0} - \kappa_{i0})^2]^2}. \end{aligned} \tag{3.25}$$

When $j = h$ (i.e., only the host is nonlinearity: $\chi_i = 0$, $\chi_h \neq 0$), the nonlinearity enhancement ratio for the Rayleigh method is

$$\begin{aligned} c = & \frac{-\beta_1^2 (f_i - 1)(\kappa_{h0} + \kappa_{i0})^2 [(f_i + 1)\kappa_{h0}^2 - 2(f_i - 1)\kappa_{h0}\kappa_{i0} + (f_i + 1)\kappa_{i0}^2]}{[\beta_1 (\kappa_{h0} + \kappa_{i0})(f_i \kappa_{h0} - f_i \kappa_{i0} + \kappa_{h0} + \kappa_{i0}) - f_i^4 (\kappa_{h0} - \kappa_{i0})^2]^2} \\ & + \frac{-2\beta_1 f_i^4 (\kappa_{h0} - \kappa_{i0})^2 [2(f_i + 1)\kappa_{h0}\kappa_{i0} + \kappa_{h0}^2 + \kappa_{i0}^2] + f_i^8 (\kappa_{h0} - \kappa_{i0})^4}{[\beta_1 (\kappa_{h0} + \kappa_{i0})(f_i \kappa_{h0} - f_i \kappa_{i0} + \kappa_{h0} + \kappa_{i0}) - f_i^4 (\kappa_{h0} - \kappa_{i0})^2]^2}. \end{aligned} \tag{3.26}$$

In the case of strong nonlinearity, i.e., $\chi_j(T + T_{\mathrm{rt}})^\alpha \gg \kappa_{j0}$, the effective conductivity takes the same form as in Eq. (3.9). Therefore, the effective nonlinear coefficient for strong nonlinearity in the Rayleigh method is

$$\chi_e = \chi_h \frac{(-\beta_1 + \beta_1 f_i + f_i^4)\chi_h^2 - 2(\beta_1 + f_i^4)\chi_h\chi_i + (-\beta_1 - \beta_1 f_i + f_i^4)\chi_i^2}{(-\beta_1 - \beta_1 f_i + f_i^4)\chi_h^2 - 2(\beta_1 + f_i^4)\chi_h\chi_i + (-\beta_1 + \beta_1 f_i + f_i^4)\chi_i^2}. \tag{3.27}$$

We can see that Eq. (3.27) has the same form as the linear case in Eq. (3.23) if we substitute χ_e, χ_h, and χ_i with κ_e, κ_h, and κ_i, respectively.

3.5. Scattering Cancellation Method

The scattering cancellation method is commonly employed to calculate the appropriate material parameters to minimize scattering from spherical objects. This technique can be used to design a thermal cloak, making small objects completely invisible to heat diffusion waves. A schematic of this spherical thermal cloak is provided in Fig. 3.3.

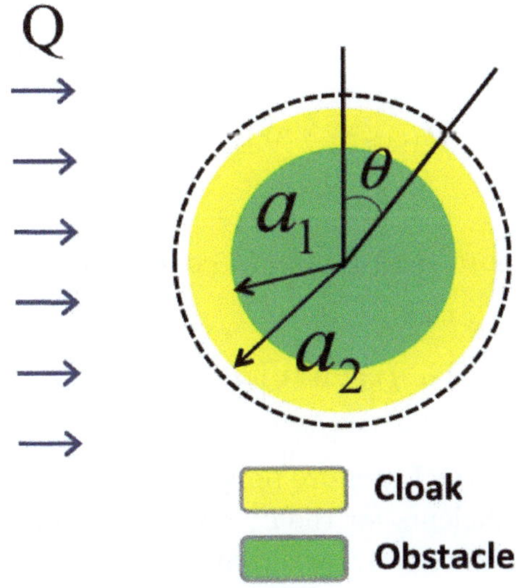

Fig. 3.3. Cross-section view of the cloaked object. Adapted from Ref. [2].

We define the parameters of this system as $[\kappa_i,\ \rho_i,\ c_i,\ k_i]_{i=0,1,2}$ representing the conductivity, fluid density, heat capacity, and wave number in the background medium ($r > a_2$), object ($r < a_1$), and shell ($a_1 < r < a_2$), respectively. This thermal conductivity problem can be effectively addressed using Fourier's law,

$$\rho c \frac{\partial T}{\partial t} = \nabla \cdot (\kappa \nabla T) + Q, \tag{3.28}$$

where ρ, c, T, t, κ and Q represent mass density, specific heat capacity, temperature, time, conductivity, and heat energy generated per unit volume per unit time, respectively. The following part discusses the scattering cancellation conditions in static and time-harmonic regimes.

3.5.1. *Static Regime*

In the static regime, the temperature distribution remains unchanged over time. Then we have $\partial T/\partial t = 0$, and Eq. (3.28) can be simplified to

$$\nabla \cdot (\kappa \nabla T) + Q = 0. \tag{3.29}$$

The temperature field T in the different regions can be expressed in spherical coordinates as

$$T(r,\theta) = \sum_{l=0}^{\infty} A_l r^l P_l(\cos\theta), \quad 0 < r < a_1, \tag{3.30}$$

$$T(r,\theta) = \sum_{l=0}^{\infty} [B_l r^l + C_l r^{-(l+1)}] P_l(\cos\theta), \quad a_1 < r < a_2, \tag{3.31}$$

$$T(r,\theta) = \sum_{l=0}^{\infty} [E_l r^l + D_l r^{-(l+1)}] P_l(\cos\theta), \quad r > a_2, \tag{3.32}$$

where P_l represents the Legendre polynomial of order l and A_l, B_l, C_l, D_l, E_l are the coefficients for the lth order. Given that the scattering effect is negligible for $l > 1$, we simplify the analysis by considering only the case with $l = 0$ and 1. We can determine all coefficients by applying the boundary conditions described in Ref. [3].

The aim of designing this thermal cloak is to ensure that the temperature distribution outside the cloak ($r > a_2$) closely resembles the scenario without any object. By comparing this solution with the general solution for the case without object, we derive the scattering cancellation condition that the coefficient D_1 must be set to zero, which is

$$D_1 = \frac{\widetilde{Q}}{2\kappa_0} \frac{(\kappa_0 - \kappa_2)[\kappa_1 + 2\kappa_2 - \gamma^3(\kappa_1 - \kappa_2)(\kappa_0 + 2\kappa_2)]}{\gamma^3(\kappa_0 - \kappa_2)(\kappa_1 - \kappa_2) - (2\kappa_0 + \kappa_2)(\kappa_1 + 2\kappa_2)} = 0. \quad (3.33)$$

$\widetilde{Q}$ represents the heat generated per unit surface and per unit time, a value influenced by the external heat source, and we denote $\gamma = a_1/a_2$.

Solving Eq. (3.33) for κ_2 allows us to determine the effective conductivity of the cloak, ensuring that the object causes no perturbation. This condition is given by

$$\kappa_2 = \left| \frac{-[(\kappa_0 - 2\kappa_1)\gamma^3 + (2\kappa_0 - \kappa_1)]}{4(\gamma^3 - 1)} \pm \frac{\sqrt{[(\kappa_0 - 2\kappa_1)\gamma^3 + (2\kappa_0 - \kappa_1)]^2 + 8\kappa_0\kappa_1(\gamma^3 - 1)^2}}{4(\gamma^3 - 1)} \right|. \quad (3.34)$$

For some cases, we assume the inner object is thermally insulating ($\kappa_1 = 0$), then Eq. (3.34) can be simplified as [10]

$$\kappa_2 = \frac{\gamma^3\kappa_0 + 2\kappa_0}{2(1 - \gamma^3)}. \quad (3.35)$$

For an analog derivation, the 2D scattering cancellation condition can also be expressed as [8]

$$\kappa_2 = \frac{1 + \gamma^2}{1 - \gamma^2}\kappa_0. \quad (3.36)$$

3.5.2. *Time-harmonic Regime*

In the time-harmonic case, the scattering coefficients connect the scattered fields to the incident ones. In a spherical coordinate system, the incident wave can be represented as an oblique plane diffusion wave in the form of $e^{ikr\cos\theta}$, where k and θ represent the wave number and the incident angle, respectively. The spherical Bessel function can expand this incident plane wave,

$$T^{\text{inc}}(r,\theta) = T_0 \sum_{l=0}^{\infty} i^l (2l+1) J_l(kr) P_l(\cos\theta), \tag{3.37}$$

where T_0 is the amplitude of the incident temperature field, J_l represents the lth order spherical Bessel function of the first kind, and P_l represents the Legendre polynomial of order l. Moreover, the scattered field can be expressed as

$$T^{\text{scat}}(r,\theta) = T_0 \sum_{l=0}^{\infty} i^l (2l+1) s_l H_l(kr) P_l(\cos\theta), \tag{3.38}$$

where s_l is the complex scattering coefficient, and H_l represents the Hankel function of the first kind. Consider a thermal cloak, as shown in Fig. 3.3, the temperature field in different regions can be expressed as

$$T(r,\theta) = T_0 \sum_{l=0}^{\infty} a_l J_l(k_1 r) P_l(\cos\theta), \quad 0 < r < a_1, \tag{3.39}$$

$$T(r,\theta) = T_0 \sum_{l=0}^{\infty} [b_l J_l(k_2 r) + c_l N_l(k_2 r)] P_l(\cos\theta), \quad a_1 < r < a_2, \tag{3.40}$$

$$\begin{aligned} T(r,\theta) &= T^{\text{inc}} + T^{\text{scat}} \\ &= T_0 \sum_{l=0}^{\infty} i^l (2l+1) [J_l(k_O r) + s_l H_l(k_O r)] P_l(\cos\theta), \quad r > a_2. \end{aligned} \tag{3.41}$$

a_l, b_l, and c_l are complex coefficients that require determination, N_l represents the lth-order Bessel function of the second kind, and k_1, k_2, and k_3 are wave numbers in different regions. It is worth noting that Eq. (3.41) differs from the other formulas, as it is represented by the incident and scattered waves.

We can define the scattering amplitude $g(\theta)$ as

$$g(\theta) = rT^{\text{scat}}/(T_0 e^{ik_0 r}) = -\frac{i}{k_0}\sum_{l=0}^{\infty}(2l+1)s_l P_l(\cos\theta)(r \to \infty). \tag{3.42}$$

The scattering cross-section ζ^{scat} can be obtained by integrating the scattering amplitude with incremental solid angle dΩ, which is

$$\zeta^{\text{scat}} = \iint d\Omega |g(\theta)|^2 = \frac{4\pi}{|k_0|^2}\sum_{l=0}^{\infty}(2l+1)|s_l|^2. \tag{3.43}$$

We can assume that only the first two orders of scattering terms contribute to the overall scattering cross-section in the quasistatic limit. Then, we can derive

$$\zeta^{\text{scat}} \approx \frac{4\pi}{|k_0|^2}(|s_0|^2 + 3|s_1|^2). \tag{3.44}$$

Considering the scattering cancellation condition, s_0 and s_1 should be zero to ensure that $\zeta^{\text{scat}} \approx 0$. These conditions depend on the material parameters, which can be determined through the boundary conditions for Eqs. (3.39)–(3.41) [2]. The requirements can be expressed as

$$\frac{\rho_2 c_2 - \rho_0 c_0}{\rho_2 c_2 - \rho_1 c_1} = \left(\frac{a_1}{a_2}\right)^3 = \gamma^3, \quad \text{for } s_0 = 0, \tag{3.45}$$

$$\frac{(\kappa_0 - \kappa_2)(\kappa_1 + 2\kappa_2)}{(\kappa_1 - \kappa_2)(\kappa_0 + 2\kappa_2)} = \left(\frac{a_1}{a_2}\right)^3 = \gamma^3, \quad \text{for } s_1 = 0. \tag{3.46}$$

In comparison to the static case, we can see that we need two distinct sets of variables (ρc and κ) to suppress the scattering wave from the spherical object in the time-harmonic case.

3.6. Dimensional Reduction Method

Passive ultra-conductive thermal metamaterials were facilitated by a vertical thermal transport channel with adjustable height [5]. The continuous effective thermal conductivity can be attained by verifying the height of each channel, enabling a wide range of conductivities. In this context, we present the derivation of the

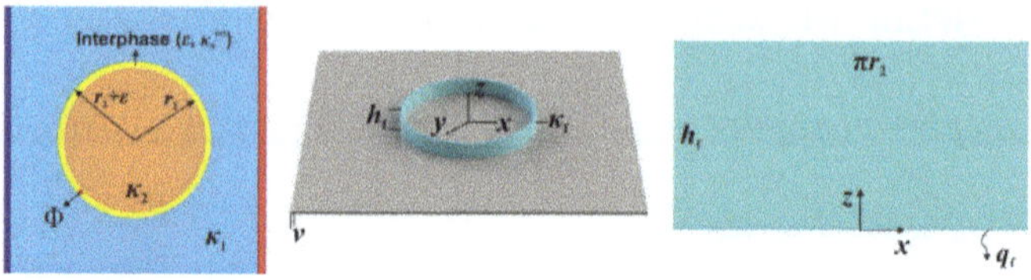

(a) (b) (c)

Fig. 3.4. (a) Schematic of embedded object and interphase. (b) 3D model of the metamaterial. (c) Unfolded channel. Adapted from supporting information of Ref. [5].

relationship between effective conductivity and height. This method can be applied to various 2D systems in pure conduction.

The vertical view of the system is depicted in Fig. 3.4a, featuring an object of conductivity κ_2 embedded within a host medium of conductivity κ_1. In the interphase region ($r_2 < r < r_2 + \varepsilon$), the conductivity is denoted as κ_0^{ori}, representing the original conductivity inside the slab. The thick blue and red lines on the left and right sides symbolize the heat sources. Therefore, the temperature gradient of the background field follows the x-direction. Fig. 3.4b shows a 3D view of this system, with a ring of height h_f located on the interphase region. The unfolded representation of the ring is shown in Fig. 3.4c. The conductivity of this channel is designated as κ_f.

With the thermal boundary conditions, the effective heat source can be expressed as

$$\Phi = \frac{q_f}{v} = M \cos\theta, \tag{3.47}$$

where q_f is the boundary heat flux shown in Fig. 3.4c, and M is a constant. The general solutions for this pseudo-2D system depicted in Fig. 3.4a is

$$T(r,\theta) = \left(-\frac{M}{3\kappa_0^{ori}} r^2 + A_0 r + B_0 r^{-1}\right) \cos\theta \, (r_2 < r \le r_2 + \varepsilon), \tag{3.48}$$

$$T(r,\theta) = (A_1 r + B_1 r^{-1}) \cos\theta \, (r > r_2 + \varepsilon), \tag{3.49}$$

$$T(r,\theta) = A_2 r \cos\theta \, (r \le r_2). \tag{3.50}$$

Now, we suppose that the thermal effect of the interphase part is equivalent to one of a passive conductivity κ_0^{eff}. The temperature solutions for this equivalent system can be written as

$$T^{\text{eff}}(r,\theta) = (C_0 r + D_0 r^{-1})\cos\theta\,(r_2 < r \leq r_2 + \varepsilon), \tag{3.51}$$

$$T^{\text{eff}}(r,\theta) = (C_1 r + D_1 r^{-1})\cos\theta\,(r > r_2 + \varepsilon), \tag{3.52}$$

$$T^{\text{eff}}(r,\theta) = C_2 r\cos\theta\,(r \leq r_2). \tag{3.53}$$

A, B, C, and D in Eqs. (3.48)–(3.53) represent the scaled components of solutions, and these values can be determined by ensuring the continuity of temperature and heat flux boundary conditions. Therefore, the primary focus of this matter is to ascertain the value of M. In comparison between Eqs. (3.49) and (3.52), the outgoing scattering wave in the background region ($r > r_2 + \varepsilon$) must be equal. The condition can be expressed as

$$B_1 = D_1. \tag{3.54}$$

Let us consider the vertical component of this system. The unfolded channel is illustrated in Fig. 3.4c. Apart from the lower boundary, the remaining boundaries are adiabatic. Consequently, the general solution for the system depicted in Fig. 3.4c is

$$T^f = \sum_{n=1}^{\infty} E_n \cos\left(\frac{n}{r_2}x\right)\cosh\left(\frac{n}{r_2}(h_f - z)\right), \tag{3.55}$$

where E_n represents the expanded coefficients, and h_f stands for the height of the interphase. Considering the temperature continuity condition between Eqs. (3.49) and (3.55) at the position of $z = 0$, the coefficients E_n can be expressed as

$$E_1 = \frac{Gr_2}{\cosh\left(\frac{h_f}{r_2}\right)}, \quad E_n = 0\,(n \neq 1), \tag{3.56}$$

and therefore we have

$$T^f = \frac{Gr_2}{\cosh\left(\dfrac{h_f}{r_2}\right)}\cos\left(\frac{x}{r_2}\right)\cosh\left(\frac{h_f - z}{r_2}\right). \tag{3.57}$$

The equation $q_f = -\kappa_f(\partial T^f/\partial z|_{z=0})$ obtains the boundary heat flux. Combining with Eq. (3.47), the constant M can be calculated as

$$M = -\frac{G\kappa_f}{v}\tanh\left(\frac{h_f}{r_2}\right). \tag{3.58}$$

Finally, by combining Eqs. (3.54), (3.58), and the boundary conditions of Eqs. (3.48)–(3.53), the effective thermal conductivity κ_0^{eff} is determined as follows,

$$\kappa_0^{\mathrm{eff}} = -\frac{Cp_1 + p_2 + \sqrt{C^2 p_3 + Cp_4 + p_5}}{p_6}, \tag{3.59}$$

with the parameters of

$$\begin{aligned} C &= -[\tanh(h_f/r_2)\kappa_f]/(v\kappa_0^{ori} r_2),\\ p_1 &= d_4 e_1, \quad p_2 = d_4 e_2 + e_3 e_5, \quad p_3 = d_4^2 e_1^2,\\ p_4 &= 4d_3 e_1 e_3 e_4 + 2d_4^2 e_1 e_2 + 2d_4 e_1 e_3 e_5,\\ p_5 &= 4d_3 e_2 e_3 e_4 + d_4^2 e_2^2 + 2d_4 e_2 e_3 e_5 + e_3^2 e_5^2,\\ p_6 &= 2d_3 e_3. \end{aligned} \tag{3.60}$$

The functions of e_1-e_5 in Eq. (3.60) are

$$\begin{aligned} e_1 &= (\kappa_0^{ori})^2 d_1 + \kappa_0^{ori}\kappa_2 d_2, \quad e_2 = 3(\kappa_0^{ori})^2 d_3 + 3\kappa_0^{ori}\kappa_2 d_4,\\ e_3 &= -3\kappa_0^{ori} d_4 - 3\kappa_2 d_3, \quad e_4 = \kappa_2 d_3, \quad e_5 = \kappa_2 d_4. \end{aligned} \tag{3.61}$$

The functions of d_1–d_4 are expressed as

$$\begin{aligned} d_1 &= 4r_2^3(r_2+\varepsilon) - 3r_2^2(r_2+\varepsilon)^2 - (r_2+\varepsilon)^4,\\ d_2 &= -2r_2^3(r_2+\varepsilon) + 3r_2^2(r_2+\varepsilon)^2 - (r_2+\varepsilon)^4,\\ d_3 &= -r_2^2 + (r_2+\varepsilon)^2, \quad d_4 = r_2^2 + (r_2+\varepsilon)^2. \end{aligned} \tag{3.62}$$

Up to this point, we establish a relationship between the height of the interphase channel and its effective thermal conductivity. This method introduces a strategy for efficient and convenient modulation of material conductivities in higher dimensions.

3.7. Convective Method

This section introduces the effective-medium method for systems incorporating moving devices. Thermal convection is introduced in these systems, increasing the efficiency and flexibility of designing

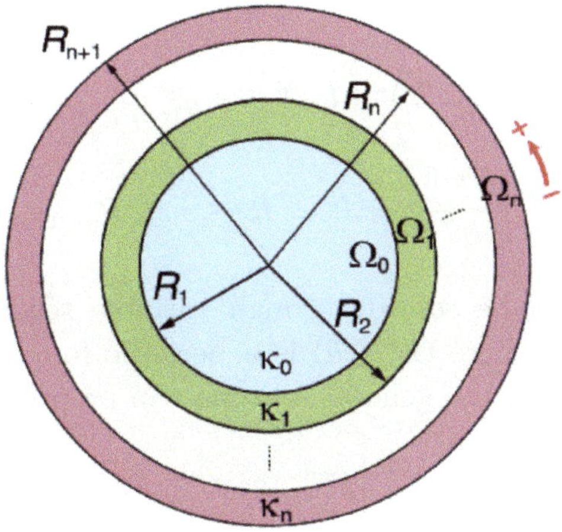

Fig. 3.5. Schematic of a core (blue) surrounded by a shell with n layers. Adapted from Ref. [9].

metamaterials. This approach achieves effects that purely thermal conduction cannot. Here, we demonstrate the effective-medium analysis of the systems with mechanically rotating structures from one layer to multilayers. Thermal convection enriches the interpretation of material parameters and offers expanded possibilities for designing metamaterials.

We investigate the general case of a core-shell structure, as illustrated in Fig. 3.5. This structure consists of n layers, with the kth layer ($k = 1, 2, \ldots, n$) possessing the following properties: rotation angular speed Ω_k, thermal conductivity κ_k, diffusivity D_k, interior radius R_k, and exterior radius R_{k+1}. Due to the rotational motion, the effective conductivity of this structure exhibits anisotropy. The thermal conductivity of this multilayered structure should be expressed as a tensor,

$$\left[\kappa_{1\sim n}^{\text{eff}}\right] = \begin{bmatrix} \kappa_{1\sim n}^{rr} & \kappa_{1\sim n}^{r\theta} \\ \kappa_{1\sim n}^{\theta r} & \kappa_{1\sim n}^{\theta\theta} \end{bmatrix}. \tag{3.63}$$

Assuming that $T_{1\sim n}^{\text{eff}}$ is the temperature field on the nth layer shell with its conductivity tensor $[\kappa_{1\sim n}^{\text{eff}}]$, the heat transfer equation becomes

$$\begin{aligned}&\kappa^{rr}_{1\sim\mathrm{n}}\frac{1}{r}\frac{\partial}{\partial r}\left(r\frac{\partial T^{\mathrm{eff}}_{1\sim\mathrm{n}}}{\partial r}\right)+\left(\kappa^{r\theta}_{1\sim\mathrm{n}}+\kappa^{\theta r}_{1\sim\mathrm{n}}\right)\frac{1}{r}\frac{\partial^2 T^{\mathrm{eff}}_{1\sim\mathrm{n}}}{\partial r\partial\theta}\\&\quad+\kappa^{\theta\theta}_{1\sim\mathrm{n}}\frac{1}{r^2}\frac{\partial^2 T^{\mathrm{eff}}_{1\sim\mathrm{n}}}{\partial\theta^2}=0,\qquad R_1\le r\le R_{n+1}.\end{aligned}\tag{3.64}$$

The equation can be solved through variable separation: $T^{\mathrm{eff}}_{1\sim n} = F^{\mathrm{eff}}_{1\sim n}(r)G^{\mathrm{eff}}_{1\sim n}(\theta)$, where $G^{\mathrm{eff}}_{1\sim n}(\theta)$ is a periodic function. We can set $G^{\mathrm{eff}}_{1\sim n}(\theta) = \mathrm{e}^{i\theta}$, then the general solution can be presented as

$$T^{\mathrm{eff}}_{1\sim n}(r,\theta) = T_{const} + \frac{1}{2}\left[F^{\mathrm{eff}}_{1\sim n}(r)e^{i\theta} + \overline{F^{\mathrm{eff}}_{1\sim n}(r)e^{-i\theta}}\right]. \tag{3.65}$$

Substituting Eq. (3.65) into Eq. (3.64), we can obtain the general solution for the r-dependent part,

$$\begin{aligned}F^{\mathrm{eff}}_{1\sim n}(r) &= z_{1\sim n}r^{u_{1\sim n}} + z_{1\sim n}r^{-\overline{u_{1\sim n}}},\ u_{1\sim n}\\&=\sqrt{\frac{\kappa^{\theta\theta}_{1\sim n}}{\kappa^{rr}_{1\sim n}}-\left(\frac{\kappa^{r\theta}_{1\sim n}+\kappa^{\theta r}_{1\sim n}}{2\kappa^{rr}_{1\sim n}}\right)^2}-i\frac{\kappa^{r\theta}_{1\sim n}+\kappa^{\theta r}_{1\sim n}}{2\kappa^{rr}_{1\sim n}},\end{aligned}\tag{3.66}$$

where $z_{1\sim n}$ represents constant coefficients. The temperature field inside the core (κ_0) and outside background (κ_b) have general solutions of $F_0(r) = z_0 r$ and $F_b(r) = Ar + z_b/r$, respectively. Here, A, z_0, and z_b are the constant coefficients. By canceling the intermediate variables through the application of boundary conditions, we can deduce the matching conditions as

$$\begin{aligned}&(AR_{n+1}+z_bR_{n+1}^{-1})\begin{bmatrix}1\\ \kappa_b\alpha(R_{n+1})\end{bmatrix}\\&\quad=\frac{z_0R_1}{X+\overline{X}}\begin{bmatrix}\overline{X}Y+X\overline{Y}^{-1} & \kappa_0^{-1}(Y-\overline{Y}^{-1})\\ \kappa_0X\overline{X}(Y-\overline{Y}^{-1}) & XY+\overline{XY}^{-1}\end{bmatrix}\begin{bmatrix}1\\ \kappa_0\end{bmatrix},\end{aligned}\tag{3.67}$$

where $X = (\kappa^{rr}_{1\sim n}u + i\kappa^{r\theta}_{1\sim n})/\kappa_0$ and $Y = (R_{n+1}/R_1)^{u_{1\sim n}}(R_{n+1}/R_1)^{u_{1\sim n}}(R_{n+1}/R_1)^{u_{1\sim n}}$ with $\alpha(R_{n+1})$ representing the thermal scattering effect on the outer boundary ($r = R_{n+1}$) of the object. Alternatively, we can represent the effective conductivity using a single value (rather than a tensor). In this scenario, the temperature field is consistent with the rotating model only in the external region, while the internal temperature distribution is different. In this effective-medium strategy, the matching condition can also be calculated and expressed as [9]

$$(AR_{n+1} + z_b R_{n+1}^{-1}) \begin{bmatrix} 1 \\ \kappa_b \alpha(R_{n+1}) \end{bmatrix} = \prod_{k=1}^{n} \frac{z_0 R_1}{t_k} \begin{bmatrix} q_k & -\kappa_k^{-1} p_k \\ \kappa_k s_k & -r_k \end{bmatrix} \begin{bmatrix} 1 \\ \kappa_0 \end{bmatrix}, \tag{3.68}$$

with the parameters of

$$\begin{aligned} p_k &= M(\Omega_k, R_{k+1})N(\Omega_k, R_k) - M(\Omega_k, R_k)N(\Omega_k, R_{k+1}), \\ q_k &= R_k[M(\Omega_k, R_{k+1})N'(\Omega_k, R_k) - M'(\Omega_k, R_k)N(\Omega_k, R_{k+1})], \\ r_k &= R_{k+1}[M'(\Omega_k, R_{k+1})N(\Omega_k, R_k) \\ &\quad - M(\Omega_k, R_k)N'(\Omega_k, R_{k+1})], \\ s_k &= R_k R_{k+1}[M'(\Omega_k, R_{k+1})N'(\Omega_k, R_k) \\ &\quad - M'(\Omega_k, R_k)N'(\Omega_k, R_{k+1})], \\ t_k &= R_k[M(\Omega_k, R_k)N'(\Omega_k, R_k) - M'(\Omega_k, R_k)N(\Omega_k, R_k)]. \end{aligned} \tag{3.69}$$

The functions of $M(\Omega,R)$ and $N(\Omega,R)$ are the r-dependent solutions of the heat transfer equation. These two functions can be expressed as

$$\begin{aligned} M(\Omega, r) &= \begin{cases} I_1\left(\sqrt{i\Omega/D}r\right) & \Omega \neq 0 \\ r & \Omega = 0 \end{cases}, \\ N(\Omega, r) &= \begin{cases} K_1\left(\sqrt{i\Omega/D}r\right) & \Omega \neq 0 \\ 1/r & \Omega = 0 \end{cases}, \end{aligned} \tag{3.70}$$

where I_1 and K_1 represent the first-order modified Bessel function of the first and second kind, respectively. The detailed derivation can be found in Ref. [9]. Both Eqs. (3.67) and (3.68) provide the matching conditions for the same system. Therefore, their solutions for the external temperature field must be identical. By comparing these two functions, we can solve for the variables X and Y as

$$X = \left[(\overline{\lambda}\mu - \lambda\overline{\mu}) \mp \sqrt{(\overline{\lambda}\mu - \lambda\overline{\mu})^2 + 4(\lambda\overline{\lambda} - 1)(\mu\overline{\mu} - 1)}\right] / [2(\lambda\overline{\lambda} - 1)],$$
$$Y = \left[(\overline{\lambda}\mu - \lambda\overline{\mu} + 2) \pm \sqrt{(\overline{\lambda}\mu - \lambda\overline{\mu})^2 + 4(\lambda\overline{\lambda} - 1)(\mu\overline{\mu} - 1)}\right] / [2(\overline{\lambda} + \overline{\mu})], \tag{3.71}$$

where λ and μ are calculated by

$$\begin{bmatrix} \lambda \\ \kappa_0\mu \end{bmatrix} = \prod_{k=1}^{n} \frac{1}{t_k} \begin{bmatrix} q_k & -\kappa_k^{-1}p_k \\ \kappa_k s_k & -r_k \end{bmatrix} \begin{bmatrix} 1 \\ \kappa_c \end{bmatrix}. \tag{3.72}$$

Therefore, the effective-medium tensor of the thermal conductivity in Eq. (3.63) can be finally deduced as

$$\begin{aligned} \kappa_{1\sim n}^{rr} &= \kappa_0 \frac{X + \overline{X}}{\ln Y\overline{Y}} \ln(R_{n+1}/R_1) & \kappa_{1\sim n}^{r\theta} &= i\kappa_0 \frac{\overline{X}\ln Y + X\ln\overline{Y}}{\ln Y\overline{Y}} \\ \kappa_{1\sim n}^{\theta r} &= i\kappa_0 \frac{X\ln Y + \overline{X}\ln\overline{Y}}{\ln Y\overline{Y}} & \kappa_{1\sim n}^{\theta\theta} &= \kappa_0 \frac{X + \overline{X}}{\ln Y\overline{Y}} \frac{(\ln Y)(\ln\overline{Y})}{\ln(R_{n+1}/R_1)} \end{aligned}. \tag{3.73}$$

The conductivity tensor depends on the core parameter κ_0 but remains independent of the background parameter κ_b. In other words, the thermal scattering effect is influenced by the material inside the shell.

References

1. Dai, G. and Huang, J. (2020). Nonlinear thermal conductivity of periodic composites. *Int. J. Heat Mass Transf.*, 147, pp. 118917.
2. Farhat, M., Chen, P.-Y., Bagci, H., Amra, C., Guenneau, S., and Alù, A. (2015). Thermal invisibility based on scattering cancellation and mantle cloaking. *Sci. Rep.*, 5, pp. 9876.
3. Gömöry, F., Solovyov, M., Šouc, J., Navau, C., Prat-Camps, J., and Sanchez, A. (2012). Experimental realization of a magnetic cloak. *Science*, 335, pp. 1466–1468.
4. Gu, G., Yu, K. W., and Hui, P. M. (1998). First-principles approach to conductivity of a nonlinear composite. *Phys. Rev. B*, 58, pp. 3057–3062.
5. Guo, J., Xu, G., Tian, D., Qu, Z., and Qiu, C.-W. (2022). Passive ultra-conductive thermal metamaterials. *Adv. Mater.*, 34, pp. 2200329.

6. Gong, L., Wang, Y., Cheng, X., Zhang, R., and Zhang, H. (2014). A novel effective medium theory for modelling the thermal conductivity of porous materials. *Int. J. Heat Mass Transf.*, 68, pp. 295–298.
7. Huang, J. (2020). *Theoretical Thermotics.* Springer.
8. Han, T., Bai, X., Gao, D., Thong, J. T. L., Li, B., and Qiu, C.-W. (2014). Experimental demonstration of a bilayer thermal cloak. *Phys. Rev. Lett.*, 112, pp. 054302.
9. Li, J., Li, Y., Wang, W., Li, L., and Qiu, C.-W. (2020). Effective medium theory for thermal scattering off rotating structures. *Opt. Express*, 28, pp. 25894.
10. Xu, H., Shi, X., Gao, F., Sun, H., and Zhang, B. (2014). Ultrathin three-dimensional thermal cloak. *Phys. Rev. Lett.*, 112, pp. 054301.

Part II

Time-Harmonic Heat Transfer

Chapter 4

Scattering of Thermal Fields

4.1. Introduction

Recently, it was proposed that heat transfer systems could be a unique platform for studying non-Hermitian physics [1–4], originally used to describe dissipative wave systems [5, 6]. Under this perspective, the effective Hamiltonian is a new tool for designing heat transfer systems and realizing unconventional functionalities. Despite the progress, this method only applies to the time evolution of isolated systems. In many cases, the thermal response of a particular structure to an external excitation needs to be analyzed. The relationship between the input and output thermal fields requires a quantitative description, so many novel applications in wave physics, such as isolators and gyrators, can meet their counterparts in diffusion physics. Therefore, to study the response of a system to external heat sources, another important tool — the scattering theory — is needed.

This chapter presents the thermal scattering theory, a novel method developed to describe the heat transport behavior between the input and output ports of thermal systems. We begin by discussing the general definitions and properties of the thermal scattering matrix. Then, we demonstrate an example where thermal scattering theory can offer valuable insights into the analysis of thermal nonreciprocity. By employing the Green Function approach, we have classified thermal reciprocity into different types and analyzed the feasibility of various methods for controlling heat flow in a preferred direction. Furthermore, we introduce a three-port thermal meta-device based on the concepts discussed earlier, which incorporates both dynamic and steady-state nonreciprocity within a unified framework. After that, we delve into the non-Hermitian scattering symmetry observed in diffusive channels. This analysis reveals the presence of symmetric and symmetry-broken scattering phases, manifesting as the collapse and revival of temperature oscillations. Finally, we demonstrate the thermal analog of coherent perfect absorption, highlighting the fascinating phenomena that arise in thermal scattering. These findings represent a significant advancement in our understanding of thermal phenomena. By integrating these properties, we can achieve enhanced functionality and control in thermal systems.

4.2. General Properties of the Thermal Scattering Matrix

The scattering investigation is of great prominence in physics research because it could reveal fundamental properties about the scatterer, such as its reflectivity, transmittance, and reciprocity. Take the signal transmission process between two points in a physical system. The scattering behavior can be described by a two-order matrix which connects the input signals $(A_1, A_2)^{\mathrm{T}}$ and output signals $(B_1, B_2)^{\mathrm{T}}$, namely $(B_1, B_2)^{\mathrm{T}} = \boldsymbol{S}(A_1, A_2)^{\mathrm{T}}$. In the scattering matrix $\boldsymbol{S} = [S_{11}, S_{12}; S_{21}, S_{22}]$, the off-diagonal components S_{12} and S_{21} represent the transmission coefficients, while the diagonal components S_{11} and S_{22} represent the reflection coefficients.

4.2.1. *Symmetry and Unitarity of the Scattering Matrix*

Nonreciprocity refers to the characteristic of the signal response of the device changing after swapping the source and receiver. As a widely studied property in various physical systems, nonreciprocity offers numerous potential applications, including topologically protected one-way edge states [7, 8], isolations [9, 10], and rectifications [11, 12]. With the scattering matrix, the reciprocal transmission of a multiple-port system can be demonstrated by the symmetry of $\boldsymbol{S}$.

In the wave community, a common assumption is that the waves could travel through the system without loss, which implies unitarity for scattering matrix $\boldsymbol{S}$. Limited by this relation, it is demonstrated that the off-diagonal coefficients of the lossless two-port system can only differ in phase rather than amplitude, namely, the system can only be nonreciprocal in the transmission phase but never in amplitude. Therefore, to create a device with nonreciprocal amplitude transmission for waves, the assumption of unitarity must be relaxed, where the absorption losses should be considered in the physical process. From this perspective, heat transfer describes the extreme case of field evolution in a purely lossy medium, namely the diffusion process. Starting from the parabolic form of Fourier's function, the mathematical solution that describes heat conduction in the 1D case can always be rewritten into its exponential form,

$$T(x) = (A\mathrm{e}^{ikx} + B\mathrm{e}^{-ikx})\mathrm{e}^{-i\omega t} + T_0 \tag{4.1}$$

where Ae^{ikx} and Be^{-ikx} are the incoming and outgoing parts of the thermal field, respectively. Based on this, a group of ill-defined concepts in traditional heat transfer can be revived, and the scattering method can be developed to study heat diffusion behavior between different material points in a system.

Hereto, the general properties of the scattering matrix and the potential applications, such as the analysis of reciprocity, have already been introduced. One must note that, though reciprocity can be characterized as the symmetry of the scattering matrix, it is necessary to distinguish it from the time-reversal symmetry of the physical process. In other words, there is no equivalence between time-reversal symmetry and reciprocity. Take the macroscopic heat conduction process as an example. Its time-reversal case is a process with an increasing temperature gradient, which violates the second law of thermodynamics and is asymmetrical in time. However, although the purely dissipative heat conduction equation is not invariant under time reversal, the fundamental property of reciprocity remains unchanged in linear materials because heat diffusion remains symmetric in space. Such a constraint is believed to be due to the microscopic reversibility or the time-reversal symmetry of heat carrier dynamics, based on which the Onsager-Casimir relation requires that the thermal conductivity is a symmetric tensor.

4.2.2. *Green's Function Method and Global Nonreciprocity*

As mentioned above, a typical application of scattering theory is to analyze the reciprocity between different system ports. Therefore, based on the thermal scattering theory, this section describes using Green's function to determine the global reciprocity of thermal systems.

In general, the scattering coefficients of a system can be obtained using Green's function. By integrating Green's function of the point source, the superposition of the contributions of all the fields at the target point can be obtained accordingly. Then, the amplitude of the outgoing field is also obtainable, which acts as the scattering coefficients. To begin with, we first need to clarify the difference between global reciprocity and local reciprocity. Generally, Green's function for the equation governing heat diffusion in linear materials must be symmetric [13]. Namely, if one swaps the positions of a point heat source and the target point, the corresponding Green's function remains the same.

We refer to this property as the local reciprocity of heat diffusion, as the reference point can be any local point inside the material (Fig. 4.1a). In contrast, global reciprocity describes the transport relationships between system ports (Fig. 4.1b), where the scattering parameters directly correlate the output and input signals. Therefore, from the standpoint of practical application, global reciprocity is much more important than local reciprocity, and Green's function relationship that it satisfies is the main focus of this section.

For wave propagation, the reciprocity of such processes is often defined as the symmetry of the scattering matrix. Similarly, we can study the reciprocity in terms of the relation between field amplitudes at different ports of a diffusive system. Consider a two-port heat transfer system as in Fig. 4.1b. Region V is thermally insulated except for two channels C_1 and C_2 that have uniform and linear

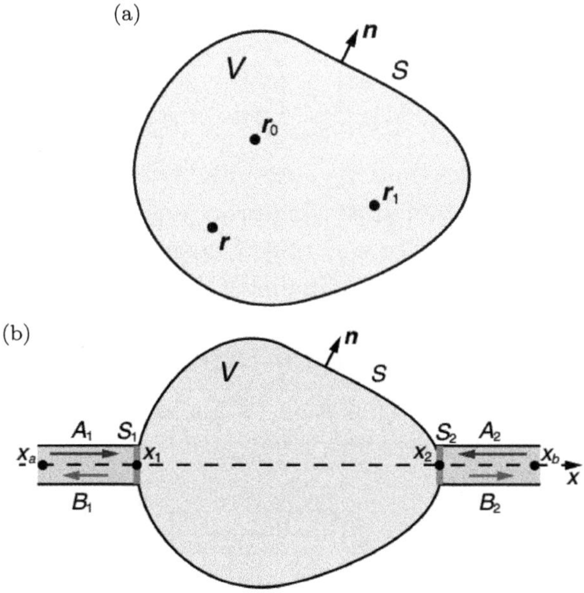

Fig. 4.1. (a) Local reciprocity in heat transfer. (b) Global reciprocity of heat transfer through a two-port system. Adapted from Ref. [26].

material properties. It is also assumed that both channels are long, narrow, and aligned in the lateral x direction, with their upper and lower boundaries insulated. At oscillatory frequency ω, the supported modes are

$$T_c(x,\omega) = e^{\pm ik_c x}, \quad k_c = (1-i)\sqrt{\frac{\omega}{2D_c}} \tag{4.2}$$

where D_c is the diffusivity of the channels. If the interfaces between the left and right ports and the system are at x_1 and x_2, then we can define the fields in the channels as

$$T_c(x,\omega) = \begin{cases} A_1 e^{-ik_c(x-x_1)} + B_1 e^{ik_c(x-x_1)}, x \leq x_1 \\ A_2 e^{ik_c(x-x_2)} + B_2 e^{-ik_c(x-x_2)}, x \geq x_2 \end{cases}. \tag{4.3}$$

Note that the field evolution is obtained by multiplying $T(x,\omega)$ with $e^{i\omega t}$, so A_1 and A_2 are the amplitudes of the incident fields while B_1 and B_2 are the amplitudes of the outgoing fields. They are related through the scattering matrix S,

$$\begin{pmatrix} B_1 \\ B_2 \end{pmatrix} = \mathrm{S} \begin{pmatrix} A_1 \\ A_2 \end{pmatrix} = \begin{pmatrix} r_{11} & t_{12} \\ t_{21} & r_{22} \end{pmatrix} \begin{pmatrix} A_1 \\ A_2 \end{pmatrix} \tag{4.4}$$

where r_{11} and r_{22} (t_{12} and t_{21}) are the reflection (transmission) coefficients. Reciprocity of scattering processes through the system is defined as a symmetric S or $t_{12} = t_{21}$. Consider Green's function for the entire system plus the two channels ($V \cup C_1 \cup C_2$). It satisfies the Neumann boundary condition except for the two ends at $x = \pm L$, and must have a similar form as Eq. (4.3) in the two channels

$$G(x|x_a,\omega) = \begin{cases} A_1(x_a,\omega)\dfrac{e^{-ik_c(x-x_1)} - e^{ik_c x_1}}{1-e^{ik_c x_1}} \\ \quad + B_1(x_a,\omega)\dfrac{e^{ik_c(x-x_1)} - e^{-ik_c x_1}}{1-e^{-ik_c x_1}}, & x_a \le x \le x_1 \\ B_2(x_a,\omega)\dfrac{e^{-ik_c(x-x_2)} - e^{-ik_c(L-x_2)}}{1-e^{-ik_c(L-x_2)}}, & x \ge x_2 \end{cases}$$

$$G(x|x_b,\omega) = \begin{cases} B_1(x_b,\omega)\dfrac{e^{ik_c(x-x_1)} - e^{-ik_c(L+x_1)}}{1-e^{-ik_c(L+x_1)}}, & x \le x_1 \\ A_2(x_b,\omega)\dfrac{e^{ik_c(x-x_2)} - e^{-ik_c x_2}}{1-e^{-ik_c x_2}} \\ \quad + B_2(x_b,\omega)\dfrac{e^{-ik_c(x-x_2)} - e^{ik_c x_2}}{1-e^{ik_c x_2}}, & x_2 \le x \le x_b \end{cases} \tag{4.5}$$

where $x_a \le x_1$ and $x_b \ge x_2$. We adopt this special form to ensure a meaningful linear distribution in the limit of $\omega(k_c) \to 0$, instead of the zero fields for Eq. (4.3). The boundary condition is $G(\pm L|x_a,\omega) = G(\pm L|x_b,\omega) = 0$ for $L \to \infty$. Therefore, we have

$$\begin{aligned} t_{12} &= \frac{B_1(x_b,\omega)}{A_2(x_b,\omega)} = \frac{G(x_1|x_b,\omega)}{A_2(x_b,\omega)} \\ t_{21} &= \frac{B_2(x_a,\omega)}{A_1(x_a,\omega)} = \frac{G(x_2|x_a,\omega)}{A_1(x_a,\omega)} \end{aligned}. \tag{4.6}$$

When the sources are close to the interfaces: $x_a \to x_1$ and $x_b \to x_2$, Eq. (4.6) is

$$\mathrm{t}_{12} = \frac{G(x_1|x_2,\omega)}{A_2(x_2,\omega)}, \quad \mathrm{t}_{21} = \frac{G(x_2|x_1,\omega)}{A_1(x_1,\omega)}. \tag{4.7}$$

The amplitude of an incident field only depends on the source since the reflections at $x = \pm L \to \pm\infty$ are negligible. It is thus reasonable to assume that $A_1(x_1,\omega) = A_2(x_2,\omega) = A(\omega)$, and the reciprocity of the two-port system is equivalent to

$$G(x_1|x_2,\omega) = G(x_2|x_1,\omega). \tag{4.8}$$

Furthermore, it can be proved that the condition of local reciprocity is almost identical to that of global reciprocity, except that the positions of source and target can be any local points inside the material. We note that in the limit of $\omega(k_c) \to 0$, one cannot distinguish the incident, reflected, and transmitted parts of a field since there is no directionality of field propagation. However, the definition of reciprocity in Eq. (4.8) is still meaningful as an extension of the cases with nonzero frequency.

4.2.3. *Dynamic Reciprocity and Steady-State Reciprocity*

For wave propagation, a nonreciprocal two-port system is almost the same as a diode, and both refer to an asymmetric transmission. It is worth noting that for a diode, the incident and outgoing waves under consideration are generally required to be at equal frequency, while it is correct to state that the transmission between two modes at different frequencies is nonreciprocal. Here, we focus on the equal-frequency transmission. In diffusive processes, a diode is usually defined differently. For example, a thermal diode refers to a system in contact with two thermal reservoirs at temperatures T_1 and T_2, and the heat fluxes passing through it must be asymmetric after swapping the values of T_1 and T_2, as shown in Fig. 4.2. Naturally, one would ask whether a nonreciprocal heat transfer system is still equivalent to a thermal diode. We could answer the question by fixing the temperatures of the interfaces S_1 and S_2 at T_1 (T_2).

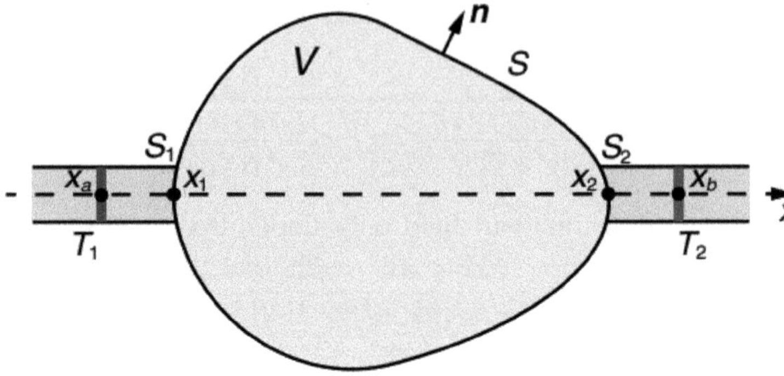

Fig. 4.2. Schematic diagram of a thermal diode. Adapted from Ref. [26].

Only the fields at zero frequency or steady state are of interest for such setups, so we consider the steady-state Green's function. It must be linear in the channels,

$$G(x|x_a,0)=\begin{cases}C_1(x_a)(x+L),\ -L\le x\le x_a\\ D_1(x_a)(x-x_a)+C_1(x_a)(x_a+L), & x_a\le x\le x_1\\ C_2(x_a)(x-L), x_2\le x\le L\end{cases}$$

$$G(x|x_b,0)=\begin{cases}C_1(x_b)(x+L),\ -L\le x\le x_1\\ D_2(x_b)(x-x_b)+C_2(x_b)(x_b-L), & x_2\le x\le x_b\\ C_2(x_b)(x-L), x_b\le x\le L\end{cases} \tag{4.9}$$

where $x=\pm L$ are the positions of the left and right ends of the channels. Again $x_a\le x_1$ and $x_b\ge x_2$. The function satisfies the Dirichlet boundary condition $G(\pm L|x_a,0=G(\pm L|x_b,0=0$. Assuming that the total heat generation $H=\int h(\boldsymbol{r})dV$ in V is independent of the conditions outside, the energy conservation requires that

$$\begin{aligned}-\kappa_c D_1(x_a)\sigma&=-\kappa_c C_2(x_a)\sigma+H\\ -\kappa_c C_1(x_b)\sigma&=-\kappa_c D_2(x_b)\sigma+H\\ -\kappa_c[C_1(x_a)-C_2(x_a)]\sigma&=-\kappa_c[C_1(x_b)-C_2(x_b)]\sigma=Q_0+H\end{aligned} \tag{4.10}$$

where κ_c is thermal conductivity, and σ is the cross-section area of the channels. If the differential operator on the field in region V is linear, then the linear combination $F(x) = f_a G(x|x_a, 0) + f_b G(x|x_b, 0)$ is also a solution to the equation in region V. If we require $F(x_a) = T_1$ and $F(x_b) = T_2$, the coefficients satisfy

$$G\begin{pmatrix} f_a \\ f_b \end{pmatrix} = \begin{pmatrix} G(x_a|x_a, 0) & G(x_a|x_b, 0) \\ G(x_b|x_a, 0) & G(x_b|x_b, 0) \end{pmatrix}\begin{pmatrix} f_a \\ f_b \end{pmatrix} = \begin{pmatrix} T_1 \\ T_2 \end{pmatrix}. \tag{4.11}$$

In the limit $L \to \infty$, the third equation of Eq. (4.10) gives

$$G_{11} + G_{12} = G_{22} + G_{21}. \tag{4.12}$$

The heat fluxes q_1 and q_2 in C_1 and C_2 near the interfaces S_1 and S_2 are

$$\begin{pmatrix} q_1 \\ q_2 \end{pmatrix} = \kappa_c \begin{pmatrix} D_1(x_a) & C_1(x_b) \\ C_2(x_a) & D_2(x_b) \end{pmatrix}\begin{pmatrix} f_a \\ f_b \end{pmatrix} = Q\begin{pmatrix} f_a \\ f_b \end{pmatrix}. \tag{4.13}$$

In the limit $L \to \infty$, we have

$$\begin{aligned} Q_{12} &= \frac{\kappa_c G_{21}}{L} \\ Q_{21} &= -\frac{\kappa_c G_{12}}{L} \end{aligned}. \tag{4.14}$$

The first two equations of Eq. (4.10) give

$$Q_{11} + Q_{22} = Q_{12} + Q_{21}. \tag{4.15}$$

Now, we can swap the values of T_1 and T_2 to solve another set of coefficients,

$$G\begin{pmatrix} \tilde{f}_a \\ \tilde{f}_b \end{pmatrix} = \begin{pmatrix} T_2 \\ T_1 \end{pmatrix} = \begin{pmatrix} 0 & 1 \\ 1 & 0 \end{pmatrix}\begin{pmatrix} T_1 \\ T_2 \end{pmatrix}. \tag{4.16}$$

Therefore, we have

$$\begin{pmatrix} \tilde{f}_a \\ \tilde{f}_b \end{pmatrix} = G^{-1}\begin{pmatrix} 0 & 1 \\ 1 & 0 \end{pmatrix} G \begin{pmatrix} f_a \\ f_b \end{pmatrix}. \tag{4.17}$$

The corresponding heat fluxes are

$$\begin{pmatrix} \tilde{q}_1 \\ \tilde{q}_2 \end{pmatrix} = Q\begin{pmatrix} \tilde{f}_a \\ \tilde{f}_b \end{pmatrix} = QG^{-1}\begin{pmatrix} 0 & 1 \\ 1 & 0 \end{pmatrix} GQ^{-1}\begin{pmatrix} q_1 \\ q_2 \end{pmatrix}. \tag{4.18}$$

Note that when $H = 0$, Q is singular, but the following results can be similarly obtained. If the system is not a diode, then it requires

$$\begin{pmatrix} \widetilde{q}_1 \\ \widetilde{q}_2 \end{pmatrix} = -\begin{pmatrix} q_2 \\ q_1 \end{pmatrix} = -\begin{pmatrix} 0 & 1 \\ 1 & 0 \end{pmatrix}\begin{pmatrix} q_1 \\ q_2 \end{pmatrix}. \tag{4.19}$$

Substituting into Eq. (4.18) and considering that the temperature values are arbitrarily chosen, we have

$$\boldsymbol{Q}\boldsymbol{G}^{-1}\begin{pmatrix} 0 & 1 \\ 1 & 0 \end{pmatrix}\boldsymbol{G}\boldsymbol{Q}^{-1} + \begin{pmatrix} 0 & 1 \\ 1 & 0 \end{pmatrix} = 0. \tag{4.20}$$

Assuming that the system is reciprocal, $G_{12} = G_{21}$. In the limit $L \to \infty$, Eq. (4.12) gives $G_{11} = G_{22}$. From Eqs. (4.14) and (4.15), we see that Q is anti-symmetric and traceless. Using all the properties, we find that Eq. (4.20) is satisfied, and the system is not a diode. However, if Eq. (4.20) is satisfied, then we have

$$\begin{aligned} (\boldsymbol{G}_{12} - \boldsymbol{G}_{21})(\boldsymbol{G}_{12} + \boldsymbol{G}_{21} - \boldsymbol{G}_{11} + \boldsymbol{Q}_{11}L/\kappa_c) &= 0 \\ (\boldsymbol{G}_{12} - \boldsymbol{G}_{21})(\boldsymbol{G}_{12} - \boldsymbol{G}_{21} + \boldsymbol{G}_{11} + \boldsymbol{Q}_{11}L/\kappa_c) &= 0. \end{aligned} \tag{4.21}$$

The only physical solution is $\boldsymbol{G}_{12} = \boldsymbol{G}_{21}$. We prove the equivalence between steady-state global reciprocity and the absence of the diode effect, or steady-state global nonreciprocity and being a diode. It is worth mentioning that with heat generation inside the system, there are multiple choices to define a thermal diode, except for the violation of Eq. (4.19). For example, it can be defined based on the average heat flux $(q_1 + q_2)/2$. However, as we have shown above, Eq. (4.19) is directly related to global reciprocity, thereby reflecting the essential properties of the system.

From another aspect, when the time scale of observation goes to infinity, the so-called steady state is nothing but a brief moment. Based on this viewpoint, the steady-state heat transfer can be understood as an instant of the time-harmonic heat transfer whose wave frequency ω approaches zero (Fig. 4.3). However, if we allow this process to happen spontaneously, the temperature field of the system will be uncertain due to the randomness of the time phase ωt. To eliminate the unwanted ambiguity, here we set the time phase of the signal as $\omega t = 2\pi N$ while ω approaches zero.

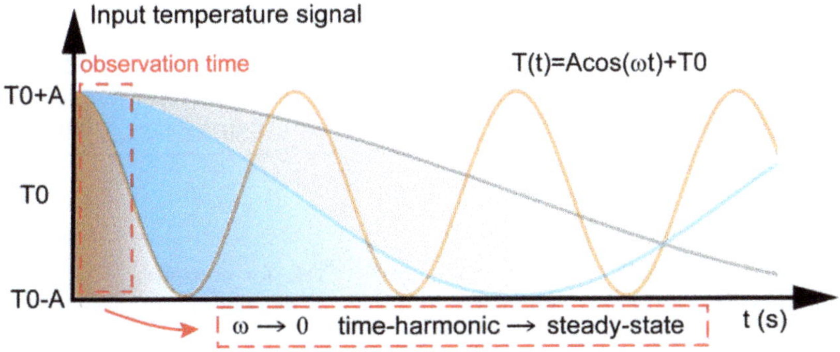

Fig. 4.3. Relationship between dynamic and steady-state nonreciprocity. Adapted from Ref. [51].

Note that under this circumstance, the time-harmonic heat signal will transform from $Ae^{-i\omega t} + T_0$ to $T_h = A + T_0$, and the previous amplitude of the heat signal will become part of the quasi-steady-state temperature field at time $t = 2\pi N/\omega$ for an infinitely long period of oscillation.

So far, the relevant concepts of the thermal scattering matrix have been elaborated, which can be used to analyze the global reciprocity of the system in combination with Green's function method. Although the above process is a qualitative analysis of the reciprocity of the system, the function of the thermal scattering theory does not stop there. Through mathematical modeling of the device and providing the physical coupling conditions, thermal scattering theory can quantitatively describe the heat transfer response characteristics of the system accurately. The specific content will be introduced in the following sections.

4.3. Diffusive Nonreciprocity and Thermal Diode

The device that transfers heat asymmetrically in opposite directions is called a thermal diode [14, 15] or thermal rectifier [16], which plays an important role in temperature management and thermal information processing. Combining linear materials is insufficient to build a thermal diode because heat diffusion is always symmetric.

Table 4.1. A summary of non-reciprocity under various mechanisms.

Potential Mechanisms	Local Nonreciprocity	Global Nonreciprocity	Make a Thermal Diode?
Nonlinearity with Asymmetric Geometry	Meaningless		Yes
Asymmetric Thermal Conductivity Tensor	Yes (for Neumann boundary conditions)	Yes (for nonzero frequency)	No
Thermal Convection	Yes	Yes (for nonzero frequency)	No
Heat Generation (Temperature-independent)	Yes	Yes	Yes
Heat Generation (Temperature-dependent)	No	No	No

Here, we provide an in-depth study of the local and global reciprocities in heat diffusion. The scattering theory for heat diffusion through ports is established by considering time-harmonic situations. The corresponding conditions for preserving and breaking global reciprocity are formulated, with meaningful results in the zero frequency limit when the fields cannot be decomposed into input and output parts. We further prove that any device that preserves global reciprocity cannot be a thermal diode, while that with global nonreciprocity is a thermal diode. Based on our findings, methods to break diffusive reciprocity and realize a thermal diode are summarized, with some typical examples illustrated.

4.3.1. *Nonlinearity*

Combining nonlinearity with asymmetry is a well-known strategy for building a thermal diode. It has been shown that simply using nonlinearity is not enough [17]. For 1D heat diffusion, it is also required that the space and temperature dependence of the thermal conductivity should be nonseparable [18]. Detailed analytical studies on the rectification ratio using temperature-dependent thermal conductivities have been performed recently [19]. However, when the governing differential equation is nonlinear due to the temperature-dependent thermal conductivity or structure, its solution cannot be obtained through Green's function. In such cases, local and global reciprocity is not very meaningful, but a diode effect can be expected.

4.3.2. *Heat Generation*

In our analysis, heat generation could be inside the system, so its scope is broader than many other works on thermal diode and rectification. As proved in the above section, when the total heat generation H is independent of the outside thermal conditions, the equivalence between steady-state global reciprocity and diode effect is still valid. In addition, it is easy to see that when the distribution of heat generation density $h(\boldsymbol{r})$ in the system is inhomogeneous and asymmetric, it can break the steady-state global reciprocity without relying on other effects. Therefore, inhomogeneous heat generation is a convenient mechanism to make a thermal diode.

Here, we further consider the case when a heat source in the system depends on the temperatures in the channels. Equation (4.10) and the following derivations are incorrect as H is temperature-dependent. Such a case can be easily realized using a setup, as shown in Fig. 4.4a.

The system is rectangular of size 10 cm $\times$ 5 cm (thickness 1 mm) with thermal conductivity $\kappa = 100\,\mathrm{W/(m \cdot K)}$. A square of size 1 cm $\times$ 1 cm is made inside. The center of the square is 1 cm left from the center of the rectangle. Two channels of size 2 cm $\times$ 1 cm with thermal conductivity $\kappa_c = 400\,\mathrm{W/(m \cdot K)}$ are attached to the left and right boundaries of the system. The boundaries of the square are maintained at a fixed temperature $T_0 = 300\,\mathrm{K}$ to introduce a temperature-dependent heat source. The local and global reciprocities are preserved since this is a Dirichlet boundary condition. The global reciprocity is confirmed (Fig. 4.4a) using a heat source $Q = 5\,\mathrm{W}$ to a small spot of radius 0.3 mm at the center of channel 1.

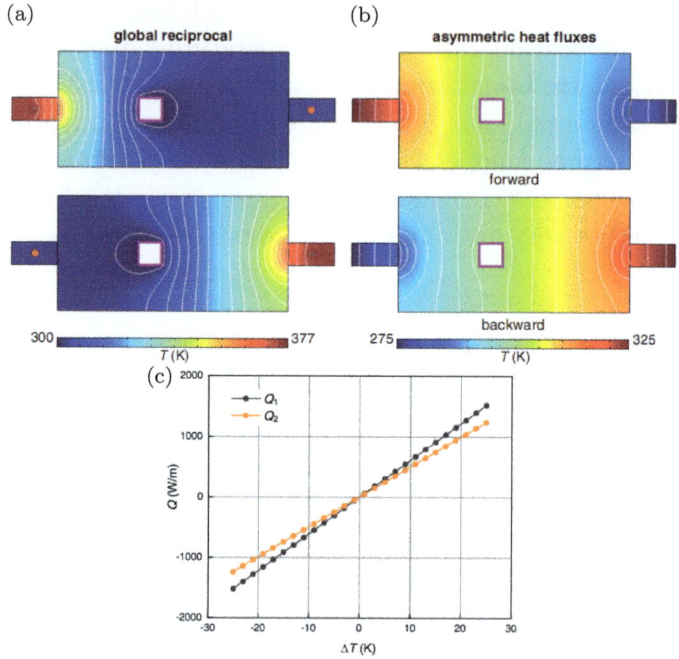

Fig. 4.4. Effects of temperature-dependent heat source. (a) Temperature profiles on a two-port system with an interior hole at fixed temperature generated by a point heat source with all boundaries thermally insulated. The global reciprocity is preserved. (b) Steady-state temperature profiles with the left and right boundaries at fixed temperatures. The forward ($T = 25$ K) and backward ($T = -25$ K) cases are asymmetric. (c) Heat (per unit thickness) across the system. Boundaries with fixed temperatures are indicated with purple lines. The target positions where temperatures are measured are indicated by orange dots. Adapted from Ref. [26].

The temperature at the center of channel 2 is 303.33 K. The value remains the same after swapping the heat source and target positions.

Despite steady-state global reciprocity, the system violates Eq. (4.19) (Fig. 4.4b) because the heat generation inside depends on the temperature difference between T_0 and $T_1(T_2)$. As the hole is closer to channel 1, the difference between T_0 and T_1 contributes more. Therefore, in the forward (backward) case, the heat generation is negative (positive) as T_0 is lower (higher) than T_1, namely, $Q_1(\Delta T) - Q_2(\Delta T) > 0$ for $\Delta T > 0$ and $Q_1(\Delta T) - Q_2(\Delta T) < 0$

for $\Delta T < 0$, where Q_1 and Q_2 are the heat in channels 1 and 2. The linearity and symmetric boundary conditions also imply $Q_1(-\Delta T) = -Q_1(\Delta T)$ and $Q_2(-\Delta T) = -Q_2(\Delta T)$. It follows that $Q_1(\Delta T) + Q_2(-\Delta T) \neq 0$ or the average heat $[Q_1(\Delta T) + Q_2(\Delta T)]/2 \neq -[Q_1(-\Delta T)+Q_2(-\Delta T)]/2$ (Fig. 4.4c). Strictly speaking, the system is three-port with the central hole in contact with some thermal reservoir, so it does not meet the rigorous definition of a thermal diode. Yet, it is still a representative counterexample for the equivalence between Eq. (4.19) and steady-state global reciprocity.

Based on the frequency domain Green's function, we have provided a reasonable definition of the scattering matrix and global reciprocity of diffusion, which extends to the zero-frequency limit. We also proved that for linear systems without temperature-dependent heat generation, steady-state global reciprocity indicates the absence of the diode effect, while breaking the global reciprocity makes a thermal diode. We discussed several potential mechanisms to break the reciprocity and make a thermal diode. Nonlinearity with asymmetric geometry makes a thermal diode. Asymmetric thermal conductivity tensor breaks local and global reciprocity at nonzero frequencies. However, the steady-state global reciprocity is preserved for a two-port system, so the asymmetric thermal conductivity tensor does not make a thermal diode. Typical systems with heat convection and temperature-dependent heat generation are also studied. They also preserve steady-state global reciprocity, but the forward and backward heat fluxes could be asymmetric in the latter.

4.3.3. *Asymmetric Thermal Conductivity Tensor*

When the thermal conductivity tensor is asymmetric, such as the thermal Hall effect in magnetic fields, we can get the following Green's function,

$$\int_V \nabla G(\boldsymbol{r}|\boldsymbol{r}_1,\omega)\cdot(\kappa-\kappa^{\mathrm{T}})\cdot\nabla G(\boldsymbol{r}|\boldsymbol{r}_0,\omega)dV \qquad (4.22)$$

which could, but not necessarily, be nonzero. We rewrite Eq. (4.22) using the property of $\boldsymbol{\kappa}-\boldsymbol{\kappa}^{\mathrm{T}}$, which is an antisymmetric tensor in 3D.

In the Cartesian coordinate system, it can be expressed as

$$\left[\boldsymbol{\kappa}-\boldsymbol{\kappa}^{\mathrm{T}}\right]=\begin{bmatrix}0 & -a_3 & a_2\\ a_3 & 0 & -a_1\\ -a_2 & a_1 & 0\end{bmatrix}. \tag{4.23}$$

Therefore, we can define a vector $\boldsymbol{a}$, whose three Cartesian components are a_1, a_2, and a_3. Equation (4.22) becomes

$$\begin{aligned}&\int_V \nabla G(\boldsymbol{r}|\boldsymbol{r}_1,\omega)\cdot a\times\nabla G(\boldsymbol{r}|\boldsymbol{r}_0,\omega)dV\\ &=\int_V \nabla\cdot[G(\boldsymbol{r}|\boldsymbol{r}_1,\omega)a\times\nabla G(\boldsymbol{r}|\boldsymbol{r}_0,\omega)]dV.\\ &-\int_V \nabla G(\boldsymbol{r}|\boldsymbol{r}_1,\omega)\nabla\cdot[a\times\nabla G(\boldsymbol{r}|\boldsymbol{r}_0,\omega)]dV\end{aligned} \tag{4.24}$$

If only the Dirichlet boundary condition is applied, then the first term is zero based on Gauss's law. In the second term, we have

$$\begin{aligned}&\nabla\cdot[a\times\nabla G(\boldsymbol{r}|\boldsymbol{r}_0,\omega)]\\ &=\nabla G(\boldsymbol{r}|\boldsymbol{r}_0,\omega)\cdot(\nabla\times a)-a\cdot[\nabla\times\nabla G(\boldsymbol{r}|\boldsymbol{r}_0,\omega)]\\ &=\nabla G(\boldsymbol{r}|\boldsymbol{r}_0,\omega)\cdot(\nabla\times a).\end{aligned} \tag{4.25}$$

As a result, the additional term is still zero for uniform materials, which also applies to 2D cases. When Neumann or mixed boundary condition exists, the first term in Eq. (4.24) does not vanish. We thus show that the local reciprocity in a uniform material with asymmetric thermal conductivity tensor is preserved when all the boundaries are at fixed temperatures.

Next, we study the global reciprocity at a steady state to determine whether the system could become a thermal diode. At zero frequency, simply using an asymmetric thermal conductivity tensor does not break the global reciprocity of a two-port system, so a thermal diode cannot be built this way. We use the same setup as in Fig. 4.2b to prove this. An integral of the system gives

$$\begin{aligned}\int_V \{G(\boldsymbol{r}|\boldsymbol{r}_1,\omega)\boldsymbol{\nabla}\cdot[\boldsymbol{\kappa}\cdot\boldsymbol{\nabla}G(\boldsymbol{r}|\boldsymbol{r}_0,\omega)] \\ - G(\boldsymbol{r}|\boldsymbol{r}_0,\omega)\boldsymbol{\nabla}\cdot[\boldsymbol{\kappa}\cdot\boldsymbol{\nabla}G(\boldsymbol{r}|\boldsymbol{r}_1,\omega)]\}dV \\ = \int_V [G(\boldsymbol{r}|\boldsymbol{r}_0,\omega)Q_0\delta(\boldsymbol{r}-\boldsymbol{r}_1) \\ - G(\boldsymbol{r}|\boldsymbol{r}_1,\omega)Q_0\delta(\boldsymbol{r}-\boldsymbol{r}_0)]dV = 0.\end{aligned} \tag{4.26}$$

Now that r_0 and r_1 are outside the system in the channels, the integral vanishes. We can also integrate the system plus the two channels:

$$\begin{aligned}\int_{V\cup C_1\cup C_2} \{G(\boldsymbol{r}|\boldsymbol{r}_1,\omega)\boldsymbol{\nabla}\cdot[\boldsymbol{\kappa}\cdot\boldsymbol{\nabla}G(\boldsymbol{r}|\boldsymbol{r}_0,\omega)] \\ - G(\boldsymbol{r}|\boldsymbol{r}_0,\omega)\boldsymbol{\nabla}\cdot[\boldsymbol{\kappa}\cdot\boldsymbol{\nabla}G(\boldsymbol{r}|\boldsymbol{r}_1,\omega)]\}dV \\ = \int_{V\cup C_1\cup C_2} [G(\boldsymbol{r}|\boldsymbol{r}_0,\omega)Q_0\delta(\boldsymbol{r}-\boldsymbol{r}_1) \\ - G(\boldsymbol{r}|\boldsymbol{r}_1,\omega)Q_0\delta(\boldsymbol{r}-\boldsymbol{r}_0)]dV = 0 \\ = \boldsymbol{Q_0}[G(\boldsymbol{r_1}|\boldsymbol{r}_0,\omega) - G(\boldsymbol{r_0}|\boldsymbol{r}_1,\omega)].\end{aligned} \tag{4.27}$$

Using Eq. (4.26), the left-hand side (l.h.s.) is simplified as

$$\begin{aligned}\text{l.h.s.} = \int_{C_1\cup C_2} \{G(\boldsymbol{r}|\boldsymbol{r}_1,\omega)\boldsymbol{\nabla}\cdot[\boldsymbol{\kappa_c}\cdot\boldsymbol{\nabla}G(\boldsymbol{r}|\boldsymbol{r}_0,\omega)] \\ - G(\boldsymbol{r}|\boldsymbol{r}_0,\omega)\boldsymbol{\nabla}\cdot[\boldsymbol{\kappa_c}\cdot\boldsymbol{\nabla}G(\boldsymbol{r}|\boldsymbol{r}_1,\omega)]\}dV \\ = \int_{\partial C_1\cup C_2} \kappa_c[G(\boldsymbol{r}|\boldsymbol{r}_1,\omega)\boldsymbol{\nabla}G(\boldsymbol{r}|\boldsymbol{r}_0,\omega) \\ - G(\boldsymbol{r}|\boldsymbol{r}_0,\omega)\boldsymbol{\nabla}G(\boldsymbol{r}|\boldsymbol{r}_1,\omega)]\cdot\boldsymbol{n}dS\end{aligned} \tag{4.28}$$

on the two interfaces S_1 and S_2 between the channels and the system. At zero frequency ($\omega = 0$), energy conservation requires

$$\begin{aligned}\int_{S_1+S_2} \kappa_c\nabla G(\boldsymbol{r}|\boldsymbol{r}_0,\omega)\cdot\boldsymbol{n}dS = 0 \\ \int_{S_1+S_2} \kappa_c\nabla G(\boldsymbol{r}|\boldsymbol{r}_1,\omega)\cdot\boldsymbol{n}dS = 0.\end{aligned} \tag{4.29}$$

Combining with the other boundary conditions of C_1 and C_2, the surface integral in Eq. (4.28) vanishes. Therefore, the right-hand side of Eq. (4.27) equals zero. We thus prove that the steady-state global reciprocity is preserved even for a system with an asymmetric thermal conductivity tensor, which could be non-uniform. Based on the equivalence shown in Section 4.2, it cannot become a thermal diode. Of course, we do not have Eq. (4.29) at nonzero frequencies, so the global reciprocity is broken.

To confirm the above results, we perform finite-element simulations on a system with the same shape as in Fig. 4.5. Two channels of the size 2 cm × 1 cm with thermal conductivity $\kappa_c = 400\,\mathrm{W/(m\cdot K)}$, density $\rho_c = 8900\,\mathrm{kg/m^3}$, and heat capacity $c_{pc} = 390\,\mathrm{J/(kg\cdot K)}$ (those of copper) are attached to the system. Channel 1 is connected to the left of the system. Channel 2 is connected to the bottom of the system. The thermal conductivity of the system follows Eq. (4.23). The global reciprocity is tested by applying an oscillatory point heat source, with all boundaries thermally insulated. When the point heat source is in channel 1 (5 mm from the interface), the amplitude of temperature variation in channel 2 (5 mm from the interface) is $A_1 = 38.89\,\mathrm{K}$. The value becomes $A_2 = 42.06\,\mathrm{K}$ after the positions of the heat source and target are swapped. The temperature distributions at $t = 9110$ s are shown in Fig. 4.5a.

However, the ratio A_2/A_1 decreases toward unity as the oscillating frequency ω approaches zero, as shown in Fig. 4.5b. It confirms our theoretical prediction. We also apply fixed boundary conditions on the left end of channel 1 and the lower end of channel 2, whose temperatures $T_1 = T_0 + T$ and $T_2 = T_0 - T$ are set the same as for the nonlinear cases. As shown in Fig. 4.5c, the temperature distributions for the forward and backward cases are symmetric. As expected, the heat is linearly proportional to T, so $Q(T) = -Q(-T)$, and the system is not a thermal diode (Fig. 4.5d).

4.3.4. *Thermal Convection*

On a macro scale, thermal conduction can be described by Fourier's function $\rho c_p \frac{\partial T(\boldsymbol{r},t)}{\partial t} = \kappa \nabla^2 T(\boldsymbol{r}, t)$, where ρ is density, c_p is specific heat capacity at constant pressure, and κ is thermal conductivity. In its

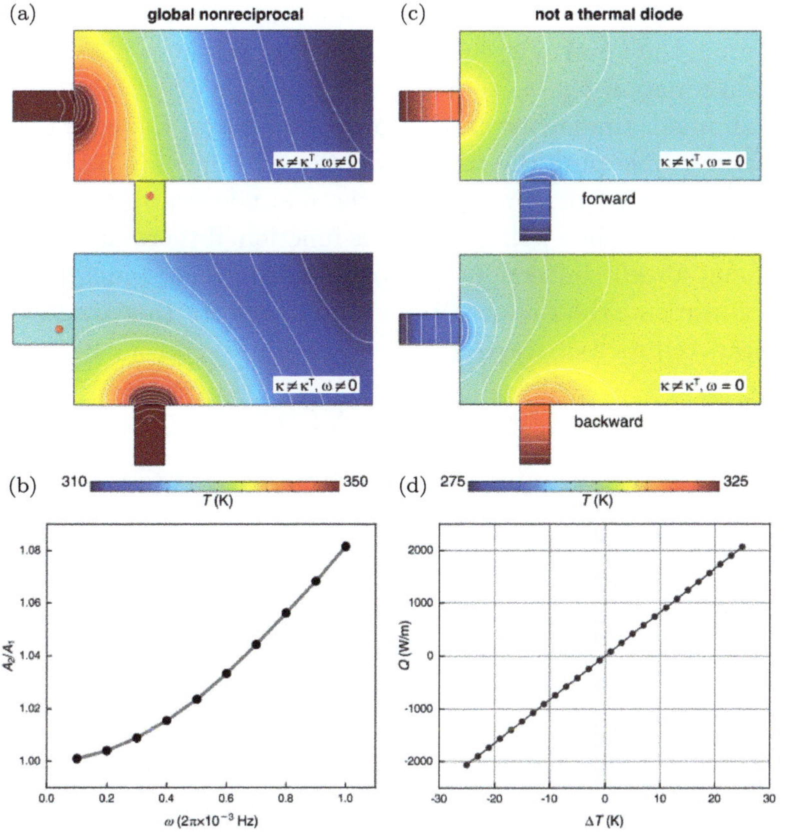

Fig. 4.5. Effects of asymmetric thermal conductivity tensor: global reciprocity and diode effect. (a) Temperature profiles at $t = 9110$ s generated by a time-harmonic point heat source with all boundaries thermally insulated. The global reciprocity is broken. (b) As the oscillating frequency $\omega$of the heat source approaches zero, the ratio between the two amplitudes approaches one, indicating a preserved global reciprocity at a steady state. (c) Steady-state temperature profiles with the left and lower boundaries at fixed temperatures. The forward ($T = 25$ K) and backward ($T = -25$ K) cases are symmetric, so the system is not a thermal diode. (d) Heat (per unit thickness) across the system. The target positions where temperatures are measured are indicated by orange dots. Adapted from Ref. [26].

simplest form, when these three parameters are assumed to be independent of time t and position vector $\boldsymbol{r}$, diffusion of the temperature field $T(\boldsymbol{r}, t)$ is a spatially symmetric physical process. Consequently, thermal reciprocity is an inevitable result in any general media.

Convection with velocity $\boldsymbol{v}$ is introduced to change this, which transforms the governing equation into the well-known convection-diffusion equation $\rho c_p \frac{\partial T(\boldsymbol{r},t)}{\partial t} = \kappa \nabla^2 T(\boldsymbol{r},t) - \rho c_p \boldsymbol{v} \nabla T(\boldsymbol{r},t)$. As an odd function under time reversal, the convection makes the heat transfer process resemble pure thermal conduction in a moving background, which sheds new light on exploring thermal nonreciprocity.

From the standing point of Green's function, if thermal convection (assuming forced convection with temperature-independent velocity field) is present, then an additional term appears that could break the local reciprocity,

$$\begin{aligned} -\int_V &[G(\boldsymbol{r}|\boldsymbol{r}_1,\omega)\rho c_p \boldsymbol{v} \cdot \nabla G(\boldsymbol{r}|\boldsymbol{r}_0,\omega) \\ &- G(\boldsymbol{r}|\boldsymbol{r}_0,\omega)\rho c_p \boldsymbol{v} \cdot \nabla G(\boldsymbol{r}|\boldsymbol{r}_1,\omega)]dV. \end{aligned} \tag{4.30}$$

Such a convection-induced nonreciprocity has been studied for mechanically rotating structures, whose effective thermal conductivities tensor are shown to be asymmetric. The nonreciprocity is reflected in a rotated temperature field around the structure. Interestingly, the reciprocity can be restored by combining counter-rotating parts, thereby disguising a convective system as a purely conductive [20]. It should be noted that this reciprocity is between any two arbitrary directions, so it is a local one. In terms of a two-port setup, the result is quite different.

Based on the effective thermal conductivity [21], most asymmetric thermal conductivity tensor results are also valid for heat convection through rotating structures. Namely, the local reciprocity is broken; the global reciprocity for a two-port system is broken for nonzero frequencies but preserved at zero frequency based on a similar argument for Eq. (4.27); the system cannot become a thermal diode. Equation (4.30) cannot be manipulated like Eq. (4.22), so the local reciprocity is broken even for pure Dirichlet boundary conditions.

All predictions are numerically demonstrated for a 1-mm-thick solid ring with inner and outer radii 1.5 cm and 3 cm rotating at angular speed $\Omega = 2\pi \times 0.002$ rad/s. Thermal conductivity, density, and heat capacity are $\kappa = 16\,\mathrm{W/(m \cdot K)}$, $\rho = 8000$ kg/m^3, and $c_p = 500\,\mathrm{J/(kg \cdot K)}$ (those of stainless steel). For local reciprocity, all the boundaries are thermally insulated. In a polar coordinate system (r, θ) whose origin is built at the center of the system, a heat source $Q = 2\cos(\omega t)$ (W) is placed at $[r_0] = [2.25\,\text{cm}\ 5\pi/4]^{\mathrm{T}}$. For $\omega = 2\pi \times 0.001\,\text{rad/s}$, the amplitude of temperature variation at $[\boldsymbol{r}_1] = [2.25\,\text{cm}\ 7\pi/4]^{\mathrm{T}}$ is $A_1 = 52.23\,\text{K}$. The value becomes $A_2 = 37.46\,\text{K}$ after the positions of $\boldsymbol{r}_0$ and $\boldsymbol{r}_1$ are swapped. The local reciprocity is broken, as shown in Fig. 4.6a. The same is confirmed for fixed boundary conditions. The temperature fields are not shown as they are not very informative: only the vicinity of the heat source is heated up.

For the global reciprocity, two channels are attached to the ring at $\theta = 5\pi/4$ and $7\pi/4$ as in Fig. 4.6b. The size and material parameters of the channels are the same as used in the above simulations. The heat source Q is placed at $[\boldsymbol{r}_0] = [3.5\,\text{cm}\ 5\pi/4]^{\mathrm{T}}$. The amplitude of temperature variation at $[\boldsymbol{r}_1] = [3.5\,\text{cm}\ 7\pi/4]^{\mathrm{T}}$ is $A_1 = 43.28\,\text{K}$. The value becomes $A_2 = 31.17\,\text{K}$ after the positions of $\boldsymbol{r}_0$ and $\boldsymbol{r}_1$ are swapped, indicating global nonreciprocity. As $\omega \to 0$, the ratio A_1/A_2 approaches unity, which confirms the steady-state global reciprocity. The system is thus not a thermal diode, as shown in Fig. 4.6c, where the outer end of channel 1 (2) is maintained at a fixed temperature $T_1 = T_0 + \Delta T\ (T_2 = T_0 - \Delta T)$, same as for the nonlinear cases. The temperature profiles for forward and backward cases are symmetric. Since there is no heat generation in the system, the amount of heat through both channels should be equal. We thus choose to calculate the heat $Q(T)$ as an integral over a cross-section in the right channel (yellow lines in Fig. 4.6c). The results for different rotation speeds of the ring are plotted in Fig. 4.6d. All of them are odd symmetric to T, indicating the absence of a diode effect.

4.4. Nonreciprocal Heat Circulation Metadevices

Several mechanisms that may be promising in achieving thermal nonreciprocity are introduced above, and the global reciprocity of the model is analyzed using the frequency domain Green's function method. However, constrained by their inherent mechanisms, these ways of thermal nonreciprocity can hardly be effective under dynamic

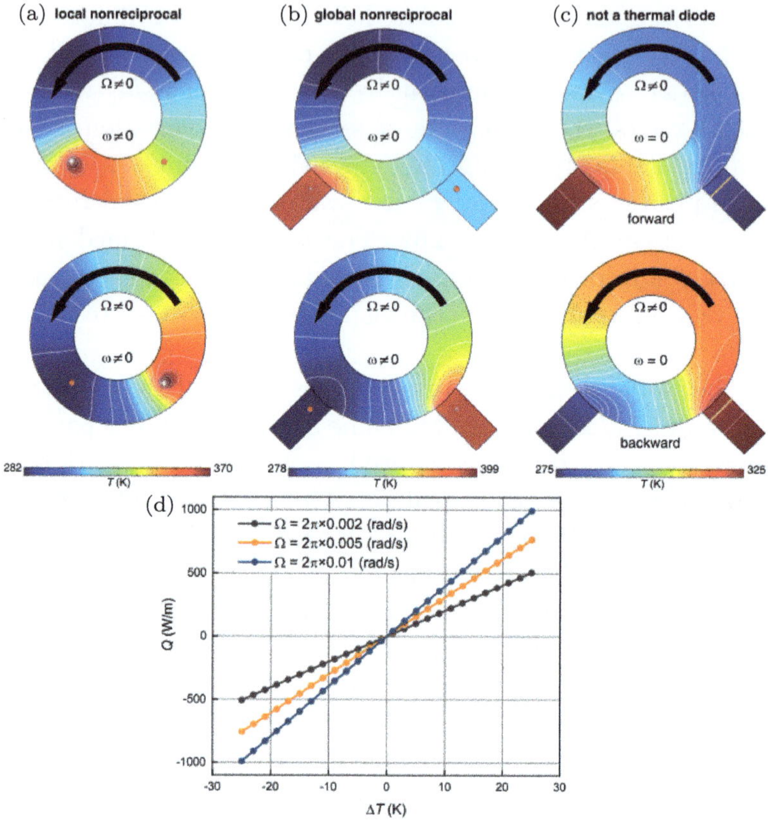

Fig. 4.6. Effects of heat convection. (a) Temperature profiles on a rotating ring (with rotation speed) at $t = 9110$ s generated by a time-harmonic point heat source (at angular frequency ω) with all boundaries thermally insulated. The local reciprocity is broken. (b) Temperature profiles on a rotating ring attached with two channels at 9080 s generated by a time-harmonic point heat source with all boundaries thermally insulated. The global reciprocity is broken. (c) Steady-state temperature profiles with the left and lower boundaries at fixed temperatures. The forward ($T = 25$ K) and backward ($T = -25$ K) cases are symmetric, so the system is not a thermal diode. The target positions where temperatures are measured are indicated by orange dots. (d) Heat (per unit thickness) across the system at different rotation speeds. Adapted from Ref. [26].

and static excitations [22, 24]. For example, previous works have theoretically proved that the bias of magnetic field or convection cannot break the steady-state reciprocity of a two-port thermal system [25, 26]. This section will focus on whether an external bias can help realize steady-state thermal nonreciprocity. Furthermore, we will show the potential of thermal scattering theory for quantitative analysis of thermal responses.

Here, we develop a convection-based thermal meta-device where nonreciprocity is realized under dynamic and steady-state temperature excitations. The dynamic nonreciprocity is attributed to an asymmetric scattering matrix. Counterintuitively, this matrix approaches a symmetric limit at the steady state but still conveys significant steady-state nonreciprocity. Moreover, we reveal a unique nonreciprocal mechanism where the boundary-induced multiple scattering counterbalances the diminishing asymmetry and experimentally verifies this effect.

4.4.1. *Scattering Matrix of the Thermal Circulator*

Generally, there are two kinds of nonreciprocity: nonreciprocal energy transfer and nonreciprocal phase transfer. Therefore, it must be noted that in this chapter, when we mention nonreciprocity, we always mean nonreciprocal energy transfer, which the amplitudes of the heat signal or heat flux can measure. To examine this feasibility, we consider a thermal meta-device composed of a center solid ring and three ports symmetrically placed at 120° intervals, known as a thermal circulator (Fig. 4.7). We establish a comprehensive framework to explore thermal scattering under the time-harmonic excitation $T(t) = A\cos(\omega t) + T_0$, then judiciously extend this method to the zero-frequency case.

Here, we start by considering a periodical temperature profile both in the channels and the central ring of the thermal circulator,

$$T = T(x,\omega)e^{-i\omega t} + T_0 \tag{4.31}$$

where $T(x,\omega), \omega, T_0$ are the complex amplitude, thermal wave frequency, and reference temperature. Among them, the complex amplitude $T(x,\omega)$ can be further expressed in the form of a plane wave solution $T(x,\omega) = Ae^{ikx} + Be^{-ikx}$, where A and B are the real amplitude of the thermal signal, and $k = \beta + i\gamma$ denotes the

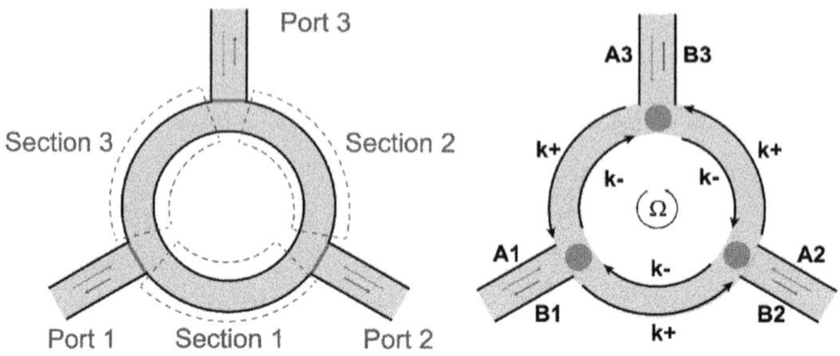

Fig. 4.7. Scheme of the thermal circulator. Adapted from Ref. [51].

wavenumber of the thermal wave. β is phase constant, and γ is attenuation constant. Therefore, the instantaneous value of the wave-like temperature field in the circulator is

$$T(x,t) = (Ae^{ikx} + Be^{-ikx})e^{-i\omega t} + T_0. \tag{4.32}$$

Substitute Eq. (4.32) into the 1D convection-diffusion equation $\frac{\partial T}{\partial t} = D\nabla^2 T - \boldsymbol{v}\nabla T$, then the relationship among thermal wave frequency ω, wavenumber k, convective velocity $\boldsymbol{v}$, and the thermal diffusivity D of the material can be deduced as $\omega = k\boldsymbol{v} - ik^2 D$. With $\pm\sqrt{a+ib} = \pm\frac{\sqrt{2}}{2}\left[\left(\sqrt{\sqrt{a^2+b^2}+a} + i\,\mathrm{sign}(b)\sqrt{\sqrt{a^2+b^2}-a}\right)\right]$, k can be written as an explicit function of parameters $\omega, \boldsymbol{v}, D$,

$$k(\omega, \boldsymbol{v}, D) = \frac{\sqrt{-2\boldsymbol{v}^2 + 2\sqrt{\boldsymbol{v}^4 + 16D^2\omega^2}}}{4D} + i\frac{\frac{\sqrt{2\sqrt{\boldsymbol{v}^4+16D^2\omega^2}+2\boldsymbol{v}^2}}{2} - \boldsymbol{v}}{2D}. \tag{4.33}$$

Therefore, the real part of k is phase constant β, and the imaginary part of k is attenuation constant γ,

$$\begin{cases} \beta = \dfrac{\sqrt{-2\boldsymbol{v}^2 + 2\sqrt{\boldsymbol{v}^4 + 16D^2\omega^2}}}{4D} \\ \gamma = \dfrac{\dfrac{\sqrt{2\sqrt{\boldsymbol{v}^4 + 16D^2\omega^2} + 2\boldsymbol{v}^2}}{2} - \boldsymbol{v}}{2D} \end{cases}. \tag{4.34}$$

γ can be further written as $\gamma = \frac{\omega - \boldsymbol{v}\beta}{2D\beta} = \frac{\frac{\omega}{\beta} - \boldsymbol{v}}{2D} = \frac{v_p - \boldsymbol{v}}{2D}$, where v_p denotes the phase velocity of the wave. Phase velocity v_p is always larger than convection velocity $\boldsymbol{v}$, so attenuation constant

γ is always positive, which is consistent with the diffusive nature of the thermal wave. Then, consider that the convection velocity can be positive or negative. Accordingly, attenuation constant γ can be along or against the convection. Therefore, three different attenuation constants appear in this model,

$$\begin{cases} \gamma_0 = \sqrt{\dfrac{\omega}{2D}} \\ \gamma_+ = \dfrac{v_p - \boldsymbol{v}}{2D} \\ \gamma_- = \dfrac{v_p + \boldsymbol{v}}{2D} \end{cases} \tag{4.35}$$

where γ_0 denotes the attenuation constant in three channels, γ^+ denotes the downstream attenuation constant in the rotating ring, and γ^- denotes the upstream attenuation constant in the rotating ring. To sum up, wavenumbers in the channel and the ring can be derived accordingly, which help describe the evanescent characteristics of thermal waves,

$$\begin{cases} k_0 = \dfrac{\sqrt{2}}{2}\left(\sqrt{\dfrac{\omega}{D}} + i\sqrt{\dfrac{\omega}{D}}\right) \\ k_+ = k(|\boldsymbol{v}|) = \dfrac{\sqrt{-2v^2 + 2\sqrt{v^4 + 16D^2\omega^2}}}{4D} \\ \qquad + i\dfrac{\dfrac{\sqrt{2\sqrt{v^4 + 16D^2\omega^2} + 2v^2}}{2} - |\boldsymbol{v}|}{2D} \\ k_- = k(-|\boldsymbol{v}|) = \dfrac{\sqrt{-2v^2 + 2\sqrt{v^4 + 16D^2\omega^2}}}{4D} \\ \qquad + i\dfrac{\dfrac{\sqrt{2\sqrt{v^4 + 16D^2\omega^2} + 2v^2}}{2} + |\boldsymbol{v}|}{2D} \end{cases} . \tag{4.36}$$

For $\boldsymbol{v} \neq 0$, the wavenumbers are split into k_+ and k_-, depending on whether it travels with or against the convection. The wavenumber's real part is an even velocity function, while the imaginary part produces a split. This feature means that the direction of convection does not change the phase constant of the thermal wave, but it does affect the attenuation rate of the thermal wave, which lays a foundation for the realization of nonreciprocity. Consequently, we can decompose the oscillatory temperature fields into counter-propagating parts like the wave frame, where $T(x,\omega) = Ae^{ik_+(x-x_0)} + Be^{-ik_-(x-x_0)}$. By this definition, the ingoing and outgoing signals (A_i and B_i) can be related through the scattering matrix $\boldsymbol{S}$, where $(B_1, B_2, B_3)^{\mathrm{T}} = \boldsymbol{S}(A_1, A_2, A_3)^{\mathrm{T}}$. Due to the rotation-symmetry property of the circulator, the elements of the scattering matrix $\boldsymbol{S}$ obey the rule $S_{31} = S_{23} = S_{12}$. Therefore, by assuming port 1 to be the input port and ports 2 and 3 to be the output ports, then the strength of nonreciprocity can be defined as $\Gamma = |S_{21}/S_{31}|$.

Hereto, the derivation method of the wavenumber has been elaborated. In the following part, we will focus on the derivation of the thermal scattering matrix. The model has three fixed solid ports and one rotating solid ring. Whenever the ring passes through a port, energy exchanges between the ring and the port. Therefore, for this thermal circulator, it is necessary to divide the rotating ring into three parts to describe its temperature profile theoretically. As shown in Fig. 4.7, being symmetrically spaced 120° apart, three ports are attached tightly to the center rotation ring. Assuming the average radius of the ring to be r_0, then the temperature amplitude of the three ports can be described as

$$T_{\mathrm{port}}(r,\omega) = \begin{cases} A_1 e^{ik_0(r_0-r)} + B_1 e^{-ik_0(r_0-r)}, & r \geq r_0, \text{ port 1} \\ A_2 e^{ik_0(r_0-r)} + B_2 e^{-ik_0(r_0-r)}, & r \geq r_0, \text{ port 2} \\ A_3 e^{ik_0(r_0-r)} + B_3 e^{-ik_0(r_0-r)}, & r \geq r_0, \text{ port 3} \end{cases} \tag{4.37}$$

where r is the polar diameter in polar coordinates. By this definition, the ingoing signal amplitude A_i and outgoing signal amplitude B_i can be related through the scattering matrix

$$\begin{pmatrix} B_1 \\ B_2 \\ B_3 \end{pmatrix} = \boldsymbol{S} \begin{pmatrix} A_1 \\ A_2 \\ A_3 \end{pmatrix} = \begin{pmatrix} r_{11} & t_{12} & t_{13} \\ t_{21} & r_{22} & t_{23} \\ t_{31} & t_{32} & r_{33} \end{pmatrix} \begin{pmatrix} A_1 \\ A_2 \\ A_3 \end{pmatrix} \tag{4.38}$$

where r_{ii} and t_{ij} are the reflection and transmission coefficients.

Besides analyzing three ports, a heat-transport model of the center ring is also needed. Here, we divide the ring into three sections to describe the heat exchange process between the ring and ports. Under the assumption that only heat transport along the toroidal direction is considered, here, the temperature field in the ring can be described as

$$T_{\text{ring}}(\theta,\omega)=\begin{cases}C_1e^{ik_+r_0(\theta-\theta_1)}+D_1e^{-ik_-r_0(\theta-\theta_1)},\theta_1=-\frac{\pi}{2},\ \text{section 1}\\ C_2e^{ik_+r_0(\theta-\theta_2)}+D_2e^{-ik_-r_0(\theta-\theta_2)},\theta_2=\frac{\pi}{6},\ \text{section 2}\\ C_3e^{ik_+r_0(\theta-\theta_3)}+D_3e^{-ik_-r_0(\theta-\theta_3)},\theta_3=\frac{5\pi}{6},\ \text{section 3}\end{cases} \tag{4.39}$$

where θ is the azimuth variable in the range of $[-\frac{\pi}{3},\frac{\pi}{3}]$.

Hereto, the governing equations of the ports and ring are given. To investigate the transmission properties of this thermal metamaterial, we have to obtain the scattering matrix $\boldsymbol{S}$ by thermally intercoupling Eq. (4.37) and Eq. (4.39). Under the adiabatic boundary assumption, the ring can only gain or lose heat energy via thermal contact with three ports. Therefore, to establish the thermal scattering theory, we should start with the thermal matching of contact interfaces between the ring and ports. According to the continuity of the temperature field, we can get the following matching equations,

$$\begin{cases}T_{p1}(r_0)=T_{r1}(\theta_1-\pi/3)=T_{r3}(\theta_3+\pi/3)\\ T_{p2}(r_0)=T_{r2}(\theta_2-\pi/3)=T_{r1}(\theta_1+\pi/3).\\ T_{p3}(r_0)=T_{r3}(\theta_3-\pi/3)=T_{r2}(\theta_2+\pi/3)\end{cases} \tag{4.40}$$

It means that the temperature at the interfaces should satisfy the field equations of both the ring and the ports.

Furthermore, regarding the contact area as a node, the continuity condition of the heat flux at three nodes provides another set of matching equations, which is ruled by the energy conservation law. Thus, thermally intercoupling the ring and ports using matching conditions $T_1=T_2$ and $-\kappa\frac{\partial T_1}{\partial x}+\rho c_p\boldsymbol{v}T_1=-\kappa\frac{\partial T_2}{\partial x}+\rho c_p\boldsymbol{v}T_2$, the transmission coefficients at ports 2 and 3 can be obtained,

$$S_{21}=\left[k_0(\tau_+-\tau_+^2\tau_-)+(k_++k_-)(\tau_-^2+\tau_++\tau_+^2\tau_-)\right]/C \tag{4.41}$$

$$S_{31}=\left[k_0(\tau_--\tau_-^2\tau_+)+(k_++k_-)(\tau_+^2+\tau_-+\tau_-^2\tau_+)\right]/C \tag{4.42}$$

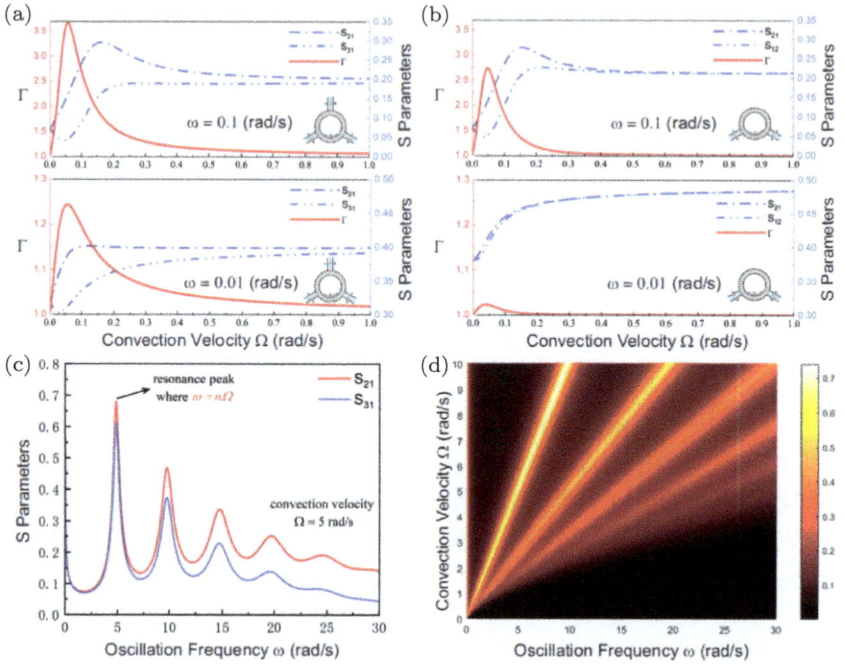

Fig. 4.8. Transmission properties revealed by the scattering matrix $\boldsymbol{S}$. (a) Effect of varying the biasing convection velocity Ω on the nonreciprocal transmission properties of the thermal circulator at different thermal wave frequencies ω. (b) Effect of varying the biasing convection velocity Ω on the nonreciprocal transmission properties of the two-port thermal metadevice at different thermal wave frequency ω. (c) Resonance of scattering coefficients S_{21} and S_{31}, which emerges when an integer multiple relation is satisfied between ω and Ω. (d) Resonance effect of scattering coefficient S_{21} of the circulator as a function of convection velocity Ω and thermal wave frequency ω. Adapted from Ref. [51].

where $e^{ik_+ 2\pi r_0/3}$ and $e^{ik_- 2\pi r_0/3}$ are represented by τ_+ and τ_-, and C is a coefficient.

In this relation, Γ varies with the thermal wave frequency ω and the convection velocity $\boldsymbol{v}$. As revealed by the red line in Fig. 4.8a, at a common oscillatory frequency ω, there exists a nontrivial relationship between Γ and circulation velocity. As $\boldsymbol{v}$ increases, nonreciprocity initially strengthens and then weakens, implicating the existence of an optimal working point in this process. Physically, the weakening in nonreciprocity at high convection velocities can be attributed to the equivalent infinite-thermal-conductivity effect, wherein the entire system restores its original reciprocity. Furthermore, we have observed a strong resonance phenomenon in the circulator. As shown in Fig. 4.8c, d, when an integer multiple relation is satisfied between thermal wave frequency ω and rotational angular velocity Ω, (e.g., $\omega = N\Omega$), the transmission coefficients S_{21} and S_{31} will be significantly enhanced, which means that with the help of the resonance effect, the thermal signal from the incident terminal can be well transmitted to the outgoing terminals.

4.4.2. *Scattering Theory Considering the Boundary Conditions*

To this end, it seems that when achieving thermal nonreciprocity through external bias, the circulator makes no significant difference compared with the two-port metadevice, and the study of the circulator is merely an incremental work following its two-port counterpart. Moreover, upon close examination of the scattering coefficients S_{21} and S_{31}, as ω approaches zero, nonreciprocity weakens and even vanishes (Fig. 4.8a, b), regardless of whether it is a three-port or a two-port circulating thermal meta-device. Although this phenomenon can be understood as the thermal signals disregarding the existence of the scattering ring when the wavelength is significantly larger than the ring size, which is a common occurrence in the wave community, it still presents a confusing problem for us inevitably. Is it time to conclude that nonreciprocity occurs at nonzero frequency but vanishes at zero frequency, which applies to both the two-port and three-port thermal meta-devices? Fortunately, after thoroughly investigating the scattering process, we find that this idea is untrue.

Generally, the scattering matrix is a physical quantity representing the transmission relationship between input and output signals, independent of boundary conditions. The mathematical model used to derive the scattering matrix $\boldsymbol{S}$ represents a system with open-boundary conditions. That is, the outgoing signals can be transmitted to the infinity of the output ports with no hindrance on the way. However, as an inherently dissipative process, the scattering behavior of heat signals is influenced by the scatterer and the loss and boundary sources of the three transmission channels, which are factors that cannot be ignored in practical models. From this viewpoint, the open-boundary condition above needs to be revised as

$$\begin{cases} T_{p1} = T_0 + Ae^{-i\omega t} \\ T_{p2} = T_0 \\ T_{p3} = T_0 \end{cases} \tag{4.43}$$

where a heat signal with amplitude A and oscillation frequency ω is added to port 1, simulating the $(1,0,0)^{\mathrm{T}}$ input scenario based on the global reference temperature T_0. By combining the boundary conditions outlined in Eq. (4.43) with the thermal matching conditions, we can obtain the signal amplitudes at ports 2 and 3 of the thermal circulator,

$$\begin{cases} B_2 = (1+\tau_0)\dfrac{(\tau_+^2\tau_- - \tau_+)}{C(k_+ + k_-)} + (\tau_0 - 1)\dfrac{(\tau_-^2 + \tau_+ + \tau_+^2\tau_-)}{Ck_0} \\ B_3 = (1+\tau_0)\dfrac{(\tau_+\tau_-^2 - \tau_-)}{C(k_+ + k_-)} + (\tau_0 - 1)\dfrac{(\tau_+^2 + \tau_- + \tau_+\tau_-^2)}{Ck_0} \end{cases} \tag{4.44}$$

with

$$\begin{aligned} C = \Bigg[& \frac{k_0}{(k_+ + k_-)^2}(1+\tau_0)^3\left(-1+\tau_+\tau_-\right)^3 \\ & + \frac{3}{(k_+ + k_-)}\left(-1+\tau_0\right)\left(1+\tau_0\right)^2\left(-1+\tau_+\tau_-\right)^2\left(1+\tau_+\tau_-\right) \\ & + \frac{3}{k_0}\left(-1+\tau_0\right)^2\left(1+\tau_0\right)\left(-1+\tau_+^3\tau_-^3\right) \\ & + \frac{k_+ + k_-}{k_0^2}(-1+\tau_0)^3(-1+\tau_+^3)(-1+\tau_-^3)]/[2\sqrt{\tau}_0(\tau_+\tau_- - 1)\Bigg] \end{aligned}$$

being unimportant in the calculation of nonreciprocity B_2/B_3 because it can be canceled out. Besides, for simplicity, $e^{ik_+2\pi r_0/3}$, $e^{ik_-2\pi r_0/3}$, and $e^{2ik_0x_0}$ are represented by τ_+, τ_- and τ_0 here. In this case, the theoretical nonreciprocity of the practical thermal circulator can be expressed as $\Gamma' = |S_{21}{}'/S_{31}'| = |B_2/B_3|$, whose explicit expression is

$$\Gamma' = \frac{\frac{\left(1+e^{2ik_0x_0}\right)}{\left(e^{2ik_0x_0}-1\right)}\frac{\left(e^{2ik_-L+4ik_+L}-e^{2ik_+L}\right)}{(k_++k_-)}+\frac{\left(e^{4ik_-L}+e^{2ik_+L}+e^{2ik_-L+4ik_+L}\right)}{k_0}}{\frac{\left(1+e^{2ik_0x_0}\right)}{\left(e^{2ik_0x_0}-1\right)}\frac{\left(e^{2ik_+L+4ik_-L}-e^{2ik_-L}\right)}{(k_++k_-)}+\frac{\left(e^{4ik_+L}+e^{2ik_-L}+e^{2ik_+L+4ik_-L}\right)}{k_0}} \tag{4.45}$$

with $L = \pi r_0/3$.

As illustrated in Fig. 4.9a, under a small frequency (e.g., $\omega = 0.01\,\mathrm{rad/s}$), an unexpected discrepancy appears between the results of Γ and Γ'. However, what is puzzling is that this phenomenon

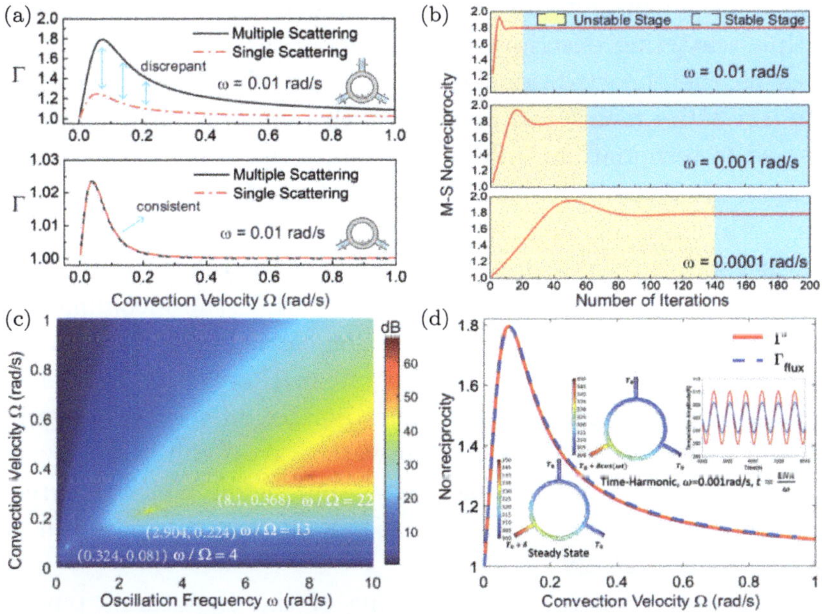

Fig. 4.9. Transmission properties of our thermal metadevices, where theory and simulations are combined for comparison. (a) The discrepant (consistent) nonreciprocity between single and multiple scattering theory of thermal circulator (two-port thermal metadevice). (b) Multiple-scattering theory describing the nonreciprocity of the circulator under zero-approaching frequency excitation. The yellow part is the unstable stage, and the blue is the stable stage. (c) The nontrivial dependence of nonreciprocity on convection velocity Ω and excitation frequency ω, where several nonreciprocal local maximum points emerge when a specific integer multiple relation is satisfied between ω and Ω. (d) A demonstration of the equivalence between thermal-scattering analysis and heat-flux analysis. Adapted from Ref. [51].

fails to appear in the two-port meta-device, where the nonreciprocity relations described by Γ and Γ' are always in high consistency.

Such problems further provoke thought: how can we explain this significant discrepancy between the nonreciprocity described by models without and with boundaries? And if Γ' is a function conforming to reality, does that mean the relation depicted by Γ is completely wrong?

4.4.3. *Multiple-Scattering Enhanced Thermal Nonreciprocity*

To gain deeper insights into this phenomenon, here we propose the multiple-scattering description of signals in thermal systems, which serves as the theory bridge between Γ and Γ'. Under the open-boundary setup mentioned previously, the outgoing heat signals are assumed to terminate at infinity. That is, although the temperature profile of the input port conforms to the form $T = Ae^{ik_0x} + Be^{-ik_0x}$ with both incoming and outgoing temperature field components, when it comes to the temperature profile of the output ports, there exists only outgoing component Be^{-ik_0x}, while the incoming term Ae^{ik_0x} is absent. As such, the information carried by $\boldsymbol{S}$ only pertains to the single-scattering case of this system. However, with a closed boundary in place, can thermal waves bounce off the boundary like electromagnetic, acoustic, and even water waves? Subsequently, the reflected signal will function as a new input signal, which undergoes a multiple-scattering process until being entirely dissipated in the thermal system [27, 28].

For a more intuitive understanding of this hypothetic physical process, we start by considering the reflection coefficients of the three-channel boundaries. Herein, we define the reflection matrix as $\boldsymbol{R}$, then the reflection process at the channel boundaries can be described as

$$\begin{pmatrix} C_1 \\ C_2 \\ C_3 \end{pmatrix} = \boldsymbol{R} \begin{pmatrix} B_1 \\ B_2 \\ B_3 \end{pmatrix} = \begin{pmatrix} \widetilde{r_1} & 0 & 0 \\ 0 & \widetilde{r_2} & 0 \\ 0 & 0 & \widetilde{r_3} \end{pmatrix} \begin{pmatrix} B_1 \\ B_2 \\ B_3 \end{pmatrix} \tag{4.46}$$

where C_1, C_2, and C_3 are the amplitudes of the reflected fields at the boundaries, and $\widetilde{r_1}$, $\widetilde{r_2}$, and $\widetilde{r_3}$ are reflection coefficients.

Such boundary conditions require that the boundary temperatures are not changed in the whole process, so we get the constraints $\sum_{i=1}^{3}\left|C_i e^{-ik_0x_0}+B_i e^{ik_0x_0}\right| = 0$. The three reflection coefficients could be derived as $\widetilde{r}_{1,2,3} = -e^{2ik_0x_0}$, and the total magnitudes of the outgoing fields can be expressed as the superposition of the signals after multiple scattering, which is an infinite sequence,

$$\begin{pmatrix} Z_1 \\ Z_2 \\ Z_3 \end{pmatrix} = \boldsymbol{S}' \begin{pmatrix} A_1 \\ A_2 \\ A_3 \end{pmatrix} = \sum_{n=1}^{+\infty} (\boldsymbol{SR})^{n-1} \boldsymbol{S} \begin{pmatrix} A_1 \\ A_2 \\ A_3 \end{pmatrix} \tag{4.47}$$

where Z_1, Z_2 and Z_3 are the output heat signals after infinite scatterings [29].

For better visualization of this physical process, the variation trend of $\Gamma^{(n)}$ with the number of iterations n is shown in Fig. 4.9b, where three examples in the zero-ω limit are given. During calculation, $|S_{21}/S_{31}|$ is taken as the initial value, which means $\Gamma^{(n)}$ equals $|S_{21}/S_{31}|$ in the case $n = 1$. With the increase of n, the multiple-scattering nonreciprocity $\Gamma^{(n)}$ will be significantly enhanced and converge to a specific value after a certain number of iterations. At this moment, the value of $\Gamma^{(n)}$ has changed from the single-scattering nonreciprocity $|S_{21}/S_{31}|$ to the multiple-scattering nonreciprocity $\Gamma' = |S'_{21}/S'_{31}|$. Our analysis reveals that as ω approaches zero, the nonreciprocity calculated by the multiple-scattering nonreciprocity Γ' requires more iterations to converge. The oscillation frequency ω and the number of scattering n are in a competitive relationship. The calculated nonreciprocity numerically satisfies the relation $\lim_{n\to+\infty}\Gamma^{(n)} = |S'_{21}/S'_{31}|$, which is in perfect agreement with our theory.

As the number of iterations approaches infinity, we can also obtain the analytical expression of the multiple scattering coefficients of the whole system, as shown below,

$$\boldsymbol{S}' = \sum_{n=0}^{+\infty} (\boldsymbol{SR})^{n} \boldsymbol{S}$$

$$= \begin{pmatrix} S'_{11} & \dfrac{\widetilde{r}S_{21}^2+S_{31}-\widetilde{r}S_{11}S_{31}}{C} & \dfrac{\widetilde{r}S_{31}^2+S_{21}-\widetilde{r}S_{11}S_{21}}{C} \\ \dfrac{\widetilde{r}S_{31}^2+S_{21}-\widetilde{r}S_{11}S_{21}}{C} & S'_{22} & \dfrac{\widetilde{r}S_{21}^2+S_{31}-\widetilde{r}S_{11}S_{31}}{C} \\ \dfrac{\widetilde{r}S_{21}^2+S_{31}-\widetilde{r}S_{11}S_{31}}{C} & \dfrac{\widetilde{r}S_{31}^2+S_{21}-\widetilde{r}S_{11}S_{21}}{C} & S'_{33} \end{pmatrix} \tag{4.48}$$

with

$$C = -[\widetilde{r}^3(S_{11}^3 + S_{21}^3 + S_{31}^3 - 3S_{11}S_{21}S_{31}) + 3\widetilde{r}^2 (S_{21}S_{31} - S_{11}^2) + 3\widetilde{r}S_{11} - 1]$$

and

$$S'_{11} = S'_{22} = S'_{33} = (\widetilde{r}^2(S_{11}^3 + S_{21}^3 + S_{31}^3 - 3S_{11}S_{21}S_{31}) + 2\widetilde{r}(S_{21}S_{31} - S_{11}^2) + S_{11})/C.$$

Here, S_{11}, S_{21}, and S_{31} are all scattering coefficients of the single-scattering case. Accordingly, the multiple-scattering nonreciprocity Γ' can be expressed as a function of the single scattering coefficients

$$\Gamma' = \left| \frac{\widetilde{r}S_{31}^2 + S_{21} - \widetilde{r}S_{11}S_{21}}{\widetilde{r}S_{21}^2 + S_{31} - \widetilde{r}S_{11}S_{31}} \right| \tag{4.49}$$

where $\widetilde{r}$ is the reflection coefficient of the boundary. To explore the properties of Eq. (4.49) in greater depth, we show its dependence on thermal wave frequency and rotation velocity in Fig. 4.9c across a broad parameter range, where a nontrivial relationship has been revealed. Interestingly, when a specific integer multiple relation is satisfied between ω and Ω, several nonreciprocal local maximum points emerge, which provides us with a meaningful reference for practical applications. Counterintuitively, though scattering coefficients S_{21} and S_{31} are almost equal in the zero-frequency limit, Γ' still conveys a significant steady-state nonreciprocity in the circulator.

Besides, it is important to note that although higher frequencies may exhibit greater nonreciprocity, it does not necessarily imply that higher frequencies are more advantageous in practical use. Higher frequencies are often associated with increased losses, a common phenomenon in wave transport. Thus, balancing the desired nonreciprocity and energy transfer efficiency becomes crucial when considering practical applications.

4.4.4. *Relationship Between Scattering and Heat-Flux Analyses*

In the above section, we studied the nonreciprocity of the system based on the scattering coefficients. That is, since S_{21} and S_{31}

quantify the heat signals transmitted from port 1 to ports 2 and 3, nonreciprocity can naturally be expressed as $|S_{21}/S_{31}|$, which measures the isolation rate of the system.

However, from a traditional heat conduction process perspective, the heat field cannot be divided into forward and backward transmission terms. Accordingly, under the classical frame where thermal scattering theory was absent, heat flux rather than scattering coefficients should quantify the extent of nonreciprocity. Therefore, considering that a new physical quantity $|S_{21}/S_{31}|$ is constructed to measure the nonreciprocity of the system, it is necessary to prove its equivalence with the classical theory; more specifically, the equivalence between the scattering analysis and heat-flux analysis.

To do this, we start from the plane wave solution $T(x,t) = (Ae^{ikx} + Be^{-ikx})e^{-i\omega t} + T_0$ of the thermal field in the scattering theory. According to the Fourier law of heat conduction, the heat flux going through a solid port is $q = -\kappa \frac{\partial T}{\partial x}$. Furthermore, we have to take the multiple scattering effect into account, and then the total heat flux in the transmission port can be expressed as

$$q = -\kappa \frac{\partial T}{\partial x} = -\kappa \sum_{n=1}^{\infty} (A^{(n)} ike^{ikx} - B^{(n)} ike^{-ikx})e^{-i\omega t}. \tag{4.50}$$

Accordingly, the nonreciprocity measured by heat flux is

$$\frac{q_2}{q_3} = \frac{\sum_{n=1}^{\infty} \left(A_2^{(n)} e^{ikx} - B_2^{(n)} e^{-ikx}\right)}{\sum_{n=1}^{\infty} \left(A_3^{(n)} e^{ikx} - B_3^{(n)} e^{-ikx}\right)}. \tag{4.51}$$

According to the mathematical relation $\sum_{n=1}^{\infty} B^{(n)} = B'$, the formula above can be written as

$$\frac{q_2}{q_3} = \frac{\sum_{n=1}^{\infty} A_2^{(n)} - B_2' e^{-2ikx_0}}{\sum_{n=1}^{\infty} A_3^{(n)} - B_3' e^{-2ikx_0}}. \tag{4.52}$$

Furthermore, the boundary-reflection assumption gives the relation $A^{(n+1} = B^{(n)}\tilde{\boldsymbol{r}}$, which means that the nth outgoing field amplitude $B^{(n)}$ is reflected by the boundary and then functions as the $(n+1)$th incident field, thus $\sum_{n=1}^{\infty} A_i^n = \sum_{n=1}^{\infty} B_i^{n-1}\tilde{\boldsymbol{r}} = \sum_{n=1}^{\infty} B_i^n \tilde{\boldsymbol{r}} = \tilde{\boldsymbol{r}} B_i'$.

Therefore, q_2/q_3 can be further simplified into the following form,

$$\frac{q_2}{q_3} = \frac{\widetilde{\boldsymbol{r}}B_2{}' - B_2' e^{-2ikx_0}}{\widetilde{\boldsymbol{r}}B_3{}' - B_3' e^{-2ikx_0}} = \frac{B_2'}{B_3'}. \tag{4.53}$$

So far, we have proved mathematically that in terms of analyzing the nonreciprocity of thermal systems, a solid equivalence exists between our scattering theory and the traditional heat-flux analysis method. As shown in Fig. 4.9d, the nonreciprocity Γ' calculated by our multiple scattering theory fits well with the nonreciprocity Γ_{flux} obtained through the traditional heat flux analysis, which lays a solid foundation for the extension of the scattering theory to the steady-state heat transfer scenario.

Based on this viewpoint, we can envision that even when the wave frequency ω is strictly equal to 0, the nonreciprocal transmission of heat signal can still be achieved in the thermal circulator. According to our scattering theory, the nonreciprocity $\Gamma'_{\omega=0}$ at this point is predicted to be

$$\Gamma'_{\omega=0} = \frac{\left(e^{\frac{-2\pi r_0 v}{3D}} - 1\right) - \frac{v}{D}x_0\left(e^{\frac{-4\pi r_0 v}{3D}} + e^{\frac{-2\pi r_0 v}{3D}} + 1\right)}{e^{\frac{-2\pi r_0 v}{3D}}\left(e^{\frac{-2\pi r_0 v}{3D}} - 1\right) - \frac{v}{D}x_0\left(e^{\frac{-4\pi r_0 v}{3D}} + e^{\frac{-2\pi r_0 v}{3D}} + 1\right)} \tag{4.54}$$

where x_0 is the length of the transmission port.

We have realized artificially tunable nonreciprocal heat transfer under dynamic and steady-state excitations. We successfully brought these two cases under the same theoretical framework by uncovering the nonnegligible multiple scattering effects in the zero-frequency limit. On a fundamental level, we bestow the thermal circulator with the ability to work properly in a far more complicated setting rather than the fixed input signal and limited working temperature range. Such a high degree of dynamical tunability without mechanical adjustment is paramount for numerous applications, such as directional heat harvesting, effective heat dissipation of integrated electronic products, and the dynamic switching of thermal topological states. Furthermore, by revealing the nonreciprocity enhancement phenomenon caused by the boundary reflection effect of thermal waves, this work bridges the theoretical gap between the dynamic thermal scattering analysis and traditional heat flux analysis. A similar concept may be extended to a broad range of heat-transfer-related problems and even other fields, such as acoustics and mechanics, if further consideration is given.

4.5. Non-Hermitian Scattering Symmetry in Purely Lossy Media

The Hamiltonian interprets how the system evolves, while the scattering coefficients describe how the system responds to inputs. In investigating scattering properties, a common belief is that the incident and scattered signals should ideally not be further modified by the scattered channels. Therefore, most previous works based their analysis on the assumption of lossless scattering channels. Inside a lossless channel, a wave field F would commonly satisfy the time-reversal (T) symmetric wave equation $\partial_t^2 F = a^2\nabla^2 F$, where a is a real parameter, and ∇^2 is the Laplacian operator. The T-symmetry is reflected in the dispersion relation as shown in Fig. 4.10a,b (black and gray lines), where all the wavenumbers are purely real. The time-reversal operation maps the state with positive wavenumber kc to that with $-kc$. With the advent of non-Hermitian physics [30–32], it was recently realized that lossy systems can exhibit intriguing novel effects, including but not limited to enhanced sensing capability [33, 34] and loss-induced lasing [35, 36]. However, the studies on non-Hermitian scattering have long been captivated by the assumption of lossless scattering channels. The significance of the non-Hermiticity of the scattering channel has thus been completely overlooked so far.

This section analyzed the interplay between the scattering channels and the scatterer, manifested in the dispersion relation. In particular, we report a distinct kind of scattering channel that is lossy and satisfies time-reversal anti-symmetry (anti-T-symmetry) rather than being T-symmetric.

4.5.1. *Time-Reversal in T-symmetric and Anti-T-symmetric Channels*

Mathematically, anti-T-symmetry requires that the channel's Hamiltonian H_C and the time-reversal operator T are anti-commutation $\{Hc, \mathrm{T}\} = 0$. The feature may appear strange, but it can be shown to

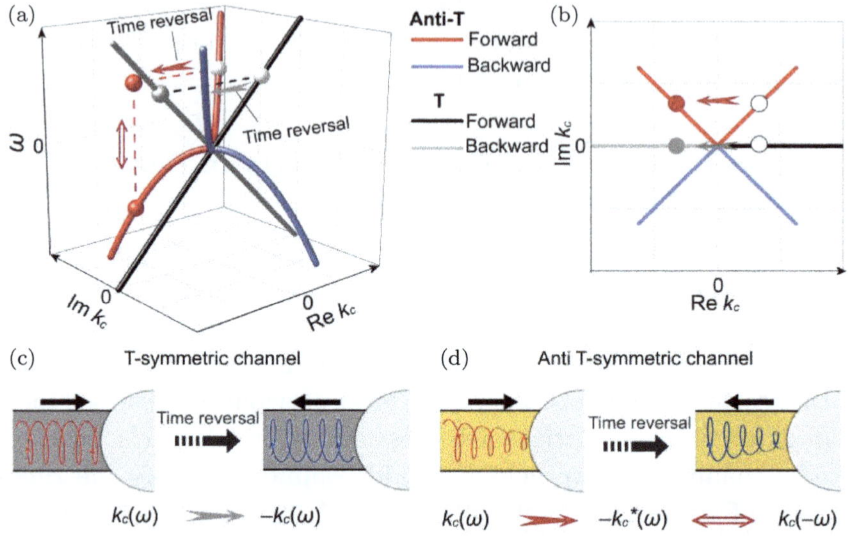

Fig. 4.10. Time-reversal in T-symmetric and anti-T-symmetric channels. (a) Dispersion of time-harmonic field in anti-T-symmetric (red or blue lines) and T-symmetric (black or gray lines) channels. The forward and backward modes are divided according to the positive and negative phase velocity $v_p = Re(\omega)/Re(k_c)$. The white balls are the initial states, and the red and gray balls are the states after the time-reversal operation. (b) Top view of the dispersion curves. (c) and (d) illustrate the physical field after time-reversal operation in the T and anti-T channels, respectively. Adapted from Ref. [52].

have a clear and fundamental physical origin. To see this, note that if we change the second-order time derivative in the wave equation to a first-order time derivative, $\partial_t^2 F \rightarrow \partial_t F$, the Hamiltonian $H_c = ia^2\nabla^2$ exactly obeys anti-T-symmetry. The resultant governing equation, $\partial_t F = a^2\nabla^2 F$, is nothing but the diffusion equation with $a^2 = D$ being the diffusivity. Therefore, as the most fundamental form of diffusive systems, heat transfer undertakes an emerging platform for studying non-Hermitian physics.

The anti-T-symmetric dispersion is shown in Fig. 4.10a,b (red and blue lines). Since the wavenumbers are complex for real frequencies, the time-reversal operation maps kc to $-kc^*$. It is worth noting that the state under time-reversal operation does not fall on the dispersion curve of the scattering channel at equal frequency, which instead goes to another branch at negative frequency [37]. The time-reversal operation on fields in anti-T symmetric channels acts as a linkage between positive and negative frequencies. Fig. 4.10c,d demonstrate the effects and differences of time-reversal operation on the physical fields in the T-channel and anti-T-channel.

Previous studies considering lossless scattering channels fail to reveal some symmetries in non-Hermitian systems. Anti-parity-time (APT) symmetry defined by $\{PT, H\} = 0$ is a typical non-Hermitian symmetry. Such Hamiltonians have aroused great interest [38–40]. However, all existing works either focus on isolated systems (Fig. 4.11a) or forcefully assume the channels to be T-symmetric (Fig. 4.11b).

We find a clear distinction between the symmetry of the Hamiltonian and the symmetry of the scattering, as shown in the scenario of Fig. 4.11a and Fig. 4.11b. That explains why previous works only use the scattering coefficient to reflect the symmetry of the scatterer's Hamiltonian, but scattering symmetry at the entire system level cannot be manifested.

We reveal a unique scattering symmetry of APT systems preserved by anti-T-symmetric or diffusive channels (Fig. 4.11c). The symmetry of scattering is essentially different from the symmetry of a Hamiltonian. The symmetric and symmetry-broken scattering phases are demonstrated with the distinguished behaviors in field evolutions. In addition, we find a specific excitation frequency at which the scattering is always in the symmetry-broken phase. Such a thresholdless symmetry-breaking feature indicates a mode decoupling under infinitesimal perturbations, a unique feature brought by the signal rather than the system's properties.

4.5.2. *Anti-parity-time Symmetry in Scattering and Hamiltonian*

Without loss of generality, we use a heat transfer system, a typical diffusive process, to study the properties of the anti-T channel. Heat transfer plays a key role in energy transport, but the non-Hermitian scattering of heat transfer has not yet been studied due to the

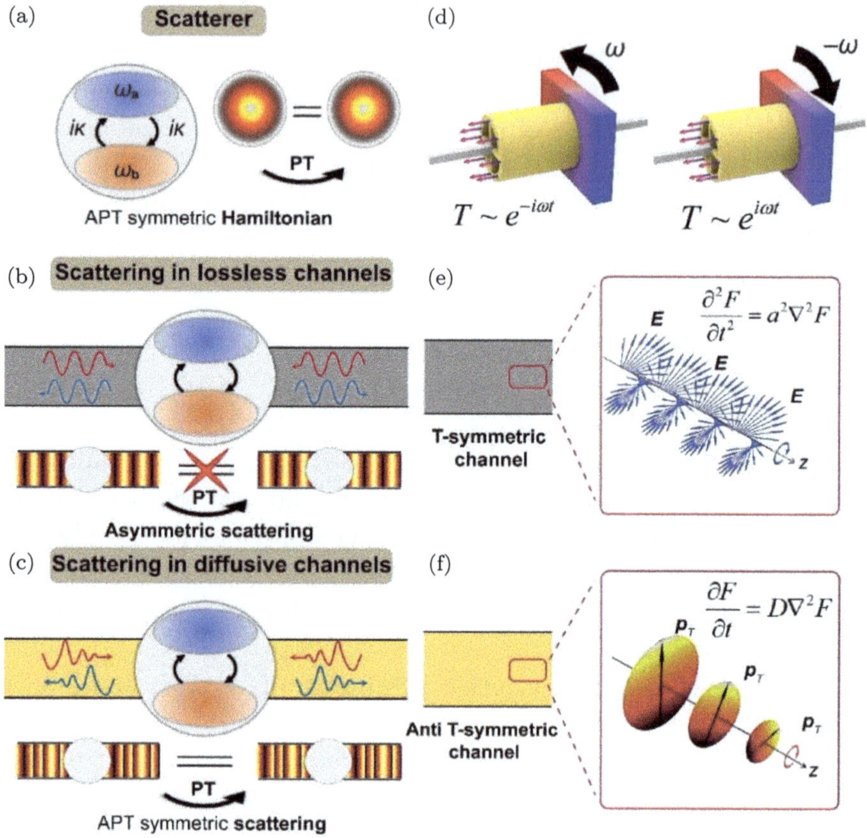

Fig. 4.11. Schematic of (a) scatterer with APT-symmetric Hamiltonian (left) and the field (right). (b) Asymmetric scattering: APT scatterer with lossless channels (up) and the field (down). (c) APT symmetric scattering: APT scatterer with diffusive channels (up) and the field (down). (d) Experimental suggestion on the realization of positive and negative frequency in the temperature field. (e) and (f) show the time-harmonic fields propagating in T-symmetric (wave system) and anti-T-symmetric channel (diffusive system). Adapted from Ref. [52].

insurmountable obstacle of decomposing the temperature field into meaningful heat signals. Inspired by the special dispersion of the anti-T channel, the heat signal with real-number frequency is constructed by the rotating heat source illustrated in Fig. 4.11d. It can be proved that the rotation directions are equivalent to the positive and negative frequencies. This treatment provides us with the possibility to define chirality in heat signals. Like the circular polarization of electromagnetic fields in T-symmetric channels (Fig. 4.11e), we could design a special kind of signal with chirality (Fig. 4.11f), which is excited by our proposed rotating chiral heat source. Furthermore, we provide a method for calculating the 2D temperature dipole. In electrodynamics, the dipole moment is defined as

$$p_e = \iint r\sigma_e dS \tag{4.55}$$

where σ_e is charge density. In the heat transfer system, we make an analogy to define the temperature dipole in the xy-plane,

$$p_T = \iint r\sigma_T dS = i\iint x\sigma_T dS + j\iint y\sigma_T dS = p_{Tx} + p_{Tx} \tag{4.56}$$

where σ_T is the temperature configuration in the cross-section. The sign change of frequency results in the flip of chirality (e.g., from LCP to RCP in Fig. 4.10c), which can directly reflect the effect of the time-inversion operation.

The general heat transfer system is generated by the fixed temperature heat source, which only provides a steady heat flow without carrying any other information about the heat signal. However, our above design introduces a time-harmonic oscillating heat source with the concept of chirality, enriching additional degrees of freedom to manipulate heat signals and investigate the propagating or scattering properties of the field in anti-T channels. When a thermally conductive cylinder is in contact with a rotating heat source, as shown in Fig. 4.11d, the temperature distribution in the cylinder satisfies the form of a time-harmonic temperature field: $T_c(z, \phi, t) = F_c(z)e^{i\phi - i\omega t}$. Here, we postulated that the medium is homogeneous along the radial direction when the cylinder's thickness is small. Then, we let the space-dependent part $F_c(z)$ have a fundamental solution of wave-like form,

$$F_c(z) = Ae^{ik_c z} + Be^{-ik_c z}. \tag{4.57}$$

By substituting $T_c(z, \phi, t)$ into the heat diffusion equation $\partial_t T_c = D\partial_z^2 T_c$ (D is thermal diffusivity), we calculate the wavenumber k_c in the anti-T channel. A and B can be complex numbers to incorporate phases in the fields. This treatment lets the temperature fields in the anti-T channel have the form of 1D wave scattering, making the heat originally thought to lack degrees of freedom come alive at the signal level. As an exemplary case to illustrate the scattering features of the anti-T channel, we consider the scatterers composed of spinning units, which are presented in the following sections.

4.5.3. *Thermal Scattering Properties of a Rotating Unit*

Here, we start our investigation from the scattering properties of a single rotating ring. Our model is depicted in Fig. 4.12. The ring rotates at speed Ω.

For thermal scattering in the z-direction, the temperature field in the lower (upper) channel T_{c1} (T_{c2}) and rotating unit T_R satisfy the governing equations for heat transfer,

$$\begin{cases} \dfrac{\partial T_{c1}}{\partial t} = D_c \dfrac{\partial^2 T_{c1}}{\partial x^2} + D_c \dfrac{\partial^2 T_{c1}}{\partial z^2}, & z \leq -b \\ \dfrac{\partial T_R}{\partial t} = D \dfrac{\partial^2 T_R}{\partial x^2} - v \dfrac{\partial T_R}{\partial x} + D \dfrac{\partial^2 T_R}{\partial z^2}, & -b \leq z \leq b \\ \dfrac{\partial T_{c2}}{\partial t} = D_c \dfrac{\partial^2 T_{c2}}{\partial x^2} + D_c \dfrac{\partial^2 T_{c2}}{\partial z^2}, & z \geq b \end{cases} \tag{4.58}$$

where D_c and D are the diffusion numbers for channels and the rotating unit, respectively, $v = \Omega R$ is the linear velocity on the rotating ring, Ω is the angular velocity of the rotating unit, R is the radius, and $2b$ is the thickness of the rotating ring.

As mentioned above, in thermal scattering, the temperature fields T_{c1}, T_R, and T_{c2} satisfy the following form,

$$\begin{cases} T_{c1} = T_0 + \left[A_1 e^{ik_c(z-z_1)} + B_1 e^{-ik_c(z-z_1)}\right] e^{ik_x x - i\omega t}, & z \leq z_1 \\ T_R = T_0 + \left(B_{R+} e^{ik_R z} + B_{R-} e^{-ik_R z}\right) e^{ik_x x - i\omega t}, & z_1 \leq z \leq z_2 \\ T_{c2} = T_0 + \left[A_2 e^{ik_c z_2} e^{ik_c(z-z_2)} + B_2 e^{-ik_c(z-z_2)}\right] e^{ik_x x - i\omega t}, & z \geq z_2 \end{cases} \tag{4.59}$$

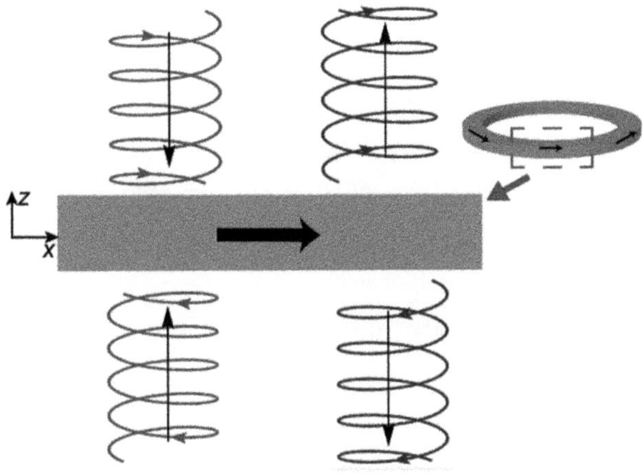

Fig. 4.12. Scattering model of the single rotation unit. The inset is a 3D model. Adapted from Ref. [52].

where k_c and k_R are the wavenumbers of the temperature field in channels and rotating units, respectively, with $z_1 = -b$ and $z_2 = b$.

By substituting Eq. (4.59) into Eq. (4.58), the dispersion relations for k_c and k_R are obtained,

$$\begin{cases} k_c^2 = -k_x^2 + \dfrac{i\omega}{D_c} \\ k_R^2 = -k_x^2 + \dfrac{i(\omega - \Omega)}{D_c} \end{cases}. \tag{4.60}$$

With the periodicity of the ring $T(x) = T(x + 2\pi R)$, we have used the relation $k_x v = 1/R \cdot \Omega R = \Omega$, regarded as a transverse wavenumber. The positive and negative k_c represent propagating parts in z and $-z$ directions, respectively.

At the interfaces between the channel and rotation unit, the matching conditions require

$$\begin{cases} T_{c1}(-b) = T_R(-b), & T_{c1}(-b) = T_R(-b) \\ \kappa_c \dfrac{\partial}{\partial z} T_{c1}(-b) = \kappa_R \dfrac{\partial}{\partial z} T_R(-b), & \kappa_R \dfrac{\partial}{\partial z} T_R(b) = \kappa_c \dfrac{\partial}{\partial z} T_{c2}(b) \end{cases} \tag{4.61}$$

where κ_c and κ_R are the thermal conductivities of the channel and rotating unit.

With Eq. (4.59), Eq. (4.61) can be reduced as

$$\begin{cases} A_1 + B_1 = B_{R+}e^{-ik_Rb} + B_{R-}e^{ik_Rb}, \\ k_c\kappa_c(A_1 - B_1) = k_R\kappa_R(B_{R+}e^{-ik_Rb} - B_{R-}e^{ik_Rb}) \\ A_2 + B_2 = B_{R+}e^{ik_Rb} + B_{R-}e^{-ik_Rb} \\ k_c\kappa_c(A_2 - B_2) = k_R\kappa_R(B_{R+}e^{ik_Rb} - B_{R-}e^{-ik_Rb}) \end{cases}. \tag{4.62}$$

Rewrite these equations into matrix form:

$$\begin{cases} \begin{pmatrix} 1 & 1 \\ 1 & -1 \end{pmatrix} \begin{pmatrix} A_1 \\ B_1 \end{pmatrix} = \begin{pmatrix} e^{-ik_Rb} & e^{ik_Rb} \\ \dfrac{k_R\kappa_R}{k_c\kappa_c}e^{-ik_Rb} & -\dfrac{k_R\kappa_R}{k_c\kappa_c}e^{ik_Rb} \end{pmatrix} \begin{pmatrix} B_{R+} \\ B_{R-} \end{pmatrix} \\ \begin{pmatrix} 1 & 1 \\ 1 & -1 \end{pmatrix} \begin{pmatrix} A_2 \\ B_2 \end{pmatrix} = \begin{pmatrix} e^{ik_Rb} & e^{-ik_Rb} \\ \dfrac{k_R\kappa_R}{k_c\kappa_c}e^{ik_Rb} & -\dfrac{k_R\kappa_R}{k_c\kappa_c}e^{-ik_Rb} \end{pmatrix} \begin{pmatrix} B_{R+} \\ B_{R-} \end{pmatrix} \end{cases}. \tag{4.63}$$

Then we could get the transfer matrix $\boldsymbol{M}$ from $(A_1, B_1)^{\mathrm{T}}$ to $(A_2, B_2)^{\mathrm{T}}$, $(A_2,\ B_2)^{\mathrm{T}} = \boldsymbol{M}(A_1, B_1)^{\mathrm{T}}$, and scattering matrix $\boldsymbol{S}$, $(B_1, B_2)^{\mathrm{T}} = \boldsymbol{S}(A_1, A_2)^{\mathrm{T}}$, where $\boldsymbol{M}$ and $\boldsymbol{S}$ are

$$\boldsymbol{M} = \begin{pmatrix} \cos 2k_Rb + \frac{1}{2}\left(\gamma + \frac{1}{\gamma}\right) i\sin 2k_Rb & \frac{1}{2}\left(\gamma - \frac{1}{\gamma}\right) i\sin 2k_Rb \\ -\frac{1}{2}\left(\gamma - \frac{1}{\gamma}\right) i\sin 2k_Rb & \cos 2k_1b - \frac{1}{2}\left(\gamma + \frac{1}{\gamma}\right) i\sin 2k_Rb \end{pmatrix} \tag{4.64}$$

$$\begin{aligned} \boldsymbol{S} &= \frac{1}{M_{22}} \begin{pmatrix} -M_{21} & 1 \\ \det M & M_{12} \end{pmatrix} \\ &= \frac{1}{\mu} \begin{pmatrix} \frac{1}{2}\left(\gamma - \frac{1}{\gamma}\right) i\sin 2k_Rb & 1 \\ 1 & \frac{1}{2}\left(\gamma - \frac{1}{\gamma}\right) i\sin 2k_Rb \end{pmatrix} \end{aligned} \tag{4.65}$$

with $\gamma = k_R\kappa_R/k_c\kappa_c$ and $\mu = \cos 2k_Rb - \frac{1}{2}\left(\gamma + 1/\gamma\right) i\sin 2k_Rb$.

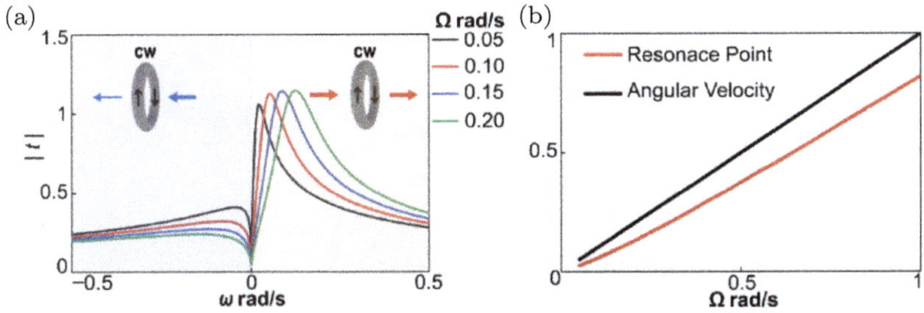

Fig. 4.13. Scattering in the single rotation unit. (a) Transmission spectra of clockwise (CW) rotating ring. (b) Resonance points of transmission spectra in (a) (red line) versus angular velocity (black line). Adapted from Ref. [52].

With the $\boldsymbol{S}$ matrix in Eq. (4.65), we find $t_{12} = t_{21} = t$ and $r_{11} = r_{22} = r$. Note that k_R and k_c are dependent on ω. Therefore, the transmission coefficient and reflection coefficient are functions of ω. In wave systems, the transmission and reflection coefficients in the scattering matrix satisfy the Stokes relation $t_{12}t_{21} + r_{11}^2 = 1$ and $r_{11} = -r_{22}$. We investigate this relation in heat transfer systems as well. From Eq. (4.65), we find that the Stokes relation is satisfied if k_R and k_c are real numbers. However, the wave numbers are complex in thermal scattering, leading to the invalidation of the Stokes relation.

With thermal scattering theory, we could obtain the transmission spectra of the scatterers, as shown in the colored lines in Fig. 4.13a. Due to the spinning structure, the transmission coefficient in negative frequency (light blue area) describes the transmission of opposite input in positive frequency.

For a clockwise rotating ring, Fig. 4.13a shows that $t(-\omega) \neq t(\omega)$, which can be explained by the structural asymmetry of this rotator. In addition, we find a resonance point in the curve that moves with different angular velocities. However, the resonance point is not equal to Ω and we infer the reason is the coupling between the spinner and channels. To illustrate this, we give Fig. 4.13b. Despite the inequivalence between resonance frequency and angular velocity, a linear relationship reflects the system's resonant features.

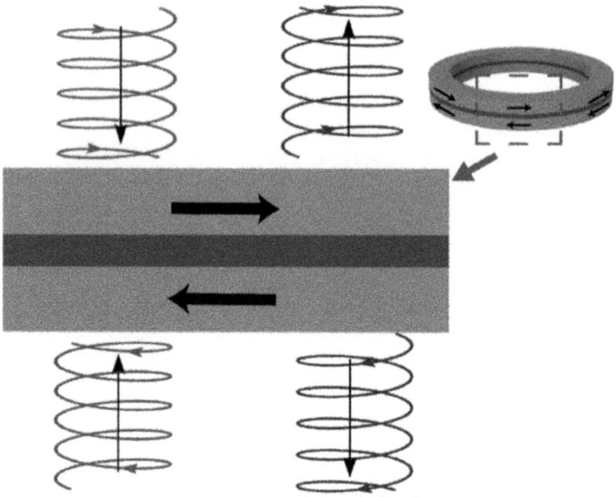

Fig. 4.14. Scattering model of APT-symmetric system. Adapted from Ref. [52].

4.5.4. *Scattering Properties in APT-Symmetric Systems*

Now, we analyze scatterers with APT symmetry. The structure is shown in Fig. 4.14, composed of two coupled rotating units with opposite rotation speeds. The static ring between them functions as a coupling layer. The APT scatterer is generated by combining the clockwise and counterclockwise spinners [1].

For thermal scattering in the z-direction, the temperature field in the lower (upper) channel T_{c1} (T_{c2}), two rings rotating in opposite directions T_1 and T_2, and isolation layer T_s satisfy the governing equations for heat transfer,

$$\begin{cases} \dfrac{\partial T_{c1}}{\partial t} = D_c\dfrac{\partial^2 T_{c1}}{\partial x^2} + D_c\dfrac{\partial^2 T_{c1}}{\partial z^2}, z \le -b \\ \dfrac{\partial T_{R1}}{\partial t} = D\dfrac{\partial^2 T_{R1}}{\partial x^2} - v\dfrac{\partial T_{R1}}{\partial x} + D\dfrac{\partial^2 T_{R1}}{\partial z^2}, -b \le z \le -a \\ \dfrac{\partial T_s}{\partial t} = D_s\dfrac{\partial^2 T_s}{\partial x^2} + D_s\dfrac{\partial^2 T_s}{\partial z^2}, -a \le z \le a \\ \dfrac{\partial T_{R2}}{\partial t} = D\dfrac{\partial^2 T_{R2}}{\partial x^2} + v\dfrac{\partial T_{R2}}{\partial x} + D\dfrac{\partial^2 T_{R2}}{\partial z^2}, a \le z \le b \\ \dfrac{\partial T_{c2}}{\partial t} = D_c\dfrac{\partial^2 T_{c2}}{\partial x^2} + D_c\dfrac{\partial^2 T_{c2}}{\partial z^2}, z \ge b \end{cases} \tag{4.66}$$

where D_c, D_s, and D are the diffusivity of the channels, isolation layer, and the two rotating rings, respectively, $v = \Omega R$ is the linear velocity on the rotating ring, Ω is the angular velocity of the rotating ring, R is the radius, $2a$ is the thickness of the isolation layer, and $d = b-a$ is the thickness of each rotating ring. Then, the scattering matrix can be obtained by a similar method in the previous section, where the transfer matrix $\boldsymbol{M}$ and scattering matrix $\boldsymbol{S}$ can be derived as

$$M = \frac{1}{2}\begin{pmatrix} \cosh m_2 + \gamma_2 \sinh m_2 & (\sinh m_2 + \gamma_2 \cosh m_2)/\eta_2 \\ \cosh m_2 - \gamma_2 \sinh m_2 & (\sinh m_2 - \gamma_2 \cosh m_2)/\eta_2 \end{pmatrix} \times \begin{pmatrix} \cosh m_s & \sinh m_s \\ \sinh m_s & \cosh m_s \end{pmatrix} \times \begin{pmatrix} \cosh m_1 + \sinh \frac{m_1}{\gamma_1} & \cosh m_1 - \sinh \frac{m_1}{\gamma_1} \\ \eta_1 \sinh m_1 + \eta_1/\gamma_1 \cosh m_1 & \eta_1 \sinh m_1 - \eta_1/\gamma_1 \cosh m_1 \end{pmatrix} \tag{4.67}$$

$$S = \begin{pmatrix} r_{11} & t_{12} \\ t_{21} & r_{22} \end{pmatrix} = \frac{1}{M_{22}} \begin{pmatrix} -M_{21} & 1 \\ \det M & M_{12} \end{pmatrix} \tag{4.68}$$

with $m_1 = \imath k_1 d$, $m_2 = \imath k_2 d$, and $m_s = 2\imath k_s a$. With Eq. (4.67), we can get the determinant of transfer matrix $\boldsymbol{M}$. $\det \boldsymbol{M} = 1$ means that the transmission coefficients in the scattering matrix satisfy $t_{12} = t_{21} = t$. $\boldsymbol{S}$ still satisfies $\boldsymbol{S} = \boldsymbol{S}^T$, and the transmission spectrum is shown in Fig. 4.15a. We give curves at different Ω, and two maximum points are observed except for the case of non-rotation.

Our calculation shows that $t(-\omega) = t(\omega)$, which is different from the case in a single rotor. Similar to the case of a single ring, resonance points are observed in the transmission spectrum of this APT scatterer. A splitting of the Hamiltonian eigenspectrum for the APT system exists, corresponding to the phase transition [41]. Thus, the peak frequencies of transmission coefficients could be predicted by the eigenfrequencies of this system, which is related to the system's Hamiltonian.

Specifically, the real parts of the Hamiltonian's eigenvalues correspond to the transmission coefficients' peaks. Here, we conducted this analysis and discovered that, despite the existence of this trend, there is no obvious splitting and phase transition, as shown in Fig. 4.15b.

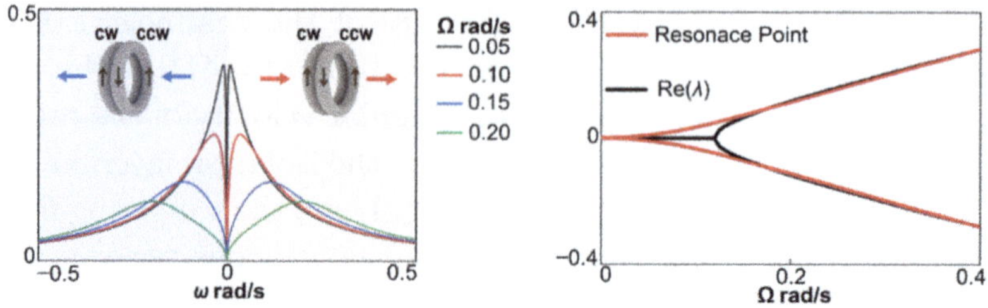

(a) (b)

Fig. 4.15. Scattering in APT-symmetric systems. (a) Transmission spectra of coupled CW and CCW rotors (APT scatterer). The light blue (red) area represents the transmission from right to left (left to right), shown as the blue (red) arrows in the inset. (b) The correspondence between eigenvalues of the scatterer's Hamiltonian and resonance points on transmission spectra in (a). The red line represents the resonance points in transmission curves. The black line indicates the real eigenspectrum to this APT scatterer's Hamiltonian. Adapted from Ref. [52].

This discrepancy was discussed in a previous work [42], and we reproduce this phenomenon in the diffusive system. However, the deeper physics behind it has not yet been revealed, and the phase transition relating to scattering needs to be further studied and discussed.

4.5.5. *Collapse and Revival of Temperature Oscillation in Scattering Phases*

For general two-port thermal scattering, the temperature field in channels has the form $T_{ci} = [A_i u_i^{\text{in}}(z,\omega) + B_i u_i^{\text{out}}(z,\omega)]$, $u_i^{\text{in}}(z,\omega)$, $u_i^{\text{out}}(z,\omega)$ are incoming and outgoing modes, and A_i, B_i are amplitudes of them, respectively. A_i and B_i are connected with scattering matrix $\boldsymbol{S}(\omega)$, $\boldsymbol{B} = \boldsymbol{S}(\omega)\boldsymbol{A}$. $A = [A_1, A_2]^{\text{T}}$ and $B = [B_1, B_2]^{\text{T}}$. By applying PT operators on both sides of T_c, we eventually have the form $(\Pi T_c)_i = [(\Pi TA)_i u_i^{\text{in}}(z,-\omega) + (\Pi TB)_i u_i^{\text{out}}(z,-\omega)]$, where $T_c = [T_{c1}, T_{c2}]^{\text{T}}$. For APT symmetric systems, if there is an eigenstate T_{ci} at ω, there must be another eigenstate $\Pi\text{T}(\boldsymbol{T}_{\text{c}})_{\text{i}}$ at $-\omega^*$. Here, since ω is real, the temperature field undergoing PT transformation should have another scattering relationship characterized by $\boldsymbol{S}(-\omega)$, where $(\Pi\text{T})\boldsymbol{B} = \boldsymbol{S}(-\omega)(\Pi\text{T})\boldsymbol{A}$. Combining the two scattering relations, we get the symmetry of the scattering matrix in APT systems, which will build a bridge linking positive and negative frequencies,

$$(\boldsymbol{PT})\boldsymbol{S}(\omega)(\boldsymbol{PT})^{-1} = \boldsymbol{S}(-\omega). \tag{4.69}$$

The symmetry of scattering in APT systems is deposited into two matrices with positive and negative frequencies. Thus, the study on genuine APT scattering is few and arduous. To demonstrate the complete symmetry of the scattering matrix, a new matrix $\boldsymbol{S}^4$ is proposed,

$$\begin{bmatrix} B_{1L} \\ B_{2L} \\ B_{1R} \\ B_{2R} \end{bmatrix} = \boldsymbol{S}^4 \begin{bmatrix} A_{1L} \\ A_{2L} \\ A_{1R} \\ A_{2R} \end{bmatrix} = \begin{bmatrix} 0 & S_{12}(-\omega) & S_{11}(-\omega) & 0 \\ S_{21}(\omega) & 0 & 0 & S_{22}(\omega) \\ S_{11}(\omega) & 0 & 0 & S_{12}(\omega) \\ 0 & S_{22}(-\omega) & S_{21}(-\omega) & 0 \end{bmatrix} \begin{bmatrix} A_{1L} \\ A_{2L} \\ A_{1R} \\ A_{2R} \end{bmatrix} \tag{4.70}$$

where the subscripts L and R represent the helicity of incoming and outgoing modes. With relation in Eq. (4.69), a symmetry has been found for $\boldsymbol{S}^4$,

$$(\boldsymbol{PT})\boldsymbol{S}^4(\omega)(\boldsymbol{PT})^{-1} = \boldsymbol{S}^4(\omega). \tag{4.71}$$

$\boldsymbol{S}^4$ is invariant under PT operation in APT systems. To reveal this property, the eigenvalues s of $\boldsymbol{S}^4$ are determined by the following formula,

$$s^2 = \mathrm{Re}(r_{11}r_{22}^*\pm)\sqrt{-\mathrm{Im}\,(r_{11}r_{22}^*)^2 + |t|^4} \tag{4.72}$$

where we have used relations in Eq. (4.69).

The eigenvalue spectrums to $\boldsymbol{S}^4$ are given in Fig. 4.16b. With the eigenspectrum, the symmetric and symmetry-broken phases for scattering are defined. In the scattering symmetric phase, the eigenvalues for $\boldsymbol{S}^4$ are real numbers, while the symmetry-broken phase for scattering implicates the appearance of imaginary parts of the matrix's eigenvalues. This nature of phase transition is uncovered in Eq. (4.72). In addition, besides ω, Ω is another modulation tool for scattering. The phase diagram in Fig. 4.16c summarizes the symmetry dependence on parameters. Surprisingly, we notice a special frequency at which the scattering symmetry is always broken regardless of the rotation speed of the system. It could be seen from Eq. (4.72) that at this special frequency, $|t|^4 = [\text{Im}(r_{11}r_{22})]^2$ when $\Omega = 0$ and $|t|^4 < [\text{Im}(r_{11}r_{22})]^2$ when $\Omega \neq 0$. The emergence of rotation is responsible for the scattering symmetry-breaking. Considering that $\boldsymbol{S}^4$ involves the integral information of this APT-symmetric scattering system, this thresholdless symmetry-broken phenomenon could be interpreted as the signal-induced decoupling of the coupled system, which is quite counterintuitive.

Because most coupled systems have a critical coupling, only when the coupling exceeds this value can the coupling be decoupled and the symmetry-breaking state be achieved.

Furthermore, we also study the eigenstates of $\boldsymbol{S}^4$. Another physical quantity named temperature amplitude ratio (TAR) is proposed to quantify the intensity relation between scattering ports, defined as $TAR = TA_r/TA_l$, which means the ratio of temperature amplitude (represents field intensity) for right and left interfaces (scattering ports) in Fig. 4.16b. When the input is the eigenstate of $\boldsymbol{S}^4$, we have the input vector: $\Psi_{\text{in}} = (s_{1L}e^{-i\omega t},\ s_{2L}e^{i\omega t},\ s_{1R}e^{i\omega t},\ s_{2R}e^{-i\omega t})^{\text{T}}$. The output vector Ψ_{out} is obtained by scattering matrix $\boldsymbol{S}^4$: $\Psi_{\text{out}} = \boldsymbol{S}^4\Psi_{\text{in}}$. Since the temperature fields for left and right ports are the combinations of respective input and output components, the amplitudes are $TA_l = |\Psi_{\text{in}1} + \Psi_{\text{in}3} + \Psi_{\text{out}1} + \Psi_{\text{out}3}|$ and $TA_r = |\Psi_{\text{in}2} + \Psi_{\text{in}4} + \Psi_{\text{out}2} + \Psi_{\text{out}4}|$. Our calculation has shown diverse features for TAR under different incident scattering eigenstates. When the incident eigenstate is in a symmetric phase, the TAR keeps a constant of 1, which reflects the symmetric scattering, as shown in Fig. 4.16d. When the incident eigenstate is in the symmetry-breaking phase, the stability of TAR is disrupted, and it presents oscillation in the time domain, as depicted in Fig. 4.16e. In addition, we investigate the phase shift of the temperature dipole in the left and right ports, $\delta\theta = \text{angle}(\boldsymbol{P}_{Tr}) - \text{angle}(\boldsymbol{P}_{Tl})$, to represent phase information. We find that it shows different behaviors in different phases as well.

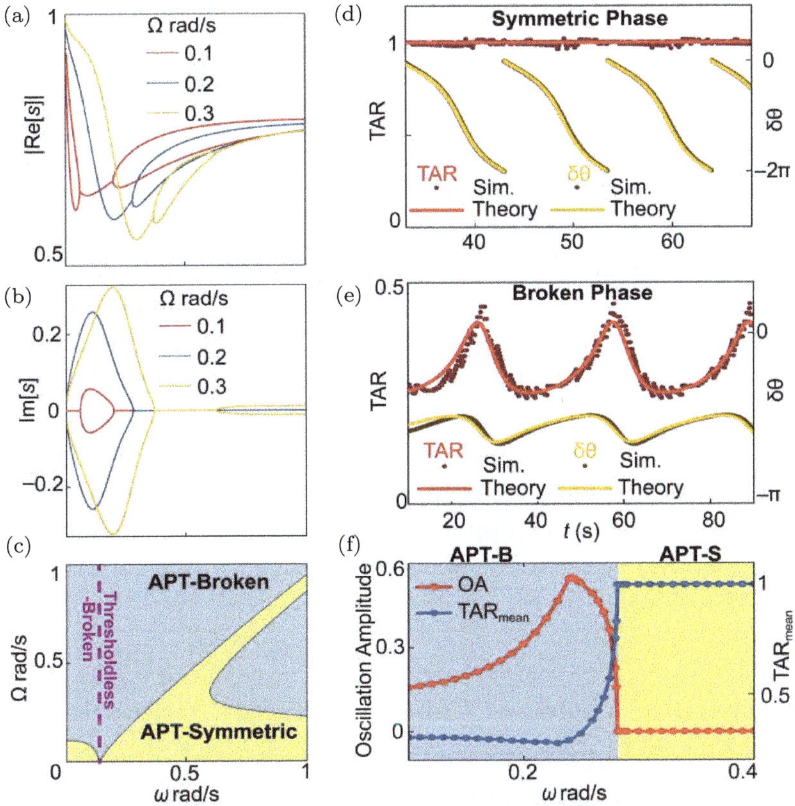

Fig. 4.16. Symmetry in scattering and the phase transition. (a,b) Eigenspectra of the scattering matrix $\boldsymbol{S}^4$. (c) Phase diagram of the scattering matrix $\boldsymbol{S}^4$. Yellow and blue regions correspond to the scattering-symmetric and symmetry-broken phases, respectively. The purple dotted line represents the thresholdless-broken region. (d,e) portray the time-domain transmission amplitude ratio (TAR) and phase shift ($\delta\theta$) in a symmetric/symmetry-broken phase. ω and Ω are (0.3, 0.2) and (0.1, 0.2) respectively. (f) Lines of oscillation amplitudes in TAR (red) and average values in TAR over a period (blue). $\Omega = 0.2$ rad/s, and $\boldsymbol{\omega}$ changes from 0.1 to 0.4 rad/s. Adapted from Ref. [52].

Fig. 4.16d,e exhibits this phenomenon with numerical and simulation results. Notably, the TAR represents the intensity relation between two scattering ports. We conclude that symmetry in scattering plays a role in maintaining the intensity stability during the scattering process. Furthermore, a more vivid dynamic result of the TAR is given in Fig. 4.16f. This figure exhibits the behavior of TAR from the broken phase to the symmetric phase.

Our work considers the scattering problem of the non-Hermitian system carried by the time-reversal antisymmetric (anti-T) channel for the first time. We constructed a theoretical scattering model of the signal propagating in this anti-T channel. With the unusual feature of the anti-T scattering channel, we revealed the scattering symmetry of the APT system characterized by a scattering matrix that associates positive frequency with negative frequency, which attributes a remarkable distinction to PT systems. Furthermore, we observed different scattering phases with distinguishing phenomena corresponding to the stability of the TAR. Our work is of great significance to studying non-Hermitian physics, non-Hermitian symmetry, scattering, heat transfer, and many other related fields. It also implies tremendous applicational possibilities in manipulating various strongly dissipative phenomena such as heat transfer and charge diffusion.

4.6. Thermal Analog of Coherent Perfect Absorption

The above content shows the potential of scattering theory in analyzing thermal nonreciprocity. Another representative application of scattering theory on non-Hermitian wave systems is the coherent perfect absorption (CPA) [43–47] of electromagnetic waves on lossy materials. It is destructive interference between the scattered waves from multiple sources, which provides a convenient method to control light with light. It would be highly desirable to manipulate heat with heat by introducing additional heat sources into the system to avoid using complex and fixed structures.

For this purpose, this section proposes a method to build a correspondence between heat transfer and electromagnetic wave scattering in different dimensions. Unlike the thermal scattering theory mentioned in the previous sections, we introduce an "imitated momentum" for steady-state heat diffusion transported along a pseudo time. The thermal analog of CPA in photonics is thus discovered. It is further identified as the perfect absorption of the exergy flux [48, 49]. Based on it, thermal transparency can be achieved for naturally occurring materials by choosing adequate thermal inputs. This theory offers strong flexibility in heat transfer control and is expected to inspire much broader research domains in diffusive processes.

4.6.1. *Thermal Scattering Theory Based on an Auxiliary Spatial Dimension*

For heat diffusion, we are mostly interested in the steady-state temperature field $T(\boldsymbol{r})$ that follows Fourier's law,

$$\kappa \nabla^2 T = 0 \tag{4.73}$$

with a uniform κ. In 1D, the equation is simply $T''(x) = 0$, and it is obvious that there is no directionality in the solution $T(x) = Ax$ (we set the constant term T_0 to zero throughout the theoretical derivation) because it cannot be decomposed into forward and backward parts.

To tackle the problem, we must establish the correspondence between 1D electromagnetic wave scattering and 2D heat diffusion and make an analogy between them. For 1D wave scattering, the electric field $E(xt)$ satisfies

$$\frac{\partial^2 E}{\partial x^2} = \varepsilon\mu \frac{\partial^2 E}{\partial t^2} \tag{4.74}$$

where ε is the permittivity, μ is the permeability, x is the special coordinate, and t is time. We focus on the scalar field $E(x, t)$, which is the magnitude of the electric field in any certain direction.

The essential difference between Eqs. (4.73) and (4.74) comes from the time-harmonic oscillation of electromagnetic fields. Therefore, our idea is to add an auxiliary spatial dimension to Eq. (4.73) as pseudo time and impose the temperature field to be periodic on it. As a concrete example, consider the model in Fig. 4.17b, where the object is a circular ring with exterior and interior radii r_1 and r_2, respectively. The background is a rectangle with a width $L = 2r_0$. The thermal conductivities of the object and background are κ and κ_0. In the polar coordinate system (r, θ) with the origin at the center

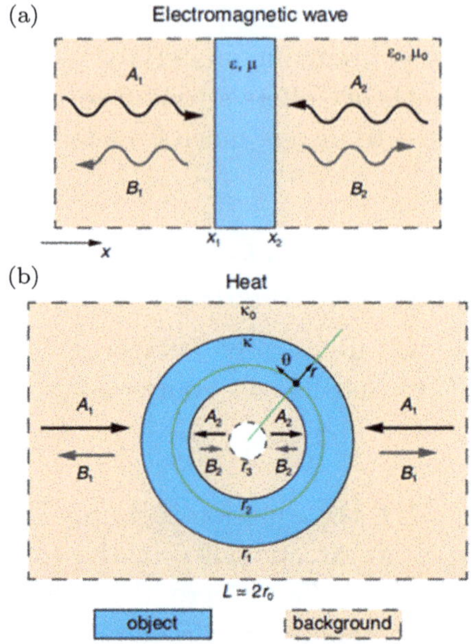

Fig. 4.17. Correspondence between two distinct processes. (a) Scattering of electromagnetic waves in 1D. (b) Steady-state heat diffusion in 2D. An object (light blue) is put in a background (beige). The black arrows represent incoming waves (fields) with amplitudes A_1 and A_2. The grey arrows are outgoing waves (fields) with amplitudes B_1 and B_2. A polar coordinate system (r, θ) is built in b (green lines). Adapted from Ref. [29].

of the object, Fourier's law is now written as

$$\kappa\left(\frac{\partial^2 T}{\partial r^2}+\frac{1}{r}\frac{\partial T}{\partial r}+\frac{1}{r^2}\frac{\partial^2 T}{\partial \theta^2}\right)=0. \tag{4.75}$$

For 2D heat diffusion, by setting $r = e^x$, the temperature field $T(r, \theta)$ above can be transformed into the following form,

$$\frac{\partial^2 T}{\partial x^2}=-\frac{\partial^2 T}{\partial \theta^2} \tag{4.76}$$

which can be mapped to Eq. (4.74) once $\theta \to t$ and $-1 \to \varepsilon\mu$. Under this relation, the temperature field $T(x, t)$ can be mapped to the electric field $E(x, t)$. Moreover, for time-harmonic cases, the electric field $E(x, t) = \mathrm{Re}[E(x)e^{i\omega t}]$ is periodic in time, where ω is frequency. On the other hand, the temperature field $T(x, \theta) = \mathrm{Re}[F(x)e^{im\theta}]$

is also periodic on the angular coordinate θ, where m is an integer. Therefore, it is reasonable to regard θ as a "pseudo time" in our case, with $m \to \omega$.

We consider the 2D heat diffusion process. As shown in Fig. 4.17b, a circular scatterer is put in the background. The temperature in the center and both sides can be written as

$$F(x) = Ae^{ikx} + Be^{-ikx}, x_2 \le x \le x_1 \tag{4.77a}$$

$$F_1(x) = A_1 e^{ik_0(x-x_1)} + B_1 e^{-ik_0(x-x_1)}, x \ge x_1 \tag{4.77b}$$

$$F_2(x) = A_2 e^{-ik_0(x-x_2)} + B_2 e^{ik_0(x-x_2)}, x \le x_2 \tag{4.77c}$$

where the amplitudes A, $B, A_{1,2}$, and $B_{1,2}$ can be complex numbers that incorporate phases in the fields. Focusing on a common case with $m = 1$, a key operation here is to introduce a variable change $r = e^{ikx} = e^x$ with a "wavenumber" $k = -i$, an imaginary number. Such an extended definition of wavenumber is commonly used in wave physics [50] to describe evanescent waves. By doing this, Eqs. (4.77a)–(4.77c) can be further written as Eqs. (4.78a)–(4.78c), which has the same form as for 1D wave scattering,

$$F(r) = Ar + B/r, r_2 \le r \le r_1 \tag{4.78a}$$

$$F_1(r) = A_1 r/r_1 + B_1 r_1/r, r \ge r_1 \tag{4.78b}$$

$$F_2(r) = A_2 r_2/r + B_2 r/r_2, r \le r_2. \tag{4.78c}$$

The two components of the field $F(x)$ appear to carry imitated momentums in the x and $-x$ directions. For 2D heat diffusion, the continuity of the temperature field and the heat flux require the following matching conditions,

$$F_{1,2}(x_{1,2}) = F(x_{1,2}) \tag{4.79a}$$

$$\kappa_0 F'_{1,2}(x_{1,2}) = \kappa F'(x_{1,2}) \tag{4.79b}$$

with $x_{1,2} = \ln r_{1,2}$. Equations (4.79a) and (4.79b) have the same form as the 1D electromagnetic wave scattering matching conditions. Therefore, it is natural for us to regard the original problem as 1D scattering in the r-direction, for which the r- and r^{-1}- components are the "forward" and "backward" parts, while θ is the pseudo time

(Fig. 4.17b). Moreover, it can be shown that the two components carry exergy fluxes in the $-r$ and r directions. The exergy is a thermodynamic quantity defined as the maximum useful work a system can do by bringing it into thermodynamic equilibrium with the environment [48]. In our case, the useful work comes from the temperature difference between any local point in the system and the environment [49], meaning that one can extract work by putting a heat engine between them. Therefore, our decomposition of the temperature field gives important information about how the potentially useful thermal energy is distributed and transferred in the system.

We can study the thermal scattering problem through the transfer matrix $\boldsymbol{M}$, which is defined based on the amplitudes of the forward and backward fields,

$$\begin{pmatrix} B_2 \\ A_2 \end{pmatrix} = \boldsymbol{M} \begin{pmatrix} A_1 \\ B_1 \end{pmatrix}. \tag{4.80}$$

The transfer matrix can be calculated by substituting Eqs. (4.77a)–(4.77c) into Eqs. (4.79a) and (4.79b),

$$\boldsymbol{M} = \begin{pmatrix} 1 & 1 \\ \kappa_0 & -\kappa_0 \end{pmatrix}^{-1} \begin{pmatrix} \cosh\Delta x & \frac{1}{\kappa}\sinh\Delta x \\ \kappa\sinh\Delta x & \cosh\Delta x \end{pmatrix} \begin{pmatrix} 1 & 1 \\ \kappa_0 & -\kappa_0 \end{pmatrix} \tag{4.81}$$

where κ_0 and κ are the thermal conductivity of the background and object, with $\Delta x = x_2 - x_1$. By comparing Eq. (4.81) with the transfer matrix for the electric fields in wave scattering, the thermal conductivity κ in heat diffusion corresponds to the admittance $Y = (\varepsilon/\mu)^{1/2}$ (ε is permittivity, and μ is permeability) for electromagnetic waves. A rearrangement of Eq. (4.81) gives the scattering matrix $\boldsymbol{S}$ between the incoming and outgoing fields,

$$\begin{pmatrix} B_1 \\ B_2 \end{pmatrix} = \boldsymbol{S} \begin{pmatrix} A_1 \\ A_2 \end{pmatrix} = \frac{1}{\boldsymbol{M}_{22}} \begin{pmatrix} -\boldsymbol{M}_{21} & 1 \\ \det\boldsymbol{M} & \boldsymbol{M}_{12} \end{pmatrix} \begin{pmatrix} A_1 \\ A_2 \end{pmatrix} = \begin{pmatrix} r_{11} & t_{12} \\ t_{21} & r_{22} \end{pmatrix} \begin{pmatrix} A_1 \\ A_2 \end{pmatrix}. \tag{4.82}$$

The entries of $\boldsymbol{S}$ are the reflection (r_{11} and r_{22}) and transmission (t_{12} and t_{21}) coefficients, which can be determined through the entries of $\boldsymbol{M}$. With $\boldsymbol{M}_{12} = -\boldsymbol{M}_{21}$, the reflection coefficients of the scattering matrix **S** are equal ($r_{11} = r_{22}$). Despite the imaginary wavenumber $k = -i$, we still have det $\boldsymbol{M} = \cosh^2 \Delta x - \sinh^2 \Delta x = 1$. Therefore, the heat transfer is reciprocal ($t_{12} = t_{21}$). Together, the system preserves a parity symmetry, meaning that the temperature field is unchanged if we swap the exterior and interior parts of the object ($A_1 \leftrightarrow A_2$, $x \leftrightarrow x_1 + x_2 - x$). This symmetry is unobvious because it is based on the logarithmic coordinate x. We explicitly write out $\boldsymbol{S}$ as

$$\boldsymbol{S} = \begin{pmatrix} \tilde{r} & \tilde{t} \\ \tilde{t} & \tilde{r} \end{pmatrix} = (\text{coth}\Delta x - \text{cosh}\gamma)^{-1} \begin{pmatrix} \text{sinh}\gamma & \text{csch}\Delta x \\ \text{csch}\Delta x & \text{sinh}y \end{pmatrix} \tag{4.83}$$

with $\gamma = \ln(\kappa/\kappa_0)$.

4.6.2. *Physical Meanings of Perfect Absorption in Heat Diffusion*

For the 1D wave scattering in Fig. 4.17a, when the material of the scatterer is lossless, it is easy to verify that the scattering matrix $\boldsymbol{S}_{\text{EM}}$ satisfies $|\det \boldsymbol{S}_{\text{EM}}| = 1$, which is a result of energy conservation. When the material is lossy, n is a complex number, and $|\det \boldsymbol{S}_{\text{EM}}| < 1$. Therefore, it is possible to add a suitable amount of loss into the material to meet the CPA condition det $\boldsymbol{S}_{\text{EM}} = 0$.

For heat transfer, the scattering matrix in Eq. (4.83) naturally satisfies $|\det \boldsymbol{S}| < 1$, which provides a promising avenue for mapping the CPA effect to heat diffusion. However, the temperature field does not propagate in the r-direction, so other physical meanings are anticipated for perfect absorption in heat diffusion. Here, we provide two possible interpretations. One is based on the exergy flux, the other on the heat flux magnitude.

4.6.2.1. *Interpretation Based on Exergy Flux*

Note that we are considering the temperature difference $T - T_0$, where T_0 is the central and ambient temperature (set to be nonzero). This temperature difference is a useful source of thermal energy, and we can start by using the concept of exergy to quantify it. Exergy is the maximum useful work a system can do by bringing it into thermodynamic equilibrium with the environment. In the heat diffusion process, the local exergy density χ gives the useful thermal energy stored at the local point. It is calculated as $\chi = u-u_0-T_0(s-s_0)$, where u is the internal energy point, s is the entropy density, u_0 and s_0 are the internal energy density and entropy density of the local spot after it has reached thermodynamic equilibrium with the environment. Its governing equation is

$$\frac{\mathrm{d}\chi}{\mathrm{d}t} = -\nabla\left[\left(1-\frac{T_0}{T}\right)q\right] - T_0 q\cdot\nabla\left(\frac{1}{T}\right). \tag{4.84}$$

The second term is $-T_0$ times the local entropy generation rate, which is always positive and contributes to the dissipation of the local exergy. The exergy flow χ_f is defined through the first term as

$$\chi_f = \left(1-\frac{T_0}{T}\right)q. \tag{4.85}$$

The steady-state temperature distribution outside the object is $T_1(x,\theta) = A_1e^{x-x_1}\cos(\theta+\alpha_1) + B_1e^{-(x-x_1)}\cos(\theta+\beta_1) + T_0$. We introduce phases α_1 and β_1 for the two fields. To simplify the calculation, we reasonably assume that $T-T_0$ are much smaller than T_0,

$$\chi_f = \left(\frac{T-T_0}{T_0}\right)q. \tag{4.86}$$

Then, the exergy flux in the radial direction can be estimated as

$$\chi_f^r = \frac{(T-T_0)}{T_0}\left(-\kappa\frac{\mathrm{d}T}{\mathrm{d}r}\right) \tag{4.87}$$

which can be further written as

$$\chi_f^r = -\frac{\kappa}{T_0 r}\left[A_1^2e^{2(x-x_1)}\cos^2(\theta+\alpha_1) - B_1^2e^{-2(x-x_1)}\cos^2(\theta+\beta_1)\right]. \tag{4.88}$$

The total exergy flux through a circle is

$$Q_1 = -\frac{2\pi\kappa}{T_0}\left[A_1^2e^{2(x-x_1)} - B_1^2e^{-2(x-x_1)}\right]. \tag{4.89}$$

The amplitudes A_1 and B_2 determine the total exergy flux in the $-r$ and r directions. Similarly, the total exergy flux through a circle in the interior of the object is

$$Q_2 = -\frac{2\pi\kappa}{T_0}\left[B_2^2 e^{2(x-x_2)} - A_2^2 e^{-2(x-x_2)}\right]. \tag{4.90}$$

As can be seen above, no matter the outer boundary or the inner boundary of the object, the total exergy flux entering or leaving the object is proportional to the squared amplitude $A_1^2\,(B_1^2)$ or $A_2^2\,(B_2^2)$. Therefore, $|\det \boldsymbol{S}\,| < 1$ or $A_1^2 + A_2^2 > B_1^2 + B_2^2$ means that more exergy flux enters than leaves the object. Namely, the exergy is dissipated in the object due to the local entropy generation. Furthermore, the perfect absorption condition in heat diffusion means that no exergy flux leaves the object, which can be characterized by $B_1 = B_2 = 0$. In this case, the useful thermal energy is transferred from the heat sources into the object at maximum efficiency.

4.6.2.2. *Interpretation Based on Heat Flux Magnitude*

Another physically relevant quantity is the heat flux magnitude (outside the object),

$$\begin{aligned} |q_1| &= \kappa r^{-1}\left(T_x^2 + T_\theta^2\right)^{\frac{1}{2}} \\ &= \kappa r^{-1}\left[A_1^2 e^{2(x-x_1)} + B_1^2 e^{-2(x-x_1)} - 2A_1B_1\cos\left(2\theta + \alpha_1 + \beta_1\right)\right]^{\frac{1}{2}}. \end{aligned} \tag{4.91}$$

The heat flux magnitude is a constant when $A_1 = 0$ or $B_1 = 0$, but oscillates with θ when both fields exist. The same conclusion also applies to the interior part of the object. The perfect absorption in heat diffusion thus also corresponds to a θ-independent heat flux magnitude.

4.6.3. *Thermal Coherent Perfect Absorption and Its Maximum Outgoing Exergy Flux*

Thermal CPA ($\det \boldsymbol{S} = 0$) can be reached by tuning γ (a real number) to absorb the exergy flux entering the object completely. In addition, the magnitude of the heat flux will be θ-independent under the condition.

Setting $\det \boldsymbol{S} = 0$ gives

$$\kappa = \kappa_{\pm}^{*} = \kappa_0 \frac{r_1 \pm r_2}{r_1 \mp r_2} \tag{4.92}$$

$$s = \frac{r_1 r_2}{r_1^2 + r_2^2} \begin{pmatrix} \mp 1 & 1 \\ 1 & \mp 1 \end{pmatrix}. \tag{4.93}$$

The eigenvalues of $\boldsymbol{S}$ are $s_1 = 0$ and $s_2 = 2r_1r_2/(r_1^2+r_2^2)$. The thermal CPA condition is reached when the inputs meet the eigenvector $(A_1, \pm A_1)$ corresponding to s_1. Unlike in photonics, where loss must be introduced, the thermal counterpart is readily realizable with normal materials. The inputs are simply symmetric or anti-symmetric, demonstrating the hidden parity symmetry of the system. The next question is how to generate the "incident" fields or input-1: $A_1 r/r_1$ and input-2: $A_2 r_2/r$. Heat transfer through a conductive system is usually studied by maintaining constant temperatures at the left and right sides ($T = \pm A_1 r_0/r_1$ at $r\cos\theta = \pm r_0$) and thermally insulating the upper and lower sides. If the object is absent, the temperature field is exactly the required $A_1(r/r_1)\cos\theta$, and the boundary condition can be regarded as the source of input-1 outside the object. On the other hand, consider a circle inside the object with radius $r_3 < r_2$. We apply a constant temperature distribution $T = A_2(r_2/r_3)\cos\theta$ on it. For this boundary condition, if the object is absent and the entire system is large (so that the effects of the outer boundaries are negligible), the temperature field outside r_3 will be $A_2(r_2/r)\cos\theta$, whose radial component meets input-2. Thus, the source for the input-2 inside the object is also found.

In photonics, the other eigenvector corresponding to the nonzero eigenvalue of S will lead to constructive interference of the scattered waves. Similarly, input fields that coincide with the other eigenvector corresponding to s_2 will lead to large outgoing fields. Note that this input can be realized by inverting the orientation of input-2, where the in-phase assumption of the two incident fields is terminated. Here, we consider the CPA condition with

$$\kappa = \kappa_0 \frac{r_1 - r_2}{r_1 + r_2}, \quad \boldsymbol{S} = \frac{r_1 r_2}{r_1^2 + r_2^2} \begin{pmatrix} 1 & 1 \\ 1 & 1 \end{pmatrix}. \tag{4.94}$$

The eigenvector corresponding to the zero eigenvalue s_1 is $(A_1, A_2)^{\mathrm{T}} = (A_1, -A_1)^{\mathrm{T}}$. CPA is realized by using them as the input fields' amplitudes, as shown in Fig. 4.18a.

More generally, we can introduce a phase difference α between the two input fields in their θ-dependence (note that it is impossible to define a phase for the radial dependence because the field does not propagate in that direction) such that $(A_1, A_2)^{\mathrm{T}} = (A_1, -A_1 e^{i\alpha})^{\mathrm{T}}$. This operation can be easily done by rotating the input-2 field by $-\alpha$. When $\alpha = \pm\pi$, the input coincides with the eigenvector for s_2. The outgoing fields have amplitudes,

$$B_1 = B_2 = A_1(1 - e^{i\alpha})\frac{r_1 r_2}{r_1^2 + r_2^2} = A_1\sqrt{2(1 - \cos\alpha)}e^{i\beta}\frac{r_1 r_2}{r_1^2 + r_2^2} \tag{4.95}$$

with $\beta = \mathrm{Arg}(1 - e^{i\alpha})$. Therefore, the outgoing fields have maximized amplitudes with $\alpha = \pm\pi$ (Fig. 4.18b), where the temperature field is strongly distorted from the CPA case. It is similar to the constructive interference of the scattered fields in optical CPA, which is also achieved when the input coincides with the other eigenvector.

As shown in Fig. 4.18c, with $\alpha = -\pi/6$, a small phase difference will strongly break the CPA condition and modify the temperature field. The effect is also experimentally verified by tuning the direction of the central copper bridge. The measured temperature field is shown in Fig. 4.18d. To quantify the influence of the phase difference, we calculate $R = (|B_1|^2 + |B_2|^2)/(|A_1|^2 + |A_2|^2)$, which gives the ratio between exergy fluxes flowing out of and into the object according to the previous discussion. The result is plotted in Fig. 4.18e. We can see that more exergy flux flows out of the object as α increases. At $\alpha = \pm\pi$, the input fields meet the other eigenvector of the scattering matrix with a maximized outgoing exergy flux (around 81% of the incoming flux).

We also note that in photonics, CPA is the time-reversed lasing process. Thus, one might be interested in a thermal analog of lasing. However, the time-reversed heat diffusion process requires a material with negative thermal diffusivity as the gain medium, which has not been realized.

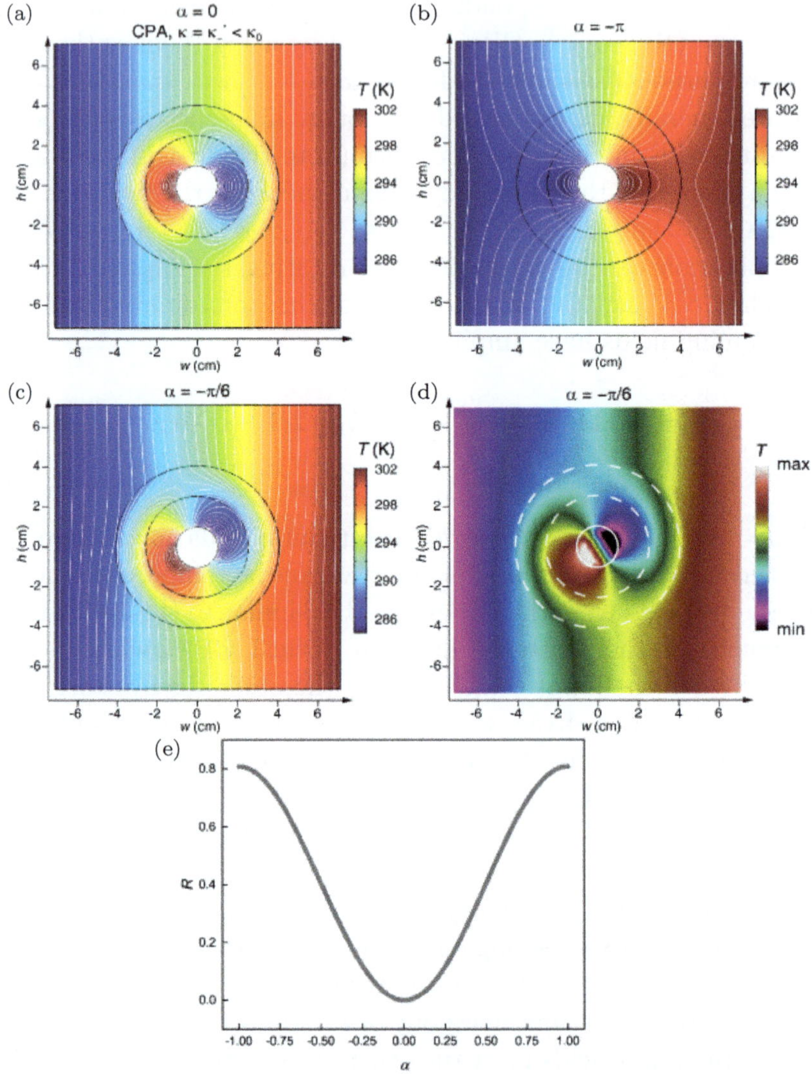

Fig. 4.18. Effects of a phase difference between the input fields. (a) Simulated temperature distribution with unrotated input-2 field. (b) Simulated temperature distribution with input-2 field rotated by $-\alpha = \pi$. (c) Simulated temperature distribution with input-2 field rotated by $-\alpha = \pi/6$. (d) Experimentally measured temperature distribution. (e) The ratio R between exergy fluxes flowing out of and into the object. Adapted from Ref. [29].

4.6.4. *One-side Coherent Perfect Absorption*

For CPA, the input fields from both sides are completely absorbed. We are often interested in the field outside the object, such as thermal transparency and cloaking. Therefore, we study the condition for no outgoing field on just one side of the object. We refer to this effect as one-side CPA, an analog of unidirectional absorption in photonics. Since B_2 does not have to be zero, one may expect that the condition for one-side CPA is $B_1 = 0$. However, the field with nonzero B_2 will be "reflected" at the boundary of heat input-2, then "transmitted" through the object to contribute to an outgoing field in region $r \geq r_1$. Thus, we must consider all outgoing fields after multiple scatterings, which requires zero summation.

To fully tackle the problem, we must obtain the reflection coefficients at the two boundaries of heat inputs. We denote another scattering matrix $\boldsymbol{D}$ as

$$\begin{pmatrix} C_1 \\ C_2 \end{pmatrix} = \boldsymbol{D} \begin{pmatrix} B_1 \\ B_2 \end{pmatrix} = \begin{pmatrix} \widetilde{r_a} & 0 \\ 0 & \widetilde{r_b} \end{pmatrix} \begin{pmatrix} B_1 \\ B_2 \end{pmatrix} \tag{4.96}$$

where C_1 and C_2 are the amplitudes of the "reflected" fields outside and inside the object. The constant boundary conditions require that the boundary temperatures are not changed by this process, so $B_1 r_1/r_0 + C_1 r_0/r_1 = 0$ and $B_2 r_3/r_2 + C_2 r_2/r_3 = 0$. For brevity, we have changed the actual boundary conditions at the left and right sides to the more symmetric condition ($T = A_1 r_0 \cos\theta/r_1$ at $r = r_0$). The reflection coefficients are $\widetilde{r_a} = -r_1^2/r_0^2$ and $\widetilde{r_b} = -r_3^2/r_2^2$. The total magnitudes Z_1 and Z_2 of the outgoing fields can be expressed as an infinite sequence,

$$\begin{pmatrix} Z_1 \\ Z_2 \end{pmatrix} = \boldsymbol{S}' \begin{pmatrix} A_1 \\ A_2 \end{pmatrix} = (\boldsymbol{S} + \boldsymbol{SDS} + \boldsymbol{SDSDS} + \cdots) \begin{pmatrix} A_1 \\ A_2 \end{pmatrix}. \tag{4.97}$$

The new scattering matrix $\boldsymbol{S}'$ can be explicitly solved by diagonalizing SD,

$$\boldsymbol{S}' = \begin{pmatrix} \widetilde{r} - \widetilde{r_b}\left(\widetilde{r}^2 - \widetilde{t}^2\right) & \widetilde{t} \\ \widetilde{t} & \widetilde{r} - \widetilde{r_a}\left(\widetilde{r}^2 - \widetilde{t}^2\right) \end{pmatrix}. \tag{4.98}$$

The result is remarkably compact. We note that $\boldsymbol{S}'$ is still symmetric, but the parity symmetry is only preserved when $\widetilde{r_a} = \widetilde{r_b}$ (i.e., $r_1/r_0 = r_3/r_2$). The requirement is a geometric mirror symmetry for variable $x = \ln r : x_0 - x_1 = x_2 - x_3$. For CPA outside the object, $A_2 = A_1$ $[\widetilde{r_b}(\widetilde{r}^2 - \widetilde{t}^2) - \widetilde{r}]/\widetilde{t}$, which is even independent of the size of the background. Another advantage of the one-side CPA condition is that it does not restrict the object's materials and background. A solution can be found for any values of κ and κ_0. The limitations of conventional thermal metamaterials can thus be avoided, especially the fixed and restricted parameters.

4.6.5. *Demonstration of Thermal and One-Side Coherent Perfect Absorption*

We build a 2D finite-element model based on the analytical results to verify and demonstrate the predicted effects. We first study the thermal CPA conditions in Eq. (4.92). The results are shown in Fig. 4.19a. A Cartesian coordinate system (w, h) is built with its origin at $r = 0$ to facilitate the display. Under thermal CPA, there should be no outgoing field on both sides of the object, such that the temperature distributions outside and inside the object should meet input-1 and input-2, respectively. The temperature distributions in Fig. 4.19a,b meet this condition for both $\kappa > \kappa_0$ and $\kappa < \kappa_0$.

For comparison, the temperature distributions with input-2 alone are simulated and plotted in Fig. 4.19c,d. The distribution for input-1 is not plotted since it is a linear profile and meets the temperature profiles outside the object in Fig. 4.19a,b. For input-2, the reflection at the outer boundary is effectively removed by enlarging the size to $L = 1\,\text{m}$, making the reflected field negligible in the displayed part. It is easy to identify the same pattern inside the object in Fig. 4.19a,b as the corresponding part of the input-2 fields in Fig. 4.19c,d (the other parts are made translucent). We verify the effects by extracting the temperature distributions on the cut line $h = 0$ across the system and plot the results in Fig. 4.19e,f (scatters). In background intervals (shaded by light orange), the numerical results overlap with the theoretical r-dependence for input-1 and input-2 (solid lines), so

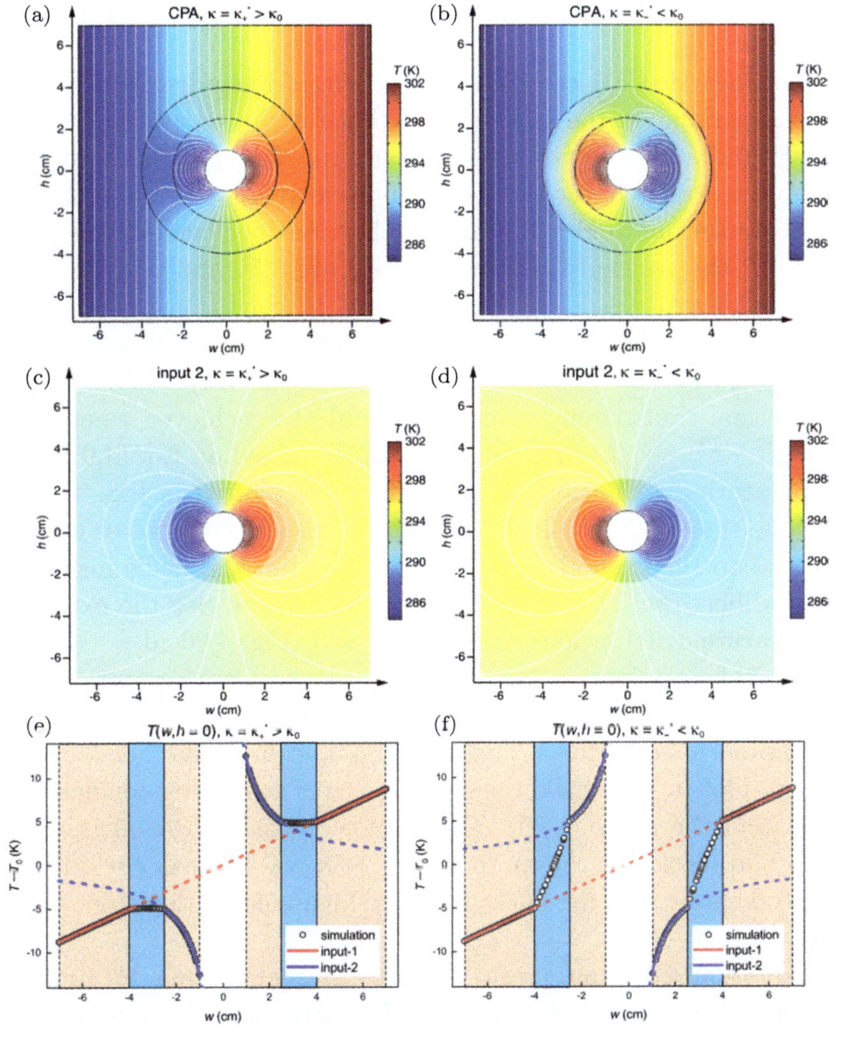

Fig. 4.19. Thermal coherent perfect absorption (CPA). The thermal conductivity κ of the ring-shaped object is (a,c,e) larger or (b,d,f) smaller than that of the background κ_0. (a,b) Temperature distributions on the entire system with isothermal lines (white). (c,d) Temperature distribution of input-2 on a pure background, much larger ($L = 1$ m) than the displayed part ($L = 14$ cm). (e,f) Temperature distributions along the line $h = 0$. The regions of the background (beige) and the object (light blue) are shaded. The incident fields input-1 (red) and input-2 (blue) are also plotted. Adapted from Ref. [29].

there is indeed no outgoing field with both incident fields perfectly absorbed.

The one-side CPA condition can be similarly studied with numerical simulations on the same model. Since it can be realized for any background and object material, we fix the thermal conductivity of the object κ and choose two representative values of $\kappa_0 = \kappa/2.4$ and $\kappa/0.3$. According to Eq. (4.83), the reflection coefficients $\widetilde{r}$ are -0.27 and 0.37; the transmission coefficients $\widetilde{t}$ are 0.56 and 0.5. The ratio $\kappa/\kappa_0 = 2.4$ and 0.3 do not match the CPA condition ($\kappa/\kappa_0 = 4.33$ and 0.23). Therefore, the scattering cannot be completely suppressed on both sides.

Thermal transparency outside the object is achieved when $A_2 = A_1[\widetilde{r}_b(\widetilde{r}^2 - \widetilde{t}^2) - \widetilde{r}]/\widetilde{t} = 2.76\,\mathrm{K}$ and $-3.52\,\mathrm{K}$, based on Eq. (4.97). The temperature distributions in these cases are plotted in Fig. 4.19a,b. The one-side CPA confirms that the fields outside the object are identical to the input-1 field in Fig. 4.19c, but the fields inside the object differ from the input-2 field in Fig. 4.19d. See the extracted temperature distributions on cutline $h = 0$ (Fig. 4.20c,d).

A highly conductive background often requires an unreachable effective κ for conventional thermal metamaterials to function, but our approach works for any values of κ_0 and κ. To further illustrate this, we plot in Fig. 4.20e the dependence of scattering coefficients $\widetilde{r}$ and $\widetilde{t}$, and the ratio A_2/A_1 required for one-side CPA on the ratio κ/κ_0. The abscissa is chosen to be $\gamma = \ln(\kappa/\kappa_0)$ to cover a wide range and to demonstrate the antisymmetric (symmetric) dependence of $\widetilde{r}$ and $\widetilde{t}$ on it.

The required input ratio is not symmetric due to multiple scatterings on the inner boundary. In particular, $A_2 = 0.1A_1 \neq 0$ for $\kappa = \kappa_0$ when the central void becomes the scatterer.

We note that the other one-side CPA condition for the field inside the object is also achievable, but one should be careful about the effects of the outer boundaries without rotational symmetry.

In this section, we propose a method to impart heat transfer processes with the concept of an "imitated momentum" and establish the theory of thermal scattering. The temperature field outside a ring-shaped normal material is found to be decomposed into two

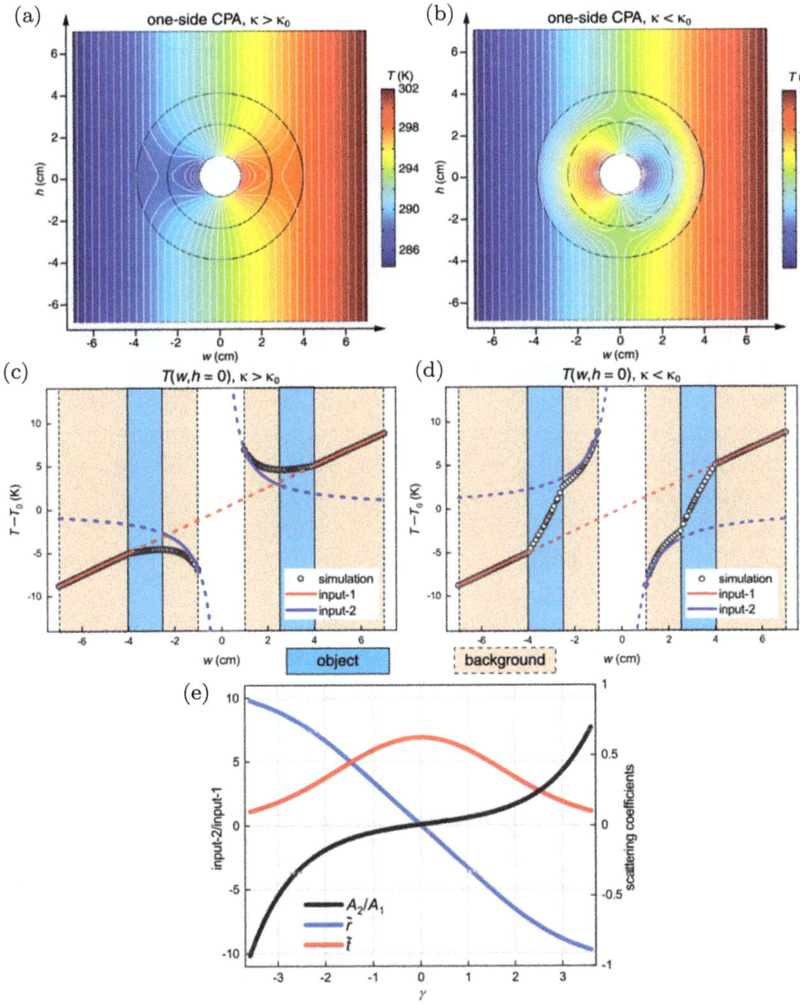

Fig. 4.20. Thermal one-side coherent perfect absorption (CPA). The ring-shaped object has an arbitrary thermal conductivity κ that is (a,c) larger or (b,d) smaller than the background κ_0. (a,b) The temperature distributions on the entire system with isothermal lines (white). (c,d) Temperature distributions along the line $h = 0$. The regions of the background (beige) and the object (light blue) are shaded. The incident fields input-1 (red) and input-2 (blue) are also plotted. For one-side CPA, only the field outside the object meets the input field. (e) The required inputs (black), the reflection (blue), and transmission (red) coefficients for different combinations of the object and background materials. Adapted from Ref. [29].

parts that carry exergy fluxes in opposite directions. The thermal scattering matrix for it is reciprocal and parity symmetric with real reflection and transmission coefficients, but the absolute value of its determinant is smaller than one due to entropy generation. It indicates the possibility of realizing the thermal analog of CPA.

References

1. Li, Y., Peng, Y.-G., Han, L., Miri, M.-A., Li, W., Xiao, M., Zhu, X.-F., Zhao, J., Alù, A., Fan, S., and Qiu, C.-W. (2019). Anti-parity-time symmetry in diffusive systems. *Science*, 364, pp. 170–173.
2. Cao, P., Li, Y., Peng, Y., Qiu, C., and Zhu, X. (2020). High-order exceptional points in diffusive systems: robust APT symmetry against perturbation and phase oscillation at APT symmetry breaking. *ES Energy & Environ.*, **7**, pp. 48–55.
3. Cao, P.-C., Li, Y., Peng, Y.-G., Qi, M., Huang, W.-X., Li, P.-Q., and Zhu, X.-F. (2021). Diffusive skin effect and topological heat funneling. *Commun. Phys.*, 4, pp. 230.
4. Xu, G., Yang, Y., Zhou, X., Chen, H., Alù, A., and Qiu, C.-W. (2022) Diffusive topological transport in spatiotemporal thermal lattices. *Nat. Phys.*, 18, pp. 450–456.
5. Kawabata, K., Shiozaki, K., Ueda, M., and Sato, M. (2019). Symmetry and topology in non-Hermitian physics. *Phys. Rev. X*, 9, pp. 041015.
6. Yan, Q., Chen, H., and Yang, Y. (2021). Non-Hermitian skin effect and delocalized edge states in photonic crystals with anomalous parity-time symmetry. *Prog. Electromagn. Res.*, 172, pp. 33–40.
7. Fang, K., Yu, Z., and Fan, S. (2012). Realizing effective magnetic field for photons by controlling the phase of dynamic modulation. *Nat. Photonics*, 6, pp. 782–787.
8. Fleury, R., Khanikaev, A. B., and Alù, A. (2016). Floquet topological insulators for sound. *Nat. Commun.*, 7, pp. 11744.
9. Lira, H., Yu, Z., Fan, S., and Lipson, M. (2012). Electrically driven nonreciprocity induced by interband photonic transition on a silicon chip. *Phys. Rev. Lett.*, 109, pp. 033901.
10. Fang, K., Yu, Z., and Fan, S. (2012). Photonic Aharonov-Bohm effect based on dynamic modulation. *Phys. Rev. Lett.*, 108, pp. 153901.
11. Wang, Y., Yousefzadeh, B., Chen, H., Nassar, H., Huang, G., and Daraio, C. (2018). Observation of nonreciprocal wave propagation in a dynamic phononic lattice. *Phys. Rev. Lett.*, 121, pp. 194301.
12. Camacho, M., Edwards, B., and Engheta, N. (2020). Achieving asymmetry and trapping in diffusion with spatiotemporal metamaterials. *Nat. Commun.*, 11, pp. 3733.

13. Mandelis, A. (2001). *Diffusion-Wave Fields: Mathematical Methods and Green Functions.* Springer.
14. Wehmeyer, G., Yabuki, T., Monachon, C., Wu, J., and Dames, C. (2017). Thermal diodes, regulators, and switches: Physical mechanisms and potential applications. *Appl. Phys. Rev.*, 4, pp. 041304.
15. Li, B., Wang, L., and Casati, G. (2004). Thermal diode: rectification of heat flux. *Phys. Rev. Lett.*, 93, pp. 184301.
16. Chang, C. W., Okawa, D., Majumdar, A., and Zettl, A. (2006). Solid-state thermal rectifier. *Science*, 314, pp. 1121–1124.
17. Wang, Y., Vallabhaneni, A., Hu, J., Qiu, B., Chen, Y. P., and Ruan, X. (2014). Phonon lateral confinement enables thermal rectification in asymmetric single-material nanostructures. *Nano Lett.*, 14, pp. 592–596.
18. Go, D. B., and Sen, M. (2010). On the condition for thermal rectification using bulk materials. *J. Heat Transf.*, 132, pp. 124502.
19. Kobayashi, W. (2020). Thermal-rectification coefficients in solid-state thermal rectifiers. *Phys. Rev. E*, 102, pp. 032142.
20. Li, J., Li, Y., Cao, P., Yang, T., Zhu, X., Wang, W., and Qiu, C. (2020). A continuously tunable solid-like convective thermal metadevice on the reciprocal line. *Adv. Mater.*, 32, pp. 2003823.
21. Li, J., Li, Y., Wang, W., Li, L., and Qiu, C.-W. (2020). Effective medium theory for thermal scattering off rotating structures. *Opt. Express*, 28, pp. 25894.
22. Maznev, A. A., Every, A. G., and Wright, O. B. (2013). Reciprocity in reflection and transmission: What is a 'phonon diode'? *Wave Motion*, 50, pp. 776–784.
23. Ju, R., Xu, G., Xu, L., Qi, M., Wang, D., Cao, P., Xi, R., Shou, Y., Chen, H., Qiu, C., and Li, Y. (2023). Convective thermal metamaterials: exploring high-efficiency, directional, and wave-like heat transfer. *Adv. Mater.*, 35, pp. 2209123.
24. Li, J., Li, Y., Cao, P.-C., Qi, M., Zheng, X., Peng, Y.-G., Li, B., Zhu, X.-F., Alù, A., Chen, H., and Qiu, C.-W. (2022). Reciprocity of thermal diffusion in time-modulated systems. *Nat. Commun.*, 13, pp. 167.
25. Li, Y. and Li, J. (2021). Advection and thermal diode. *Chin. Phys. Lett.*, 38, pp. 030501.
26. Li, Y., Li, J., Qi, M., Qiu, C.-W., and Chen, H. (2021). Diffusive nonreciprocity and thermal diode. *Phys. Rev. B*, 103, pp. 014307.
27. Lax, M. (1951). Multiple scattering of waves. *Rev. Mod. Phys.*, 23, pp. 287–310.
28. Terrón, J. M., Sánchez-Lavega, A., and Salazar, A. (2000). Multiple scattering effects of thermal waves by two subsurface cylinders. *J. Appl. Phys.*, 87, pp. 2600–2607.

29. Li, Y., Qi, M., Li, J., Cao, P.-C., Wang, D., Zhu, X.-F., Qiu, C.-W., and Chen, H. (2022). Heat transfer control using a thermal analogue of coherent perfect absorption. *Nat. Commun.*, 13, pp. 2683.
30. Miri, M.-A. and Alù, A. (2019). Exceptional points in optics and photonics. *Science*, 363, pp. 6422.
31. El-Ganainy, R., Makris, K. G., Khajavikhan, M., Musslimani, Z. H., Rotter, S., and Christodoulides, D. N. (2018). Non-Hermitian physics and PT symmetry. *Nat. Phys.*, 14, pp. 11.
32. Özdemir, Ş. K., Rotter, S., Nori, F., and Yang, L. (2019). Parity-time symmetry and exceptional points in photonics. *Nat. Mater.*, 18, pp. 783.
33. Chen, W., Özdemir, Ş. Kaya, Zhao, G., Wiersig, J., and Yang, L. (2017). Exceptional points enhance sensing in an optical microcavity. *Nature*, 548, pp. 192.
34. Hodaei, H., Hassan, A. U., Wittek, S., Garcia-Gracia, H., El-Ganainy, R., Christodoulides, D. N., and Khajavikhan, M. (2017). Enhanced sensitivity at higher-order exceptional points. *Nature*, 548, pp. 187.
35. Peng, B., Özdemir, Ş. K., Rotter, S., Yilmaz, H., Liertzer, M., Monifi, F., Bender, C. M., Nori, F., and Yang, L. (2014). Loss-induced suppression and revival of lasing. *Science*, 346, pp. 328.
36. Hodaei, H., Miri, M.-A., Heinrich, M., Christodoulides, D. N., and Khajavikhan, M. (2014). Parity-time-symmetric microring lasers. *Science*, 346, pp. 975.
37. Rousseaux, G., Mathis, C., Maïssa, P., Philbin, T. G., and Leonhardt, U. (2008). Observation of negative-frequency waves in a water tank: a classical analogue to the hawking effect? *New J. Phys.*, 10, pp. 053015.
38. Peng, P., Cao, W., Shen, C., Qu, W., Wen, J., Jiang, L., and Xiao, Y. (2016). Anti-parity-time symmetry with flying atoms. *Nat. Phys.*, 12, pp. 1139.
39. Jiang, Y., Mei, Y., Zuo, Y., Zhai, Y., Li, J., Wen, J., and Du, S. (2019). Anti-parity-time symmetric optical four-wave mixing in cold atoms. *Phys. Rev. Lett.*, 123, pp. 193604.
40. Zhang, H., Huang, R., Zhang, S.-D., Li, Y., Qiu, C.-W., Nori, F., and Jing, H. (2020). Breaking anti-PT symmetry by spinning a resonator. *Nano Lett.*, 20, pp. 7594.
41. Bergman, A., Duggan, R., Sharma, K., Tur, M., Zadok, A., and Alù, A. (2021). Observation of anti-parity-time-symmetry, phase transitions and exceptional points in an optical fibre. *Nat. Commun.*, 12, pp. 486.
42. Geng, Q. and Zhu, K.-D. (2021). Discrepancy between transmission spectrum splitting and eigenvalue splitting: a reexamination on exceptional point-based sensors. *Photon. Res.*, 9, pp. 1645.

43. Chong, Y. D., Ge, L., Cao, H., and Stone, A. D. (2010). Coherent perfect absorbers: time-reversed lasers. *Phys. Rev. Lett.*, 105, pp. 053901.
44. Wan, W., Chong, Y., Ge, L., Noh, H., Douglas Stone, A., and Cao, H. (2011). Time-reversed lasing and interferometric control of absorption. *Science*, 331, pp. 889–892.
45. Baranov, D. G., Krasnok, A., Shegai, T., Alù, A., and Chong, Y. (2017). Coherent perfect absorbers: linear control of light with light. *Nat. Rev. Mater.*, 2, pp. 17064.
46. Pichler, K., Kühmayer, M., Böhm, J., Brandstötter, A., Ambichl, P., Kuhl, U., and Rotter, S. (2019). Random anti-lasing through coherent perfect absorption in a disordered medium. *Nature*, 567, pp. 351–355.
47. Wang, C., Sweeney, W. R., Stone, A. D., and Yang, L. (2021). Coherent perfect absorption at an exceptional point. *Science*, 373, pp. 1261–1265.
48. Sciubba, E. and Wall, G. (2007). A brief commented history of exergy from the beginnings to 2004. *Int. J. Thermodyn.*, 10, pp. 1–26.
49. Choi, W., Ooka, R., and Shukuya, M. (2018). Exergy analysis for unsteady-state heat conduction. *Int. J. Heat. Mass Transf.*, 116, pp. 1124–1142.
50. Tan, S. and Tsang, L. (2019). Efficient broadband evaluations of lattice Green's functions via imaginary wavenumber components extractions. *Prog. Electromagn. Res.*, 164, pp. 63–74.
51. Ju, R., Cao, P., Wang, D., Qi, M., Xu, L., Yang, S., Qiu, C., Chen, H., and Li, Y. (2024). Nonreciprocal heat circulation metadevices. *Adv. Mater.*, 36, pp. 2309835.
52. Wang, D., Chen, H., Qiu, C., and Li, Y. (2023). Non-Hermitian scattering symmetry revealed by diffusive channels. arXiv preprint arXiv:2303.13104.

Chapter 5

Temporal Modulation

5.1. Introduction

This chapter explores the causality between temporal modulation and nonreciprocity in thermal diffusion. We introduce various temporal modulation methods and differentiate their applicability in heat transfer and wave dynamics, supported by experimental demonstrations. We clarify the necessity of boundary conditions for temperature fields in breaking reciprocity and conclude that thermal reciprocity is preserved under constant temperature boundaries when temporal mass density modulation is applied due to mass continuity. We investigate using time-harmonic heat sources to examine the thermal nonreciprocity induced by the diffusive Fizeau drag effect, explaining this phenomenon through the thermal Willis coupling mechanism. Additionally, we discuss the application of temporal modulation theory in nonreciprocal geometric heat pumping and provide feasibility suggestions for experiments. Temporal modulation theory introduces additional freedom for thermal regulation, unveiling new mechanisms for nontrivial heat transfer, which may open up new possibilities for physical effects and applications in this rapidly evolving field.

Nonreciprocity refers to unbalanced field transport in opposite directions [1], a concept fundamental to logic gates [2], and information regulation [3] in wave transport and diffusion. In thermal diffusion, non-reciprocal heat transfer could benefit the development of novel functional devices like thermal rectifiers [4] and diodes [5]. According to the Onsager reciprocal relation [6], reciprocity can be broken via three primary approaches: (1) utilizing external bias fields to break time-reversal symmetry, (2) employing nonlinear materials to break spatial translational symmetry, and (3) modulating time-varying properties. For the first approach, a typical example is the motion of electric charges in a magnetic field [7], which can be mimicked in thermal diffusion by mass fluxes passing directionally through the system, such as directional convection along or perpendicular to the heat transfer path [8]. However, this method is impractical due to integration challenges and system instability. For the second approach, phase change materials, such as oxides [4] and shape-memory alloys [5], have been widely used in developing thermal diodes. However, the harsh working conditions of phase change materials severely limit the tunability of these devices. The third approach has garnered much attention in recent years due to its high integrability and broad applicability in wave dynamics [9], but it has been less studied in thermal diffusion due to theoretical and practical challenges.

Nonreciprocal electromagnetic wave propagation can be achieved by modulating electric permittivity in a traveling wave pattern, leading to asymmetric mode conversions [10]. This approach has also been successful in thermal radiation [11], acoustic waves [12], and charge diffusion [13]. Since the governing equation for charge diffusion, Fick's law, shares the same Laplacian term as the wave equation, simultaneous modulation of capacitance and electric conductivity is necessary to achieve nonreciprocity. The governing equation of heat conduction, derived from Fourier's law, has a similar form to Fick's law. Consequently, thermal conductivity and mass density are the counterparts of electric conductivity and capacitance in thermal

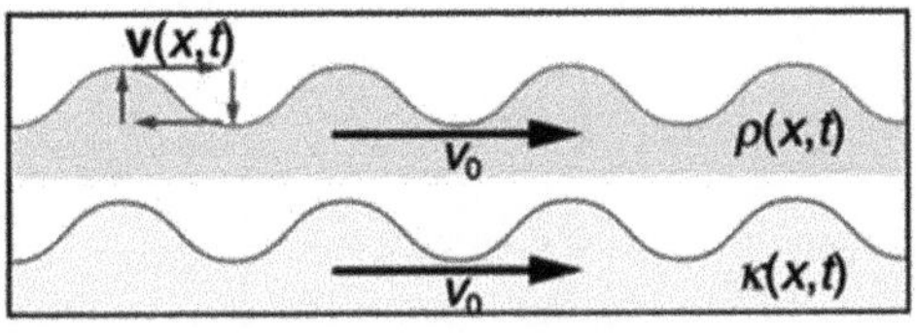

Fig. 5.1. Temporal modulation of mass density and thermal conductivity in heat transfer. Reproduced from Ref. [17], licensed under CC-BY 4.0.

diffusion (Fig. 5.1). In 2018, Torrent *et al.* theoretically predicted the possibility of inducing thermal nonreciprocity through time modulation [14]. However, experimental results by Li *et al.* demonstrated that convection-driven temporal mass motion in practice inevitably alters the governing transfer equation, making it extremely difficult to break the reciprocity of heat transfer [15].

5.2. Theoretical Basis

5.2.1. *Temporal Modulation*

Fourier's law describes the thermal diffusion process in solids: $\partial(\rho c T)/\partial t = \nabla \cdot (\kappa \nabla T)$, where ρ is mass density, c is specific heat capacity, T is temperature field, t is time, and κ is thermal conductivity. The equation has reciprocal solutions when the material is linear. If the density and thermal conductivity are dynamically modulated without introducing additional effects, Fourier's law can be written as

$$\frac{\partial[\rho(r,t)cT]}{\partial t} = \nabla \cdot [\kappa(r,t)\nabla T]. \tag{5.1}$$

According to the first law of thermodynamics, the full rate of energy accumulation E is the sum of the net heat transfer by conduction Q_{cond}, fluid flow Q_{conv}, and network W in a control volume V, where $E = \int_V \frac{\partial(\rho c T)}{\partial t} dV$, $Q_{\text{cond}} = -\int_A q \cdot n dA$, and $Q_{\text{conv}} = -\int_A \rho c T v \cdot n dA = -\int_V \nabla \cdot (\rho c T v) dV$ is the net heat transfer by fluid flow, and n is the unit normal vector at the system boundary. When we assume $W = 0$, Eq. (5.1) is changed for the dynamical modulation is driven by convection,

$$\frac{\partial[\rho c T]}{\partial t} + \nabla \cdot (\rho c T \mathrm{v}) = \nabla \cdot [\kappa \nabla T]. \tag{5.2}$$

Equation (5.2) can also be written as

$$\rho c \frac{\partial T}{\partial t} + \rho c \mathrm{v} \nabla T + c T \left[\frac{\partial \rho}{\partial t} + \nabla \cdot (\rho \mathrm{v})\right] = \nabla \cdot [\kappa \nabla T]. \tag{5.3}$$

Given the conservation of mass, we have the continuity equation of $\frac{\partial \rho}{\partial t} + \nabla \cdot (\rho v) = 0$, i.e., $\rho c \frac{\partial T}{\partial t} + \rho c v \nabla T = \nabla \cdot (\kappa \nabla T)$. Therefore, we can rewrite Eq. (5.3) as

$$\rho c \frac{DT}{Dt} = \nabla \cdot (\kappa \nabla T) \tag{5.4}$$

where D/Dt represents the "material derivative" operator $\frac{D}{Dt} = \frac{\partial}{\partial t} + v\frac{\partial}{\partial x}$.

Generally, we have two ways to achieve temporal mass density modulation without external bias: mass changes along the heat transfer path with traveling wave modulation, or mass enters/leaves the heat transfer path cyclically. In this way, no accumulated external bias or net energy flux exists in the system. Specifically, the boundary conditions are constant, and the modulation of material parameters is periodic with time to ensure a stable (time-harmonic) field.

5.2.2. *Thermal Reciprocity*

We take the two-port system as an example to study whether temporal modulation parameters can break the reciprocal in diffusion. When we swap the boundary conditions, the thermal reciprocity is defined through the equivalence of time-averaged heat fluxes $\langle q \rangle$ at different ports,

$$\begin{pmatrix} \langle q_{f,2} \rangle \\ \langle q_{f,1} \rangle \end{pmatrix} = -\begin{pmatrix} \langle q_{b,1} \rangle \\ \langle q_{b,2} \rangle \end{pmatrix} \tag{5.5}$$

where $\langle q_b \rangle$ and $\langle q_f \rangle$ represent heat fluxes before and after the exchange of boundary conditions, respectively, and the numbers 1 and 2 denote different ports.

For the first case, we modulate the mass density ρ and thermal conductivity κ along the heat transfer path spatiotemporally, and the expanded Fourier series form

$$\begin{aligned} \rho(\chi) &= \sum_l \rho_l e^{il\beta\chi}, \\ \kappa(\chi) &= \sum_l \kappa_l e^{il\beta\chi}, \end{aligned} \tag{5.6}$$

where x is the heat transfer direction, d is the spatial modulation period, $\chi = x - v_0 t$, the integer index l takes $0, \pm 1, \pm 2, \ldots$ and $\beta = 2\pi/d$. The coefficients satisfy $\rho_1 = \rho^*_{-1}$ and $\kappa_1 = \kappa^*_{-1}$ for the reality of the parameters, where the superscript star means complex conjugation. Considering the conservation of mass and the 1D continuity equation $\partial \rho/\partial t + \partial(\rho v)/\partial x = 0$, we have

$$\begin{aligned} \frac{\partial(\rho v)}{x} &= ilv \sum_l \rho_l e^{il\beta\chi}, \\ \rho v &= v_0 \sum_l \rho_l e^{il\beta\chi} + g(t) = \rho v_0 + g(t) \end{aligned} \tag{5.7}$$

where $g(t)$ is d/v_0-periodic and can be ignored when we set $g(t) = -\rho_0 v_0 + C, \rho_0$ is the average density and C is a constant. This assumption does not affect any of the conclusions.

5.3. Constant Temperature Boundary Condition

We take the average of ρv and C as zero in a period to ensure there is no accumulated net mass flux through the system. Substituting Eq. (5.7) into Eq. (5.4), the governing equation turns into

$$\rho(\chi)c\frac{\partial T}{\partial t} + \{[\rho(\chi) - \rho_0]v_0 + C\}c\frac{\partial T}{\partial x} = \frac{\partial}{\partial x}\left[\kappa(\chi)\frac{\partial T}{\partial x}\right]. \tag{5.8}$$

It is worth mentioning that additional constant mass flux passing the heat transfer path can cause direct nonreciprocity with $C \neq 0$. Equation (5.8) has a solution according to the Floquet-Bloch theorem,

$$T(\chi, \tau) = e^{i(K\chi - \Omega\tau)} f(\chi) = e^{iK\chi} e^{-i(\Omega + kv_0)t} f(\chi) \tag{5.9}$$

where $\chi = x - v_0 t$, $\tau = t$, Ω is the frequency, and K is the Bloch wavenumber. Equation (5.9) describes a temperature profile characterized by Ω and K, and modulated by a d-periodic function $f(\chi)$. In the steady state, the time-harmonic term should be a constant with $\Omega + Kv_0 = 0$. We can write Eq. (5.9) as

$$T(x,t) = e^{iKx} f(\chi) = e^{\alpha x} f(\chi). \tag{5.10}$$

For convenience, we take the Bloch wavenumber as $K = -\alpha i$ and expand the periodic function $f(\chi)$ as

$$f(\chi) = \sum_m F_m e^{im\beta\chi} \tag{5.11}$$

where m is an integer index taking $0, \pm1, \pm2, \ldots$ and $F_0 = 1$. Substituting Eq. (5.11) into Eq. (5.8), we have

$$\begin{aligned} &\kappa f'' + [2\alpha\kappa + \kappa' + (\rho_0 v_0 - C)c] f' \\ &\quad + [\alpha\kappa + \kappa' + (-\rho v_0 + \rho_0 v_0 - C)c]\alpha f = 0. \end{aligned} \tag{5.12}$$

The nth order of Eq. (5.12) is

$$\sum_m \begin{bmatrix} (\alpha + i\beta n)(\alpha + i\beta m)\kappa_{n-m} - \alpha\rho_{n-m} c v_0 \\ + \delta_{n,m}(\alpha + i\beta n)(\rho_0 v_0 - C)c \end{bmatrix} F_m = 0 \tag{5.13}$$

where $n = 0, \pm1, \pm2, \ldots$ and δ_{nm} is the Kronecker delta. For $mn \neq 0$, F_m can be solved as

$$F_m = -\sum_{n\neq 0} A_{m,n}\alpha[(\alpha + i\beta n)\kappa_n - \rho_n c v_0] F_0. \tag{5.14}$$

$[A_{m,n}]$ is a $m \times n$ matrix. It is defined from Eq. (5.13) as

$$[A_{m,n}] = \begin{bmatrix} (\alpha + i\beta n)(\alpha + i\beta m)\kappa_{n-m} - \alpha\rho_{n-m} c v_0 \\ + \delta_{n,m}(\alpha + i\beta n)(\rho_0 v_0 - C)c \end{bmatrix}^{-1}. \tag{5.15}$$

Substituting Eq. (5.14) into the 0th order of Eq. (5.13), $(n = 0)$ gives

$$\begin{aligned} \alpha(\alpha\kappa_0 - Cc) = \alpha^2 &\sum_{m\neq 0} [(\alpha + i\beta m)\kappa_{-m} - \rho_{-m} c v_0] \\ &* \sum_{n\neq 0} A_{m,n}[(\alpha + i\beta n)\kappa_n - \rho_n c v_0]. \end{aligned} \tag{5.16}$$

We take $N = 4$ to ensure the accuracy of the parameter α in Eq. (5.16). Then, F_m in Eq. (5.14) and the corresponding $f(\chi)$ can be deduced, and the temperature field in Eq. (5.10) is rewritten

$$T(x,t) = \sum_k C_k e^{\alpha_k x} f_k(\chi) = \sum_k C_k F_m^k e^{\alpha_k x + im\beta\chi} \tag{5.17}$$

where k represents the kth particular solution, and C_k is the corresponding coefficient determined by the boundary conditions at two ends. The above analysis is based on $C \neq 0$. α has a nonzero real solution represented by a convex temperature profile. Nonreciprocity can hardly be generated by temporal modulation.

For the case $C = 0$, Eq. (5.10) becomes

$$T(x,t) = \sum_k C_k e^{\alpha_k x} f_k(\chi) + C_{\#}[x + f_{\#}(\chi)] \tag{5.18}$$

where C_k and $C_{\#}$ are coefficients determined by the boundary conditions. Recalling the formula to calculate the heat flux,

$$q(x,t) = -\kappa(x,t)\frac{\partial T(x,t)}{\partial x} + [\rho(x,t) - \rho_0]v_0 c[T(x,t) - T_{ref}]. \tag{5.19}$$

When $n = 0$, the time-averaged heat flux is

$$\begin{aligned}\langle q(x)\rangle = &-\sum_{k,m} C_k[(\alpha + i\beta m)\kappa_{-m} + (\delta_{0,m} - 1)\rho_{-m}v_0 c]F_m^k e^{\alpha_k x} \\ &- C_{\#}\left\{\kappa_0 + \sum_m [(\alpha + i\beta m)\kappa_{-m} + (\delta_{0,m} - 1)\rho_{-m}v_0 c]F_m^{\#}\right\}.\end{aligned} \tag{5.20}$$

For all $\alpha_k \neq 0$, we have

$$\sum_m [(\alpha + i\beta m)\kappa_{-m} + (\delta_{0,m} - 1)\rho_{-m}v_0 c]F_m^k = 0. \tag{5.21}$$

Besides, considering Eq. (5.21), it is easy to find that for $\alpha_k = 0$, the corresponding $F_m = 0 (m \neq 0)$. Therefore, the time-averaged heat flux becomes

$$\langle q(x)\rangle = -C_{\#}\left\{\begin{array}{c}\kappa_0 + \\ \sum_m [(\alpha + i\beta m)\kappa_{-m} + (\delta_{0,m} - 1)\rho_{-m}v_0 c]F_m^{\#}\end{array}\right\}. \tag{5.22}$$

From Eq. (5.22), we can find $\langle q(x)\rangle$ is independent of x, which means the heat flux over a period is a conserved quantity, and the system will hold reciprocity.

For the 1D hypothetical case, without considering the continuity of mass, the diffusion equation is

$$\rho(\chi)c\frac{\partial T}{\partial t} = \frac{\partial}{\partial x}\left[\kappa(\chi)\frac{\partial T}{\partial x}\right]. \tag{5.23}$$

The average of heat flux over time can be deduced (the derivation is omitted)

$$\langle q(x)\rangle = -\sum_{m} C_{\#}(\alpha + i\beta m)\kappa_{-m}F_m^k e^{\alpha_k x} \tag{5.24}$$

which indicates that reciprocity is broken but is hard to implement in practice.

Previous insightful work shows the possibility of generating thermal nonreciprocity at a mathematical level in the hypothetical case [14], which can be verified in the finite-element simulation with a 1D model [15]. However, the adjustable modulation of mass density without cost is difficult to achieve. In other words, the virtual system should have an additional energy difference in the entrance and exit for nonreciprocal average heat flux. As a result, the governing equation has to be modified.

Another way to realize mass density modulation without the help of a biased field exists, i.e., the matter enters and exits the heat transfer path periodically. This idea can be achieved in a simple 2D toy model. The system is under a constant temperature boundary condition in the heat transfer path ($x = 0$ and L), with a periodic boundary condition on the upper and lower sides ($y = \pm d_y/2$). In this way, the governing equation of the 2D heat transfer model is written as

$$\rho(\xi)c\left[\frac{\partial T}{\partial t} + v_y\frac{\partial T}{\partial y}\right] = \frac{\partial}{\partial x}\left[\kappa^x(\xi)\frac{\partial T}{\partial x}\right] + \frac{\partial}{\partial y}\left[\kappa^y(\xi)\frac{\partial T}{\partial y}\right] \tag{5.25}$$

where $\zeta = x + \eta y - v_0 t$ with $\eta = d/d_y$. The mass density $\rho(x, y = 0, t)$ and thermal conductivity $\kappa(x, y = 0, t)$ are d-periodic $(x - v_0 t)$. As the mass density is modulated through mass motion along y direction (speed $v_y(x, y, t)$), $v_y(\xi, y, t) = v_{0_y}$ when the constant C is set as $\rho_0 v_{0_y}$. Consider the general form solution $T(x, y, t) = e^{\alpha x} f(\zeta)$, one can get

$$\begin{aligned}&(\kappa^x + \eta^2\kappa^y)f'' + (\alpha\kappa^x + \kappa^{x'})\alpha f \\ &\quad + [2\alpha\kappa^x + \kappa^{x'} + \eta^2\kappa^{y'} + (\rho_0 v_0 - \eta C)c]f' = 0. \end{aligned} \tag{5.26}$$

The nth order of its Fourier expansion is

$$\sum_m \begin{bmatrix} (\alpha + i\beta n)(\alpha + i\beta m)\kappa^x_{n-m} \\ -\eta^2\beta^2 nm\kappa^y_{n-m} + \delta_{n,m} i\beta n(\rho_0 v_0 - \eta C)c \end{bmatrix} F_m = 0. \tag{5.27}$$

This 2D toy model can be mapped into a practical 3D, as shown in Fig. 5.2a. The fixed and moving fan-shaped solid plates

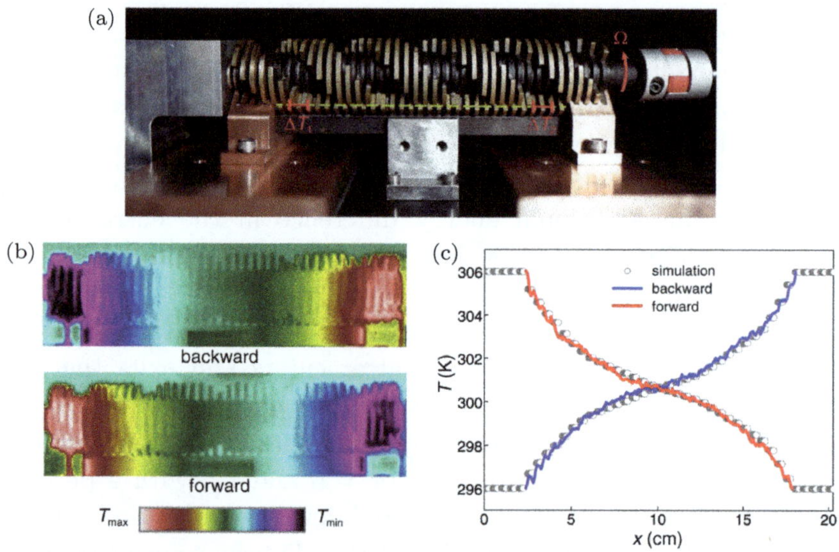

Fig. 5.2. Experimental verification of the thermal conduction reciprocity in a spatiotemporal modulated system with constant temperature boundary conditions. (a) Mass density is temporally modulated in the 3D modulation model by the rotating upper movable brass plates. (b) Captured temperature profiles with opposite ports input. (c) Measured reciprocal temperature profiles along the thermal conduction path with opposite ports input (dashed green line in (a)). The scatter points refer to simulated results considering natural air convection and interface thermal resistance. (Reproduced from Ref. [15], licensed under CC-BY 4.0).

are made of copper. The mass density ρ_A, heat capacity c_A, and thermal conductivity κ_A are $8390\,\mathrm{kg\,m^{-3}}$, $375\,\mathrm{J\,kg^{-1}\,K^{-1}}$, and $123\,\mathrm{W\,m^{-1}\,K^{-1}}$, respectively. Each plate spans $\pi/2$ with inner and exterior radius $R_1 = 1$ cm and $R_2 = 2\,\mathrm{cm}$, and thickness $\delta = 0.25\,\mathrm{cm} = d/16$ fixed on the nylon cylinder with low thermal conductivity ($0.3\,\mathrm{W\,m^{-1}\,K^{-1}}$). The constant temperatures on the left and right boundaries are $T_{\mathrm{cold}} = 296\,\mathrm{K}$ and $T_{\mathrm{hot}} = 306\,\mathrm{K}$. The heat transfer path is regarded as the portion of the system contained in the region $([R_1, R_2], [\pi/4, 3\pi/4], [0, L])$ of the cylindrical coordinate system (r, θ, x). In experiments, adjacent plates rotate at angular speed $\Omega = 0.075\pi$ rad $\mathrm{s^{-1}}$ and keep a $\pi/4$ phase lag to modulate the mass density effectively. The observed heat transfer path comprises the upper rotational and lower fixed plates. Most of the heat flux is transferred through conduction. A $0.2\,\mathrm{mm}$ gap exists between the adjacent rotational and fixed plates. To decrease the interface thermal resistance and the friction in rotation, thermal grease (thermal conductivity $4.38\,\mathrm{W\,m^{-1}\,K^{-1}}$) and silicone oil (thermal conductivity $0.16\,\mathrm{W\,m^{-1}\,K^{-1}}$) are filled in these gaps. Moreover, a vacuum chamber ($10^{-3}\,\mathrm{Pa}$) is employed to avoid the influence of natural convection-caused directional flow.

ΔT_1 and ΔT_2 with red line segments in Fig. 5.2a represent the temperature differences of measured temperature fields over time and the reference temperature. The green line denotes the heat transfer path, and the red arrow shows the rotating direction of the nylon cylinder. According to the relation between heat flux and temperature, gradient is proportional as $q_x = -\kappa \partial T/\partial x$, we can generate the reciprocity of thermal diffusion on the heat transfer path by analyzing the symmetry of temperature profiles under opposite directions input. The periodically changed mass density-caused nonreciprocity is annihilated by the conservation of mass, which is very different from the hypothetical model with only parameters artificially modulated ($\rho_A(t) = \rho_A\{1 + \cos[2\Omega t - (n-1)\pi/4]\}/2$, n is the number of the moving plates). Detailed parameters setting and simulation results of the hypothetical case can be found in Ref. [15].

Temperature fields are captured through the infrared camera on the inspection window (Fig. 5.2b). The upper picture shows the case with the left/right boundary at the temperature $T_{\mathrm{cold}}/T_{\mathrm{hot}}$, where the direction of heat flux is denoted as backward. The lower picture corresponds to the opposite case with forward heat flux direction and the left/right boundary at the temperature $T_{\mathrm{hot}}/T_{\mathrm{cold}}$. Measured temperature profiles are shown in Fig. 5.2c, and the symmetric curves show that the reciprocity of heat transfer still holds. We should note that the interference in natural convection can be ignored in the experimental settings. However, the temperature profiles are no longer linear due to thermal resistances on the interfaces. Temperature gradients at the center are much smaller. The data of temperature profiles is extracted from the heat transfer path denoted as a green line in Fig. 5.2a, which fits the simulation results well. In the simulation, thermal resistance is set as 0.2 mm thickness and $1\,\mathrm{W\,m^{-1}\,K^{-1}}$ thermal conductivity with natural air convection-induced heat exchange rate is $h = 3\,\mathrm{W\,m^{-2}\,K^{-1}}$.

The measured data of ΔT_1 and ΔT_2 in different directions over time also gives out $\langle \Delta T_{\mathrm{b},1} \rangle$, $\langle \Delta T_{\mathrm{b},2} \rangle$, $\langle \Delta T_{\mathrm{f},1} \rangle$ and $\langle \Delta T_{\mathrm{f},2} \rangle$ as $-0.874\,\mathrm{K}$, $-1.038\,\mathrm{K}$, $1.046\,\mathrm{K}$, and $0.863\,\mathrm{K}$ for the backward and forward cases, respectively. We find that the module values of $\langle \Delta T_{\mathrm{b},1} \rangle$ and $-\langle \Delta T_{\mathrm{f},2} \rangle$, $\langle \Delta T_{\mathrm{b},2} \rangle$ and $\langle \Delta T_{\mathrm{f},1} \rangle$ are nearly equal, and this symmetry relation holds for variable rotation speeds and periods. Moreover, when we change the angular velocity Ω as 0.14π rad s^{-1}, we get $\langle \Delta T_{\mathrm{b},1} \rangle$, $\langle \Delta T_{\mathrm{b},2} \rangle$, $\langle \Delta T_{\mathrm{f},1} \rangle$ and $\langle \Delta T_{\mathrm{f},2} \rangle$ as $-0.887\,\mathrm{K}$, $-0.992\,\mathrm{K}$, $0.996\,\mathrm{K}$ and $0.872\,\mathrm{K}$. We also demonstrate this reciprocity from the numerical simulation by calculating the heat flow Q from different directions, which agrees well with the expectations.

We find that thermal reciprocity is hard to break under temporal modulation for the existence of density modulation-induced convection correction. For the conservation of mass, the governing equation of thermal diffusion has to be revised by adding convection correction. This demand fundamentally differs from wave transport and charge diffusion, where the momentum or continuity equation governs these processes. It reminds us that we should pay attention to the applicability of relevant theories in heat transfer when learning from them. The thermal reciprocity of the physical system is still preserved even when the modulated parameters are asymmetrically distributed in space, which shows strong robustness. The specific heat capacity is unchanged in this experimental verification, for it is difficult to modulate in time, and the effect is the same as the mass density modulation for the relation $\partial(\rho c)/\partial t + \nabla \cdot (\rho c \mathrm{v}) = 0$. Otherwise, we can get thermal nonreciprocity by the specific heat capacity modulation with phase-change materials, but the application conditions are harsh and require external fields, so this case is not discussed here. Besides, the above theoretical analysis and experimental verification are based on the constant temperature boundary condition without external energy input over a period. We can break this thermal diffusion reciprocity by introducing directional mass/energy bias or under the periodically oscillating temperature boundary condition, which will be introduced in the following sections.

5.4. Oscillating Temperature Boundary Condition

5.4.1. *Diffusive Fizeau Drag Effect*

Oscillating temperature boundary conditions in thermal nonreciprocity are analyzed through the diffusive Fizeau drag effect and Willis coupling theory. Light travels at different speeds along and against the flow of water, which was theoretically predicted by Fresnel [16] and experimentally verified by Fizeau [17]. This phenomenon, known as Fizeau drag, is well explained by relativistic kinematics. Similar effects have been observed in other moving or spatiotemporal media. Recently, plasmonic Fizeau drag caused by the flow of electrons has been experimentally reported [18, 19], resulting from the nonlinear kinematics of drifting Dirac electrons. Diffusive systems can also exhibit wave-like behaviors [20], suggesting the possibility of diffusive Fizeau drag. However, unlike dragging photons and polaritons by momentum interaction (Fig. 5.3a,b), dragging macroscopic heat through biased advection is intrinsically challenging due to the absence of macroscopic heat momentum (Fig. 5.3c). Consequently, the forward and backward propagating speeds of temperature fields remain identical, though the amplitudes differ due to the dissipative nature of heat transfer. Thus, realizing diffusive Fizeau drag remains an extremely challenging problem.

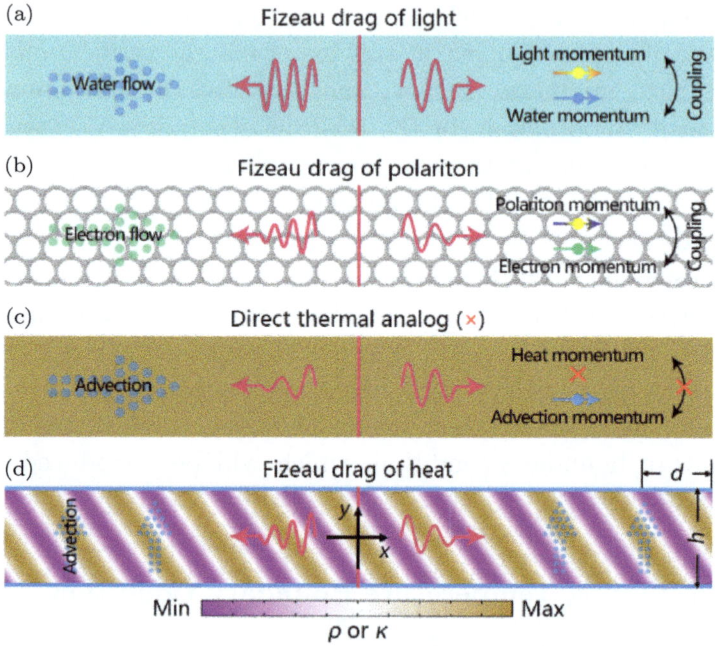

Fig. 5.3. Origin of diffusive Fizeau drag. Fizeau drag of (a) light and (b) polariton by the momentum interaction. (c) Failure of a direct thermal analog due to the lack of macroscopic heat momentum. (d) Fizeau drag of heat in a spatiotemporal thermal metamaterial by thermal Willis coupling. The red arrows contain the information on wavenumber and amplitude, indicating the forward and backward cases with (a), (b), (d) different wavenumbers, and (c) different amplitudes. Adapted from Ref. [21].

A spatiotemporal thermal metamaterial with space-related inhomogeneity and time-related advection is constructed to reveal diffusive Fizeau drag in heat transfer (Fig. 5.3d) [21]. The structure functions as a metamaterial because the characteristic length of spatiotemporal modulation is much smaller than the wavelength of wave-like temperature fields. The combined effect of periodic inhomogeneity and vertical advection induces horizontal nonreciprocity, leading to diffusive Fizeau drag.

We first explain why the direct scheme presented in Fig. 5.3c fails. Heat transfer in porous media is described by $\rho_0\partial_t T + \nabla\cdot(\phi\rho_a u T - \kappa_0\nabla T) = 0$, where ρ_0 (or ρ_a) is the product of mass density and heat capacity of the porous medium (or fluid), κ_0 is the thermal conductivity of the porous medium, ϕ is the porosity, and u is the velocity of the fluid with the horizontal and vertical components of u_x and u_y, respectively. Consider a wave-like temperature field described by $T = A\mathrm{e}^{\mathrm{i}(\beta x-\omega t)}+T_r$, where β and ω are the wavenumber and angular frequency, respectively. The term "wave-like" is used because heat transfer is governed by a diffusive rather than a wave equation. Setting the temperature field amplitude of A as 1 and the balanced temperature of T_r as 0 for brevity, and applying a periodic source with temperature $T(x=0) = \mathrm{e}^{-\mathrm{i}\omega t}$ results in a real ω and a complex β. The imaginary part of β reflects the spatial decay rate of wave-like temperature fields. The focus is on the real part of β because the propagating speed of wave-like temperature fields can be calculated by $v = \omega/\mathrm{Re}[\beta]$. The substitution of $T = \mathrm{e}^{\mathrm{i}(\beta x-\omega t)}$ with a preset real ω into the governing equation of heat transfer yields

$$\beta_{f,b} = \pm\frac{\sqrt{2}\gamma}{4\kappa_0} + \mathrm{i}\frac{-8\phi\rho_a u_x\omega\rho_0\kappa_0 \pm \sqrt{2}\gamma(2\phi^2\rho_a^2u_x^2+\gamma^2)}{16\omega\rho_0\kappa_0^2} \tag{5.28}$$

where β_f and β_b are the forward and backward wavenumbers with a definition of $\gamma = \sqrt{-\phi^2\rho_a^2u_x^2+\sqrt{\phi^4\rho_a^4u_x^4+16\omega^2\rho_0^2\kappa_0^2}}$. Since a nonzero u_x cannot generate different $|\mathrm{Re}[\beta]|$ in opposite directions, the forward and backward propagating speeds of temperature fields are identical, i.e., no diffusive Fizeau drag.

For the diffusive Fizeau drag, spatially-periodic inhomogeneity is introduced to the porous medium,

$$\rho(\xi) = \rho_0(1+\Delta_\rho\cos(G\xi+\theta)) \tag{5.29a}$$

$$\kappa(\xi) = \kappa_0(1+\Delta_\kappa\cos(G\xi)) \tag{5.29b}$$

where Δ_ρ and Δ_κ are the modulation amplitudes, $G = 2\pi/d$ is the modulation wavenumber, d is the horizontal modulation wavelength, $\xi = x+\zeta y$ is the generalized coordinate with a definition of $\zeta = d/h$, h is the vertical height, and θ is the modulation phase difference. To exclude that, the horizontal advection can generate nonreciprocal amplitudes of temperature fields, as described by the imaginary part of Eq. (5.28), we consider the upward advection with a speed of u_y, which does not contribute to the horizontal nonreciprocity.

The governing equation of heat transfer in spatiotemporal thermal metamaterials can be expressed as

$$\bar{\rho}(\xi)\frac{\partial T}{\partial t} + \phi\epsilon u_y\frac{\partial T}{\partial y} + \frac{\partial}{\partial x}\left(-D_0\bar{\kappa}(\xi)\frac{\partial T}{\partial x}\right) + \frac{\partial}{\partial y}\left(-D_0\bar{\kappa}\left(\xi\right)\frac{\partial T}{\partial y}\right) = 0 \tag{5.30}$$

with definitions of $\bar{\rho}(\xi) = \rho(\xi)/\rho_0$, $\bar{\kappa}(\xi) = \kappa(\xi)/\kappa_0$, $\epsilon = \rho_a/\rho_0$, and $D_0 = \kappa_0/\rho_0$.

Considering a wave-like temperature field with a spatially-periodic modulation,

$$T = F(\xi)\mathrm{e}^{\mathrm{i}(\beta x-\omega t)} = \left(\sum_s F_s\mathrm{e}^{\mathrm{i}sG\xi}\right)\mathrm{e}^{\mathrm{i}(\beta x-\omega t)} \tag{5.31}$$

where $F(\xi)$ is a Bloch modulation function with $s = 0, \pm1, \pm2, \ldots, \pm\infty$ and $F_0 = 1$. We can treat $\mathrm{e}^{\mathrm{i}(\beta x-\omega t)}$ as the temperature field envelope and $F(\xi)$ as local inhomogeneity. The substitution of Eq. (5.31) into Eq. (5.30) yields a series of component equations related to the order of s. For accuracy, we consider $s = 0, \pm1, \pm2, \ldots, \pm10$ and $F_{|s|>10} = 0$ to obtain 21 equations with 21 unknown numbers, including β and $F_{|s|\leq 10}$, so β can be numerically calculated.

The properties of spatiotemporal modulation are reflected in three crucial dimensionless parameters of $2\pi\Gamma = \phi\epsilon u_y d/D_0$, $\Lambda = \Delta_\rho\cos\theta/\Delta_\kappa$, and $\zeta = d/h$. The parameter of $2\pi\Gamma$ is similar to the Peclet number, which can describe the ratio of advection to diffusion. The parameters of Λ and ζ reflect the influences of modulation amplitude and wavelength, respectively. We define the speed ratio as $\eta = |v_f/v_b| = |\mathrm{Re}[\beta_b]/\mathrm{Re}[\beta_f]|$ to discuss the degree of nonreciprocity, where v_f and v_b are the forward and backward propagating speeds of temperature fields, respectively.

We first discuss Λ when $\zeta = 0.2$ (Fig. 5.4a). Since $2\pi\Gamma = 0$ and $2\pi\Gamma \to \infty$ always yield $\eta = 1$, it is necessary to introduce the vertical advection, but not the larger the better. Meanwhile, a speed difference still exists when $\Lambda = 0$ (i.e., $\Delta_\rho = 0$), so it is not necessary to modulate ρ and κ simultaneously. Two types of curves exist in Fig. 5.4a. Type I features η as always larger than 1 (the top three curves). Type II features η as first larger and then smaller than 1 (the bottom three curves). The transition between types I and II is at the critical point of $\Lambda = 1$ (the third curve from the top). When ζ changes from 0.2 to 1 (Fig. 5.4b) and 2 (Fig. 5.4c), type III curves appear,

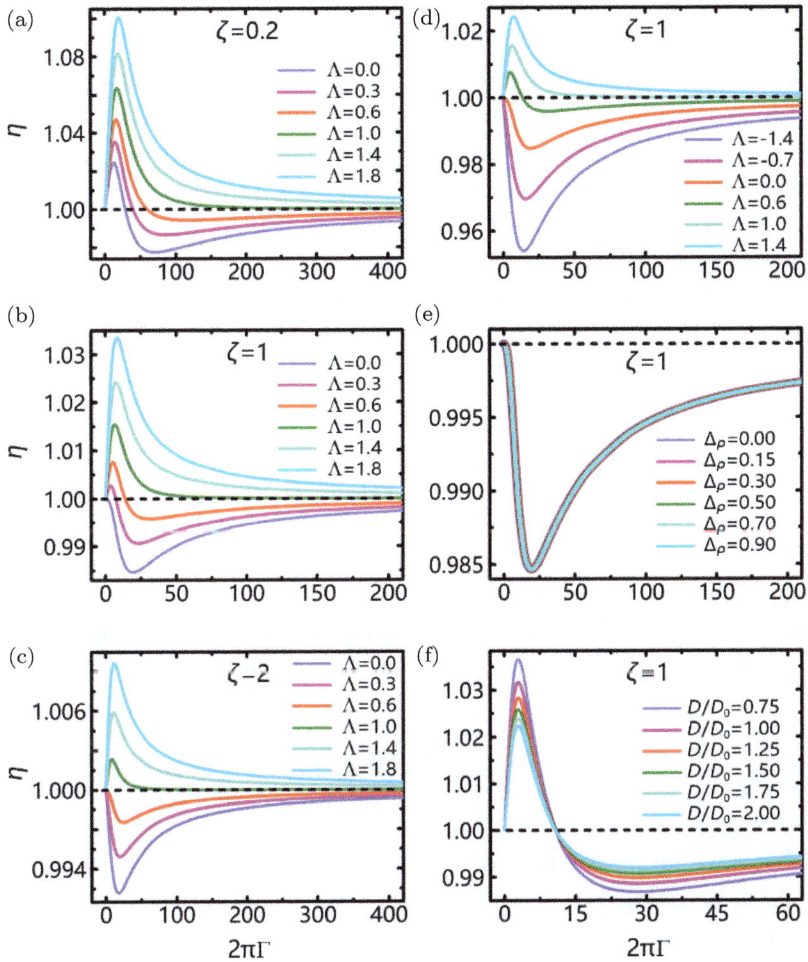

Fig. 5.4. Numerical results of the speed ratio of $\eta = |v_f/v_b|$ as a function of $2\pi\Gamma = \phi\epsilon u_y d/D_0$. $\Lambda = \Delta_\rho \cos\theta/\Delta_\kappa$ is tuned by (a–c) Δ_ρ or (d) θ. Except for the parameters presented in (a–f), the others are $\phi = 0.1$, $\epsilon = 1$, $D_0 = 5 \times 10^{-5}$ m^2/s, $d = 0.02$ m, and $\omega = \pi/10$ rad/s for (a–f); $\Delta_\rho = 0.7$ for (d); $\Delta_\rho = 0.6$ for (f); $\Delta_\kappa = 0.5$ for (a–e); $\Delta_\kappa = 0.9$ for (f); $\theta = 0$ for (a–d) and (f); and $\theta = \pi/2$ for (e). Adapted from Ref. [21].

featuring η always smaller than 1. Thus, nonreciprocal speeds can be flexibly manipulated.

We further discuss θ when $\zeta = 1$ (Fig. 5.4d), so $\Lambda = \Delta_\rho \cos\theta/\Delta_\kappa$ can be both positive and negative. The critical point of $\Lambda = 1$ still determines the transition between types I and II. Moreover, since $\theta = \pi/2$ always leads to $\Lambda = 0$, the curves in Fig. 5.4(e) are almost overlapped. We also discuss the thermal diffusivity of $D = \kappa/\rho$ (Fig. 5.4f), where κ is the balanced value of the periodic thermal conductivity, and ρ is the balanced value of the periodic product of mass density and heat capacity. The peaks of η appear at almost the same value of $2\pi\Gamma$. Meanwhile, the peak of η gets larger as the thermal diffusivity decreases, which, however, does not mean that the smaller thermal diffusivity is the better. The small thermal diffusivity is not discussed because the system becomes insulated.

The thermal dispersion is plotted in Fig. 5.5a. The thermal dispersion curve is symmetric when $2\pi\Gamma = 0$ but becomes asymmetric when $2\pi\Gamma = 8$, which is the proof of diffusive Fizeau drag. The wavenumber difference $\Delta\text{Re}[\beta] = \text{Re}[\beta_f] + \text{Re}[\beta_b]$ is shown in Fig. 5.5(b), demonstrating linear responses to ω. More intuitively, a speed difference leads to a time difference of temperature field evolution at two symmetric positions of x and $-x$ to reach the same phases. The forward phase at x is $\text{Re}[\beta_f]x - \omega t_f$, and the backward phase at $-x$ is $-\text{Re}[\beta_b]x - \omega t_b$. The same phases correspond to a time difference of $\Delta t = t_f - t_b$, which can be calculated by

$$\Delta t = \frac{\Delta\text{Re}[\beta]|x|}{\omega}. \tag{5.32}$$

Since Δt increases linearly with $|x|$, we focus on the parameter of $\Delta t/|x| = \Delta\text{Re}[\beta]/\omega$ in Fig. 5.5(c), which is almost invariant as ω changes.

Finite-element simulations are also performed. For brevity, a dimensionless temperature of $T^* = (T - T_r)/A$ and a dimensionless time of $t^* = t/t_0$ are defined, where t_0 is the time periodicity of the temperature source. When $2\pi\Gamma = 0$ (Fig. 5.5d1), the forward and backward cases are identical at $y = 0$, but a slight difference appears at $y = \pm h/4$ (Fig. 5.5d2,d3) due to the local inhomogeneity

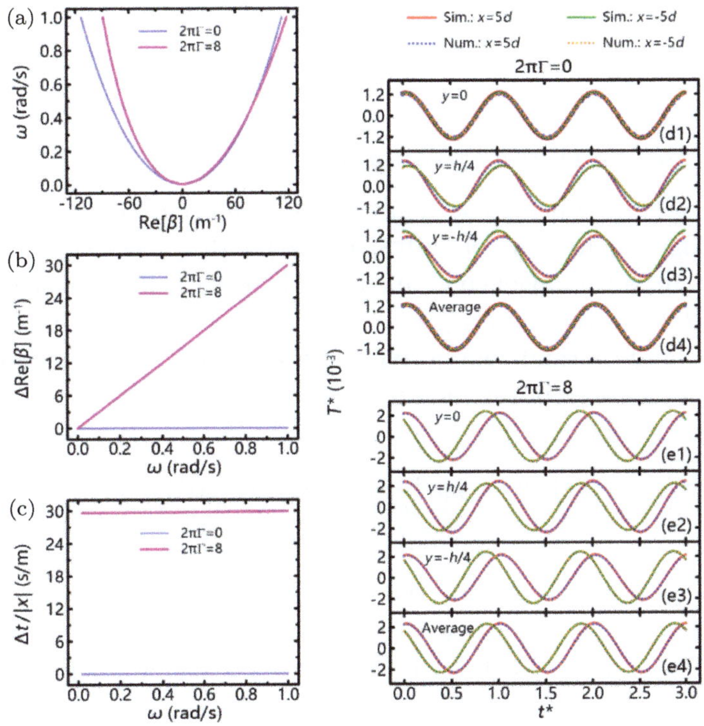

Fig. 5.5. Simulation results of the diffusive Fizeau drag. (a) Thermal dispersion. (b) Wavenumber difference $\Delta\text{Re}[\beta] = \text{Re}[\beta_f] + \text{Re}[\beta_b]$ as a function of ω. (c) Time difference per unit of distance $\Delta t/|x| = \Delta\text{Re}[\beta]/\omega$ as a function of ω. Evolution of T^* when (d1–d4) $2\pi\Gamma = 0$ or (e1–e4) $2\pi\Gamma = 8$, corresponding to $u_y = 0$ or $u_y = 0.2\,\text{m/s}$. Parameters: $\phi = 0.1$, $\epsilon = 1$, $D_0 = 5 \times 10^{-5}$ m^2/s, $\Delta_\rho = 0.9$, $\Delta_\kappa = 0.9$, $\theta = \pi$, $d = 0.02\,\text{m}$, $h = 0.02\,\text{m}$, and $t_0 = 20\,\text{s}$. The length is $30d = 0.6\,\text{m}$. The left and right boundaries are insulated. The upper and lower boundaries are set with periodic conditions. Sim.: Simulation; Num.: Numerical. Adapted from Ref. [23].

described by the $F(\xi)$ in Eq. (5.4). As long as we discuss the average temperature in the vertical direction, the effect of local inhomogeneity can be excluded, so the forward and backward cases become identical again (Fig. 5.5d4). Set $2\pi\Gamma = 8$, and the simulation results demonstrate a time difference of $\Delta t^* = 0.14$, which can be observed locally (Fig. 5.5e1–e3) and globally (Fig. 5.5e4). The numerical results predict a time difference of $\Delta t^* = 0.15$, so the numerical calculations are convincing. Meanwhile, the numerical results with dotted curves are plotted, which agree well with the simulation results.

5.4.2. *Diffusive Willis Coupling*

The governing equation is analytically homogenized to reveal the underlying mechanism of diffusive Fizeau drag. 1D thermal conduction in a homogeneous medium is governed by $\rho_0 C_0 \partial_t T + \partial_x(-\kappa_0 \partial_x T) = 0$, where ρ_0, C_0, and κ_0 are the mass density, heat capacity, and thermal conductivity of the homogeneous medium. We consider spatiotemporal modulation of mass density and thermal conductivity [22],

$$\rho(n) = \rho_0(1 + \Delta_\rho \cos(Gn + \alpha)) \tag{5.33a}$$

$$\kappa(n) = \kappa_0(1 + \Delta_\kappa \cos(Gn)) \tag{5.33b}$$

where Δ_ρ and Δ_κ denote the modulation amplitudes, α is the modulation phase difference, $G = 2\pi N/L$ represents the modulation wavenumber, L is the length, N is the period number, and $n = x - ut$ is the generalized coordinate. The diffusion equation describes classical and macroscopic heat transfer. Hence, for the validity of the diffusion equation, the spatial modulation scale should be much larger than the mean free path of heat carriers, and the temporal modulation scale should also be much longer than the relaxation time of colliding events. Then, 1D heat transfer with spatiotemporal modulation is governed by

$$\rho(n) C_0 \frac{\partial T}{\partial t} + u(\rho(n) - \rho_0) C_0 \frac{\partial T}{\partial x} + \frac{\partial}{\partial x}\left(-\kappa(n)\frac{\partial T}{\partial x}\right) = 0. \tag{5.34}$$

A local advection term of $u(\rho(n) - \rho_0) C_0 \partial_x T$ appears due to mass conservation. Nevertheless, its integration over one period is still zero; thus, the system has no macroscopically directional advection.

Considering a wave-like temperature solution with the Floquet-Bloch theorem, Eq. (5.34) can be homogenized as

$$T = \epsilon(n) e^{i(\beta x - \omega t)} + T_0 = \left(\sum_s \epsilon_s e^{isGn}\right) e^{i(\beta x - \omega t)} + T_0 \tag{5.35}$$

where β and ω are the wavenumber and frequency of the wave-like temperature field. ω is a complex number whose imaginary part indicates the temporal decay of the variation amplitude of the wave-like temperature field. Hence, the system finally goes to a constant temperature, i.e., $T = T_0$. Without loss of generality, the balanced temperature of T_0 is set at zero Kelvin, and the real part of Eq. (5.35) makes sense. $\epsilon(n)$ is the Floquet-Bloch modulation function with $s = 0, \pm 1, \pm 2, \ldots, \pm\infty$ and $\epsilon_0 = 1$. The accuracy of numerical results is ensured by considering the order of s as high as possible. Nevertheless, it is not intuitive to uncover the physical mechanism, and thus, only $s = 0, \pm 1$ is taken for analytical discussion. Assume $\beta \ll G$ to ensure a subwavelength structure, i.e., the structural unit size is far smaller than the temperature field wavelength. Hence, our structure belongs to metamaterials, and homogenization makes sense. The homogenized heat transfer equation is

$$\rho_0 C_0 \frac{\partial T_0}{\partial t} - K\frac{\partial^2 T_0}{\partial x^2} - R\frac{\partial^2 T_0}{\partial t^2} - W\frac{\partial^2 T_0}{\partial t \partial x} = 0 \tag{5.36}$$

where $T_0 = e^{i(\beta x - \omega t)}$ is the homogenized temperature. The homogenized parameters are

$$K = \kappa_0 \left(1 + \frac{-\Delta_\kappa^2 + \Delta_\rho^2 \Phi^2 + 2\Delta_\kappa \Delta_\rho \Phi^2 \cos\alpha}{2(1+\Phi^2)}\right) \tag{5.37a}$$

$$R = \frac{\kappa_0}{u^2}\frac{\Delta_\rho^2 \Phi^2}{2(1+\Phi^2)} \tag{5.37b}$$

$$W = \frac{\kappa_0}{u}\frac{\Delta_\rho^2 \Phi^2 + \Delta_\kappa \Delta_\rho \Phi^2 \cos\alpha}{1+\Phi^2} \tag{5.37c}$$

with definitions of the dimensionless parameter of $\Phi = u/(GD_0) = uL/(2\pi N D_0)$ and thermal diffusivity of $D_0 = \kappa_0/(\rho_0 C_0)$. The status of Δ_ρ and Δ_κ is, to some extent, different in Eq. (5.37c) because mass conservation should be seriously considered. Heat transfer in spatiotemporal diffusive metamaterials can no longer be homogenized within the framework of thermal conduction due to the two unexpected terms of ∂_t^2 and $\partial_t \partial_x$ in Eq. (5.36). As energy conservation indicates, the universal heat conduction equation in 1D is $\rho_0 C_0 \partial_t T_0 + \partial_x J = 0$, where J is the heat flux. Comparing the energy conservation equation and Eq. (36), the Fourier law should be modified as

$$-\frac{R}{\rho_0 C_0}\frac{\partial J}{\partial t} + J = -K\frac{\partial T_0}{\partial x} - W\frac{\partial T_0}{\partial t}. \tag{5.38}$$

The term of $\partial_t J$ is mathematically related to thermal relaxation, the term of $\partial_x T_0$ reflects the conventional contribution of thermal conduction, and the term of $\partial_t T_0$ is referred to as the thermal Willis term. Therefore, thermal Willis coupling originates from the coupling between heat flux and temperature change rate. This effect is similar to acoustic Willis coupling, i.e., acoustic propagation in inhomogeneous media cannot be described by the classical elastic wave equation [23]. Although Eq. (5.38) directly indicates thermal Willis coupling, it is obtained from two approximations of $s = 0, \pm 1$ and $\beta \ll G$. Therefore, Eq. (5.38) is not strictly accurate, but this problem is solved by exploring numerical results.

A remarkable feature distinguishing Willis thermal metamaterials from conventional thermal materials is the ability to drive asymmetric heat diffusion (Fig. 5.6a). The frequency of ω is purely imaginary in homogeneous media, making the propagation velocity of $v = \mathrm{Re}[\omega]/\beta$ zero. In contrast, Willis thermal metamaterials realize directional temperature field propagation because the real part of ω appears. Numerical results are studied by considering $s = 0, \pm 1, \pm 2, \ldots, \pm 8$ to avoid the probable inaccuracy of Eq. (5.36). Six crucial parameters are explored, which are u, D_0, N, Δ_ρ, Δ_κ, and α in Eq. (5.33), which affect the velocity of $v = \mathrm{Re}[\omega]/\beta$ and temporal decay rate of $\mu = -\mathrm{Im}[\omega]$. These six parameters are divided into two groups. One group contains u, D_0, and N, linked by the dimensionless parameter of $\Phi = u/(GD_0)$ or $2\pi\Phi = uL/(ND_0)$. Specifically, Φ has a critical point of Φ_c, crossing which the temperature field direction is reversed. The other group includes Δ_ρ, Δ_κ, and α, determining the value of Φ_c (Fig. 5.6b).

The former group related to $2\pi\Phi = uL/(ND_0)$ is first discussed. When setting $2\pi\Phi$ as the abscissa, the actual variable is u, which is flexibly controllable (Fig. 5.7a,b). All $v - 2\pi\Phi$ curves experience a process to be first positive and then negative, but the modulation

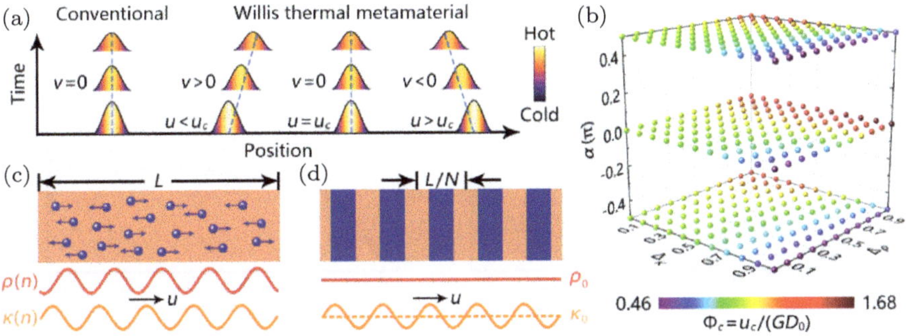

Fig. 5.6. Schematic of asymmetric heat diffusion. (a) In contrast to the conventional case with a zero velocity of temperature field propagation, i.e., $v = 0$, Willis thermal metamaterials drive directional temperature field propagation. The propagating direction is determined by a critical point of the modulation velocity of u, i.e., u_c. (b) For generality, we make u_c dimensionless by defining the dimensionless parameter of $\Phi_c = u_c/(GD_0)$. The critical point of Φ_c only depends on the modulation amplitudes of Δ_ρ and Δ_κ and the modulation phase difference of α. (c) Generalized model by modulating mass density of $\rho(n)$ and thermal conductivity of $\kappa(n)$ with $n = x - ut$. The modulation of mass density produces local advection heat flux. (d) Simplified model by modulating only thermal conductivity. Adapted from Ref. [22].

velocity of u is always positive, exhibiting unexpected direction-reversible heat diffusion. These $v - 2\pi\Phi$ curves also share the same critical point of $2\pi\Phi_c = 6.4$ (stars in Fig. 5.7a,b). Accordingly, the maximum (or minimum) values appear at the same value of $2\pi\Phi$ (vertical dashed lines in Fig. 5.7a,b). The amplitude of a $v - 2\pi\Phi$ curve is determined by the thermal diffusivity of D_0 and period number of N. Specifically, a larger D_0 (reflected in the dimensionless thermal diffusivity of $M = D_0/D_r$ where D_r is the reference thermal diffusivity) or a smaller N contributes to a larger amplitude. Since the temporal decay rate of μ mainly depends on the effective thermal diffusivity, the $\mu - 2\pi\Phi$ curves in Fig. 5.7a have different amplitudes because of the different thermal diffusivity of D_0. However, the $\mu - 2\pi\Phi$ curves in Fig. 5.7b almost overlap because the period number of N essentially does not affect the effective thermal diffusivity. We focus on the properties when $2\pi\Phi = 2.4$, corresponding to the maximum values of the $v - 2\pi\Phi$

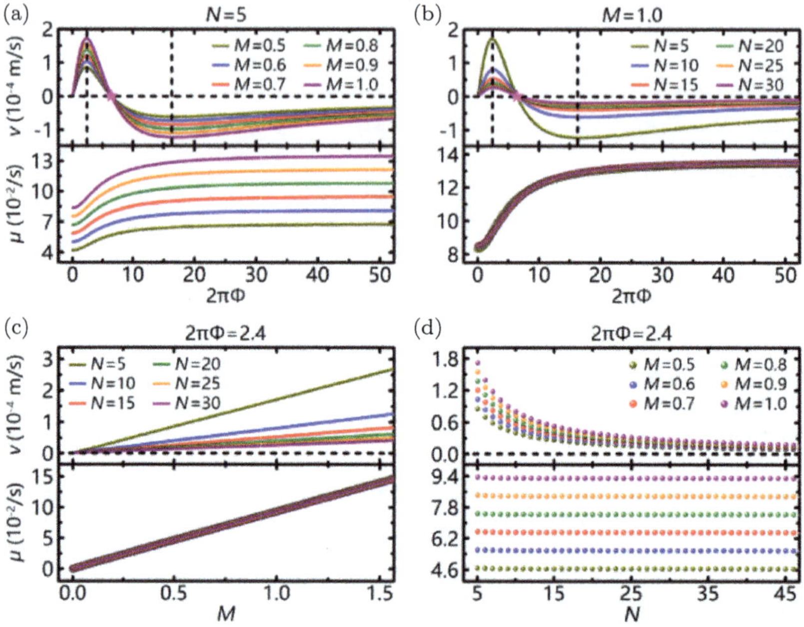

Fig. 5.7. Numerical results of asymmetric heat diffusion. Velocity of $v = \text{Re}[\omega]/\beta$ and temporal decay rate of $\mu = -\text{Im}[\omega]$ as a function of (a) and (b) dimensionless parameter of $2\pi\Phi = uL/(NMD_r)$, (c) dimensionless thermal diffusivity of $M = D_0/D_r$, and (d) period number of N, where D_r is the reference thermal diffusivity. Except for the parameters shown in each inset, the others are $\Delta_\rho = 0.5$, $\Delta_\kappa = 0.5$, $\alpha = 0$, $D_r = 10^{-4}$ m^2/s, and $L = 0.2$ m. Adapted from Ref. [22].

curves in Fig. 5.7a,b. v and μ demonstrate linear responses to the thermal diffusivity of D_0 (Fig. 5.7c). Therefore, the temperature field propagation is faster with the larger thermal diffusivity, but the temporal decay rate also becomes larger. The $v - N$ curves decrease in Fig. 5.7d, indicating that thermal Willis coupling is weakened with the decrement of inhomogeneity (i.e., the increment of N). The $\mu - N$ curves have almost no variation because the period number of N does not affect the effective thermal diffusivity. Due to the importance of inhomogeneity, the larger the modulation wavelength (i.e., the smaller N), the more obvious thermal Willis coupling in principle. However, the modulation wavelength should be much smaller than the temperature field wavelength so that homogenization makes sense. Therefore, on the premise of guaranteeing good homogenization, the larger the modulation wavelength, the better.

We further discuss the latter group related to Δ_ρ, Δ_κ, and α, determining the critical point of $2\pi\Phi_c$. As shown in Fig. 5.7a,b, the critical point for $v = 0$ is fixed at $2\pi\Phi_c = 6.4$, which is irrelevant to thermal diffusivity and period number. Then, we discuss the critical point when $\alpha = 0$, $\pi/4$, $\pi/2$, and $-\pi/2$ in Fig. 5.8. When $\Delta_\rho = 0$ or $\Delta_\kappa = 0$, α does not matter, and hence the $2\pi\Phi_c - \Delta_\rho$ curves with $\Delta_\kappa = 0$ (or the $2\pi\Phi_c - \Delta_\kappa$ curves with $\Delta_\rho = 0$) are identical in Fig. 5.8. Moreover, $\alpha = \pm\pi/2$ leads to the same patterns of $2\pi\Phi_c$, and hence, thermal Willis coupling is an even function of α. The results in Fig. 5.8 provide direct guidance for designing direction-reversible heat diffusion.

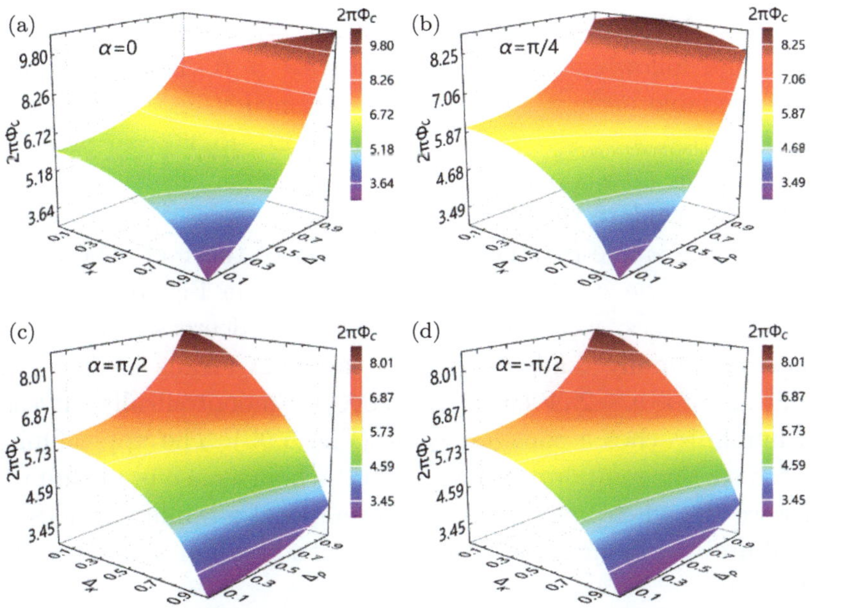

Fig. 5.8. Critical point of $2\pi\Phi_c = u_c L/(ND_0)$ as a function of the modulation amplitudes of Δ_ρ and Δ_κ. (a–d) The modulation phase difference of α takes 0, $\pi/4$, $\pi/2$, and $-\pi/2$. The critical point of $2\pi\Phi_c$ is irrelevant to specific L, N, and D_0. Adapted from Ref. [22].

Another feature of thermal Willis coupling is the freedom from modulating mass density and thermal conductivity simultaneously. Modulating two parameters simultaneously for the reported diffusive asymmetry induced by the advection-like effect is mandatory. However, the homogenized case described by Eq. (5.36) contains no advection-like term. Thermal Willis coupling between heat flux and temperature change rate realizes asymmetry, which works in transient heat transfer rather than quasi-steady heat transfer. Transient (or quasi-steady) heat transfer means that the temperature field envelope depends on (or is independent) time. Due to the essential difference, a single-parameter modulation is enough to realize asymmetric diffusion. Since the thermal conductivity of many practical materials is sensitive to external fields, a moving periodic external field can be designed to achieve the spatiotemporal modulation of thermal conductivity (Fig. 5.6d). But the inaccuracy of Eq. (5.36) also appears when $\Delta_\rho = 0$. The R in Eq. (5.37b) and W in Eq. (5.37) become zero, yielding $v = 0$. Yet the numerical results and following simulations indicate $v \neq 0$. This deviation is understandable because Eq. (5.36) is obtained using the first-order approximation, i.e., $s = 0, \pm 1$. Therefore, Eq. (5.36) can prove the absence of the advection-like term (the first-order effect) but cannot accurately calculate the high-order effects.

Then, the single-parameter modulation of thermal conductivity is discussed. The $v - 2\pi\Phi$ curves are plotted in Fig. 5.9a. As $2\pi\Phi$ (i.e., u) increases, v first increases from zero, decreases to negative values, and finally tends to zero. A larger Δ_κ leads to the larger amplitude of a $v - 2\pi\Phi$ curve. Moreover, μ monotonically increases and finally reaches a constant value (Fig. 5.9b). The upper limit is $D_0\beta^2$, i.e., the temporal decay rate in a homogeneous medium. In other words, when $2\pi\Phi \to \infty$, spatiotemporal modulation does not contribute to inhomogeneity, indicating no thermal Willis coupling and $v = 0$.

Then, finite-element simulations are performed. The thermal conductivity is set according to Eq. (5.33b) for quantitative comparison with the theory and numerical results. For brevity, a dimensionless

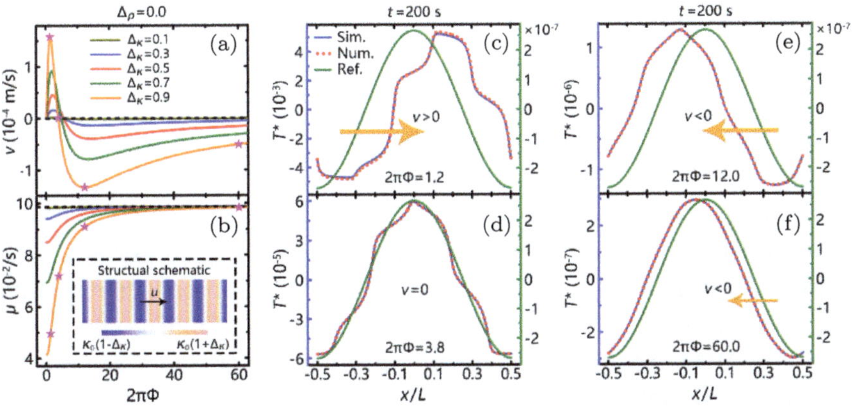

Fig. 5.9. Simulations of asymmetric heat diffusion by modulating only thermal conductivity. (a) Velocity of $v = \text{Re}[\omega]/\beta$ and (b) temporal decay rate of $\mu = -\text{Im}[\omega]$ as a function of the dimensionless parameter of $2\pi\Phi = uL/(ND_0)$. Dimensionless temperature distributions of T^* at 200 s with (c) $2\pi\Phi = 1.2$, (d) $2\pi\Phi = 3.8$, (e) $2\pi\Phi = 12.0$, and (f) $2\pi\Phi = 60.0$. The initial dimensionless temperature is $T^* = 100\cos(2\pi x/L)$. The left and right boundaries are set with periodic conditions, i.e., continuous temperatures. Parameters: $D_0 = 10^{-4}$ m^2/s, $L = 0.2$ m, and $N = 5$. Sim.: Simulation; Num.: Numerical. Adapted from Ref. [22].

temperature is defined as $T^* = (T - T_0)/(1 \text{ Kelvin})$. The wave-like temperature profile for reference is stationary and only decays. In contrast, the spatiotemporal modulation of thermal conductivity drives forward temperature field propagation (Fig. 5.9c). At the critical point of $2\pi\Phi = 3.8$, despite a nonzero modulation velocity of u, the temperature field velocity is still zero (Fig. 5.9d). The further increment of $2\pi\Phi$ leads to backward temperature field propagation (Fig. 5.9e,f). Moreover, the numerical results also agree well with the simulations, supporting the results in Figs. 5.7 and 5.8.

5.5. Nonreciprocal Geometric Heat Pumping

Under the rigorous restriction of the second law of thermodynamics, heat always flows spontaneously from higher to lower temperature in the absence of external influence in static thermal diffusion systems, leading to an increase of the entropy, so thermal nonreciprocity is unfulfillable due to the unidirectional flow. Nevertheless, as thermal metamaterials, nonlinear materials, phase-change materials, and spatiotemporal modulation develop, the nonreciprocity in thermal diffusion systems is demonstrated theoretically and experimentally [15]. Heat pumping is a device that can generate an additional thermal contribution beyond the dynamical counterpart governed by conventional thermodynamics with the aid of adiabatical and periodic modulation of several parameters of the system even under the zero bias between two thermal reservoirs, thus breaking the limitation of the law of thermodynamics and realizing the enhancement, diminution or even the reverse of the current in the presence of a finite non-zero bias. Due to the prerequisite (adiabatic evolution) and the physical nature (geometric property) of this heat-pumping effect, it is also known as geometric heat pumping, originating from the geometric phase.

The Berry (geometric) phase has attracted a great deal of interest both theoretically and experimentally since it was discovered [24], and then a series of geometric effects were explored in all aspects of physics [25–35]. A necessary but insufficient prerequisite for this geometric effect is that at least two linearly independent parameters must be modulated synchronously, periodically, adiabatically, and continuously. Then, after one cycle, a corresponding net current is completely determined by the geometric property and flux of the modulation trajectory in parameter space.

Here, we derive a general heat-pumping theory and obtain a formula as a surface integral, where the integrand, an antisymmetric tensor, is analogous to the geometric curvature. Then, we consider and calculate two cases of non-trivial zero heat flux by adjusting some parameters, which is fantastic due to the realization of non-trivial reciprocity in a nonreciprocity system. The error between the simulation and the numerical results is on the order of a thousandth under various parameter settings, which fully verify the validity and universality of our theory.

We start with the 1D classical thermal diffusion equation, which combines the third kind of boundary condition,

$$\frac{\rho c_p(x)\partial T(x,t)}{\partial t} = \frac{\partial}{\partial x}\left(\frac{\kappa(x)\partial T(x,t)}{\partial x}\right) \tag{5.39}$$

$$\begin{aligned} -\kappa(x)\tfrac{\partial T(x,t)}{\partial x} &= h_0(t)(T_0(t) - T(x,t)), \quad x = 0, \\ \kappa(x)\tfrac{\partial T(x,t)}{\partial x} &= h_1(t)(T_1(t) - T(x,t)), \quad x = L, \end{aligned} \tag{5.40}$$

where $\rho, C, \kappa(x), h_0(t), h_1(t), T_0(t), T_1(t)$ are mass density, heat capacity, thermal conductivity, temperature, and the heat transfer coefficient between two thermal reservoirs and the two ports of the system, respectively.

The general solution of this equation is

$$T(x,t) \approx G(x,t) + \sum_{i=1}^{n} c_i(t)\psi_i(x,t)\exp(-W_i(t)) \tag{5.41}$$

where $G(x,t)$ is an additional temperature term that is introduced to better match the boundary condition, $c_i(t)$ is a coefficient determined by other physical quantities, $\psi_i(x,t)$ is the eigenfunction of the Hamiltonian of the diffusion system, $\lambda_i(t)$ is the corresponding eigenvalues, and $W_i(t)$ is the antiderivative of $\lambda_i(t)$.

Obviously, $\psi_i(x,t)$ and $\lambda_i(t)$ satisfy the following relationship,

$$-\lambda_j(t)\psi_j(x,t) = \frac{1}{\rho c_p(x)}\frac{d}{dx}(\kappa(x)(\psi_j'(x,t))). \tag{5.42}$$

After some logical mathematical operations, such as taking the inner product and making some reasonable approximation, we obtain a formula of $c_i(t)$ in the form of an indefinite integral,

$$c_i(t) = -\int_0^t \left\langle \psi_i(\tau) \middle| \frac{\partial}{\partial \tau}G(x,t) \right\rangle \exp(W_i(\tau))d\tau. \tag{5.43}$$

Then, we consider the specific situation, that is, ρ, c, κ are all constants, and only the two thermal reservoirs and the heat transfer coefficients between reservoirs and the intermediate system are modulated periodically with the same period. Notably, it is the most crucial process in solving the $G(x,t)$:

$$G(x,t) = A(t)x + B(t). \tag{5.44}$$

According to the boundary conditions,

$$\begin{aligned} A(t) &= \frac{T_1(t)-T_0(t)}{\kappa\left(\frac{1}{h_0(t)}+\frac{1}{h_1(t)}\right)+L} \\ B(t) &= \frac{\kappa\left(\frac{T_0(t)}{h_1(t)}+\frac{T_1(t)}{h_0(t)}\right)+LT_0(t)}{\kappa\left(\frac{1}{h_0(t)}+\frac{1}{h_1(t)}\right)+L} \end{aligned} \tag{5.45}$$

where L is the length of the intermediate medium. Consider

$$\begin{aligned} T_1(t) - T_0(t) &= \delta T = \text{constant}, \\ T_1(t) &= T_0 + \delta T + \Delta\text{T}\cos(\omega t + \theta) \\ T_0(t) &= T_0 + \Delta\text{T}\cos(\omega t + \theta). \end{aligned} \tag{5.46}$$

We have

$$\begin{aligned} B(t) &= \frac{\kappa\left(\frac{T_0(t)}{h_1}+\frac{T_1(t)}{h_0}\right)+LT_0(t)}{\kappa\left(\frac{1}{h_0}+\frac{1}{h_1}\right)+L} = T_0(t) + \frac{\kappa A}{h_0}, \\ A(t) &= \frac{T_1(t)-T_0(t)}{\kappa\left(\frac{1}{h_0}+\frac{1}{h_1}\right)+L} = \frac{\delta T}{\kappa\left(\frac{1}{h_0}+\frac{1}{h_1}\right)+L}. \end{aligned} \tag{5.47}$$

The eigenfunctions of Hamiltonian have the form,

$$\psi_1(x,t) = A_1\left(\cos(k_1 x) + \frac{h_0(t)}{\kappa k_1}\sin(k_1 x)\right) \tag{5.48}$$

where wavenumber k_1 satisfies

$$\tan(k_i L) = \frac{\frac{h_0(t)}{\kappa k_i} + \frac{h_0(t)}{\kappa k_i}}{1 - \frac{h_0(t)}{\kappa k_i}\frac{h_0(t)}{\kappa k_i}}, \quad i = 1, 2, 3 \ldots \tag{5.49}$$

Ultimately, we obtain the formula of the temperature field and, even more importantly, the concrete formula of the geometric heat, which form the total heat together with the dynamical heat,

$$Q = -\kappa\delta T \int_0^{\text{period}} \frac{h_0(t)h_1(t)}{\kappa(h_0(t)+h_1(t)) + Lh_0(t)h_1(t)} dt$$

$$+ h_0(t)\psi_i(x,t)\left\langle \psi_i(\tau) \middle| \frac{\partial A(\tau)}{\partial \tau} x + \frac{\partial B(\tau)}{\partial \tau} \right\rangle$$

$$\times \exp \iint (W_i(\tau) - W_i(t)) d\tau dt. \quad (5.50)$$

The first term is the dynamical component, and the second is the geometric component; these contributions have a completely different nature in physics.

Using Green's theorem,

$$Q_{\text{geo}} = \oint F(\tau)\frac{\partial E(\tau)}{\partial \Gamma} d\Gamma = \oint F(\tau)\frac{\partial E(\tau)}{\partial G} dG + F(\tau)\frac{\partial E(\tau)}{\partial h} dh$$

$$= \iint \left(\frac{\partial \left(F(\tau)\frac{\partial E(\tau)}{\partial h}\right)}{\partial G} - \frac{\partial \left(F(\tau)\frac{\partial E(\tau)}{\partial G}\right)}{\partial h} \right) dGdh$$

$$= \iint \left(\frac{\partial F(\tau)}{\partial G}\frac{\partial E(\tau)}{\partial h} - \frac{\partial F(\tau)}{\partial h}\frac{\partial E(\tau)}{\partial G} \right) dGdh$$

$$= \iint \mathrm{S}_{\mathrm{G,h}}\, dGdh \quad (5.51)$$

where the parameters are

$$\Gamma(\tau) = \{G(\tau), h(\tau)\}(0 \le \tau \le \tau_p),$$

$$E(\tau) = \int_0^{\tau} \left\langle \psi_j(t) \middle| \frac{\partial}{\partial t} G(t) \right\rangle dt, \quad (5.52)$$

$$F(\tau) = \exp(W_1(\tau)) \int_{\tau}^{\tau_p} h_0(t)\psi_1(0,t)\exp(-W_1(t))dt.$$

The integrand $S_{\mathrm{G,h}}$ has the analogy with geometric curvature.

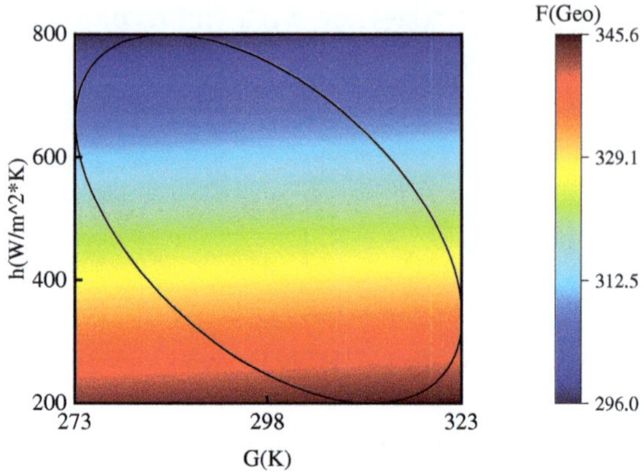

Fig. 5.10. The magnitude of the geometric curvature S_{geo} in a 2D parameter space. The solid line of the ellipse is encircled by the parameters during modulation, counterclockwise indicating the positive direction. Adapted from Ref. [36].

As shown in Fig. 5.10, the geometric heat is completely determined by the geometric property of the modulation trajectory in the parameter space and the flux on the 'surface.'

We introduce a discrete version of the geometric heat-pumping theory based on the above and Ref. [37]. First, a discretization operation on the 1D diffusion system must be made. The temperature field $T(x,t)$ is described by a column vector as

$$T_c = (T_c), \quad T_b = \begin{pmatrix} T_{b1} \\ T_{b2} \end{pmatrix} \tag{5.53}$$

where T_c and T_b are the temperature of the central system and thermal reservoirs, respectively. In contrast to the continuous theory, we are only concerned with the interaction between reservoirs and central systems but not the concrete variation of the temperature field concerning the coordinates and time. In the same way, we start with the classical diffusion equation,

$$\rho c \frac{\partial T}{\partial t} = \frac{\partial}{\partial x}\left(\kappa \frac{\partial T}{\partial x}\right), \quad J = -\kappa \frac{\partial T}{\partial x} \tag{5.54a}$$

$$-\kappa_0 T'(0) = h_0(T_0 - T(0)), \quad \kappa_1 T'(L) = h_1(T_1 - T(L)). \tag{5.54b}$$

The heat flux J and thermal conductivity κ are also vectors, and the derivative should be replaced by the difference, $\frac{\partial T}{\partial x} \to \frac{T_1 - T_2}{x_1 - x_2}$.

Under the guidance of this principle, we can divide the component of the heat flux vector into two parts,

$$J_1 = -\kappa_0 T'(0) = h_0(T_0 - T(0)) = T_{b1} - \kappa_{b1}^{eff} T_c, \tag{5.55a}$$

$$J_2 = -\kappa_1 T'(L) = h_1(T(L) - T_1)\kappa_{b2}^{eff} T_c - \kappa_{b2}^{eff} T_{b1}. \tag{5.55b}$$

The right-hand side of Eq. (5.55) reminds us to separate T_b and T_c,

$$J = \begin{pmatrix} J_1 \\ J_2 \end{pmatrix} = \mathrm{K}_b T_b + \mathrm{K}_c T_c \tag{5.56}$$

where K_b and $\mathrm{K_c}$ are conductance matrix defined as

$$\mathrm{K}_{b,c} = -\mathrm{K} \cdot D_{b,c}^T. \tag{5.57}$$

$\mathrm{K} = \mathrm{diag}(\kappa_{b1}^{eff}, \kappa_{b2}^{eff})$ is a diagonal conductance matrix, $D_{\mathrm{b,c}}$ is the incidence matrix, while $D_{\mathrm{ij}} = 1(-1)$ represents that flux into(out) of T_{b} and T_{c},

$$D_b = \begin{pmatrix} 1 & 0 \\ 0 & -1 \end{pmatrix}, \quad D_c = \begin{pmatrix} 1 & -1 \end{pmatrix}. \tag{5.58}$$

Then, the diffusion equation can be expressed by

$$\frac{\partial T_c}{\partial t} = M_b T_b + M_c T_c \tag{5.59}$$

where $M_b \equiv LK_b$, $M_c \equiv LK_c$, $L \equiv C^{-1}D$; D is the discrete version of the negative divergence matrix. The differential term in Eq. (5.59) can be considered a perturbation. Subsequently, through the application of the perturbation method, the solution of Eq. (5.59) can be solved

$$T_c = T_{ss} + M_c^{-1} \frac{\partial T_c}{\partial t} \tag{5.60}$$

where $T_{ss} \equiv -M_c^{-1} M_b T_b$ is the instantaneous steady-state temperature. The total heat during one complete period τ_p contains two different components, which are called dynamical heat and geometric heat,

$$Q_{dyn} = \int_0^{\tau_p} 1 \cdot (K_c T_{ss} + K_b T_b) dt \tag{5.61a}$$

$$Q_{geo} = \int_0^{\tau_p} 1 \cdot K_c M_c^{-1} \frac{\partial T_c}{\partial t}\, dt \tag{5.61b}$$

where '1' is a column vector able to project heat flow to the interested component J_j with $1_i = \delta_{ij}$ and δ_{ij} is the Kronecker delta.

Under the adiabatic evolution of two thermal reservoirs T_b and the thermal conductance $\kappa_{b1}^{eff}, \kappa_{b2}^{eff}$, the perturbation term is insignificant and can be neglected, so the central temperature T_c equals the instantaneous steady temperature T_{ss}:

$$Q_{geo} = \int_0^{\tau_p} K \cdot \frac{\partial T_{ss}}{\partial \lambda_\mu} \frac{\partial \lambda_\mu}{\partial t} dt = \oint K \cdot \frac{\partial T_{ss}}{\partial \lambda_\mu} d\lambda_\mu \tag{5.62}$$

where $K = 1 \cdot K_c M_c^{-1}$ is the row vector mapping the variation of T_{ss} into the interested heat current and is in the dual space of T_{ss}. We also express Q_{geo} as a surface integral by using Green's formula:

$$Q_{geo} = \iint F_{\mu\nu}(\lambda) dS_{\mu\nu} \tag{5.63a}$$

$$F_{\mu\nu}(\lambda) = \frac{\partial K}{\partial \lambda_\mu} \cdot \frac{\partial T_{ss}}{\partial \lambda_\nu} - \frac{\partial K}{\partial \lambda_\nu} \cdot \frac{\partial T_{ss}}{\partial \lambda_\mu}. \tag{5.63b}$$

The $A_\mu = K \cdot \frac{\partial T_{ss}}{\partial \lambda_\mu}$ and the antisymmetric tensor $F_{\mu\nu}(\lambda)$ are reminiscent of the Berry connection and Berry curvature, respectively, and $dS_{\mu\nu}$ is the area element of the integral surface enclosed by the driving parameters during evolution.

Some other heat-pumping effects induced by phase-changed materials in thermal diffusion [38] and thermal radiation [39] have also somewhat realized thermal nonreciprocity. Interested readers can refer to the original paper.

5.6. Conclusion

Over the past decades, temporal modulation has been widely investigated for its ability to induce non-reciprocal wave transport and charge diffusion. However, these relevant theories met challenges when they were introduced into thermal diffusion. The reciprocity of heat transfer is hard to break only through temporal modulation of mass density, and the external energy flow input is necessary. In this way, thermal Willis coupling and related diffusive Fizeau drag effect are discussed, as well as the non-reciprocal geometric heat pumping. Although theoretical advancements in this area have been considerable, many challenges remain to be addressed. Firstly, the current heat flow analysis method is more direct for the two-port systems in this chapter but seems complicated for complex structures such as multi-port systems. We can study the interaction of temperature fields with the scattering method and the effective Hamiltonian theory. Relevant practices have successfully studied thermal reciprocity [39] and symmetry [20, 41–43]. Secondly, the physical effects of spatiotemporal modulation heat transfer can be more extensive rather than focusing on reciprocity. We can extend it to the thermal time crystal system to study the impedance matching and time interface reflection [44]. Finally, the progress of relevant experiments is still limited due to the difficulty of temporal modulation of thermal conductivity and other parameters. We can use multiple physical fields such as phase change materials, external magnetic fields, thermoelectric materials, and microfluidics to enrich the means of regulation and its physical effects [45].

References

1. Potton, R. J. (2004). Reciprocity in optics. *Rep. Prog. Phys.*, 67, pp. 717–754.
2. Yu, Z. and Fan, S. (2009). Complete optical isolation created by indirect interband photonic transitions. *Nat. Photonics*, 3, pp. 91–94.
3. Chen, Z. and Segev, M. (2021). Highlighting photonics: looking into the next decade. *eLight*, 1, pp. 2.
4. Zhu, J., Hippalgaonkar, K., Shen, S., Wang, K., Abate, Y., Lee, S., Wu, J., Majumdar, A., and Zhang, X. (2014). Temperature-gated thermal rectifier for active heat flow control. *Nano Lett.*, 14, pp. 4867–4872.
5. Li, B., Wang, L., and Casati, G. (2004). Thermal diode: rectification of heat flux. *Phys. Rev. Lett.*, 93, pp. 184301.

6. Li, Y., Li, W., Chen, T., Zheng, X., Li, J., Li, B., Fan, S., and Qiu, C.-W. (2021). Transforming heat transfer with thermal metamaterials and devices. *Nat. Rev. Mater.*, 6, pp. 488–507.
7. Tishin, A. M. and Spichkin, Y. I. (2016). *The Magnetocaloric Effect and its Applications.* CRC Press.
8. Li, Y., Li, J., Qi, M., Qiu, C.-W., and Chen, H. (2021). Diffusive nonreciprocity and thermal diode. *Phys. Rev. B*, 103, pp. 014307. oxide thermal rectifier. *Appl. Phys. Lett.*, 95, pp. 171905. energy-free maintenance of constant temperatures as ambient temperature gradients change. *Phys. Rev. Lett.*, 117, pp. 055501.
9. Lira, H., Yu, Z., Fan, S., and Lipson, M. (2012). Electrically driven nonreciprocity induced by interband photonic transition on a silicon chip. *Phys. Rev. Lett.*, 109, pp. 033901.
10. Sounas, D. L. and Alù, A. (2017). Non-reciprocal photonics based on time modulation. *Nat. Photonics*, 11, pp. 774–783.
11. Hadad, Y., Soric, J. C., and Alù, A. (2016). Breaking temporal symmetries for emission and absorption. *Proc. Natl. Acad. Sci. USA*, 113, pp. 3471–3475.
12. Fleury, R., Khanikaev, A. B., and Alù, A. (2016). Floquet topological insulators for sound. *Nat. Commun.*, 7, pp. 11744.
13. Camacho, M., Edwards, B., and Engheta, N. (2020). Achieving asymmetry and trapping in diffusion with spatiotemporal metamaterials. *Nat. Commun.*, 11, pp. 3733.
14. Torrent, D., Poncelet, O., and Batsale, J.-C. (2018). Nonreciprocal thermal material by spatiotemporal modulation. *Phys. Rev. Lett.*, 120, pp. 125501.
15. Li, J., Li, Y., Cao, P.-C., Qi, M., Zheng, X., Peng, Y.-G., Li, B., Zhu, X.-F., Alù, A., Chen, H., and Qiu, C.-W. (2022). Reciprocity of thermal diffusion in time-modulated systems. *Nat. Commun.*, 13, pp. 167.
16. Fresnel, A. (1818). Lettre d'Augustin Fresnel à François Arago sur l'influence du mouvement terrestre dans quelques phénomènes d'optique. *Ann. Chim. Phys.*, 9, pp. 57.
17. Fizeau, H. (1851). Sur les hypothèses relatives à l'éther lumineux. *C. R. Acad. Sci.*, 33, pp. 349.
18. Dong, Y., Xiong, L., Phinney, I. Y., Sun, Z., Jing, R., McLeod, A. S., Zhang, S., Liu, S., Ruta, F. L., Gao, H., Dong, Z., Pan, R., Edgar, J. H., Jarillo-Herrero, P., Levitov, L. S., Millis, A. J., Fogler, M. M., Bandurin, D. A., and Basov, D. N. (2021). Fizeau drag in graphene plasmonics. *Nature*, 594, pp. 513.
19. Zhao, W. Y., Zhao, S. H., Li, H. Y., Wang, S., Wang, S. X., Utama, M. I. B., Kahn, S., Jiang, Y., Xiao, X., Yoo, S., Watanabe, K., Taniguchi, T., Zettl, A., and Wang, F. (2021). Efficient Fizeau drag from Dirac electrons in monolayer graphene. *Nature*, 594, pp. 517.

20. Li, Y., Peng, Y.-G., Han, L., Miri, M.-A., Li, W., Xiao, M., Zhu, X.-F., Zhao, J. L., Alù, A., Fan, S. H., and Qiu, C.-W. (2019). Anti–parity-time symmetry in diffusive systems. *Science*, 364, pp. 170.
21. Xu, L., Xu, G., Huang, J., and Qiu, C.-W. (2022). Diffusive Fizeau drag in spatiotemporal thermal metamaterials. *Phys. Rev. Lett.*, 128, pp. 145901.
22. Xu, L., Xu, G., Li, J., Li, Y., Huang, J., and Qiu, C.-W. (2022). Thermal Willis coupling in spatiotemporal diffusive metamaterials. *Phys. Rev. Lett.*, 129, pp. 155901.
23. Willis, J. R. (1981). Variational principles for dynamic problems for inhomogeneous elastic media. *Wave Motion*, 3, pp. 1.
24. Berry, M. V. (1984). Quantal phase factors accompanying adiabatic changes. *Proc. R. Soc. Lond. A*, 392, pp. 45.
25. Schmid, C. P., Weigl, L., Grössing, P., Junk, V., Gorini, C., Schlauderer, S., Ito, S., Meierhofer, M., Hofmann, N., Afanasiev, D., Crewse, J., Kokh, K. A., Tereshchenko, O. E., Güdde, J., Evers, F., Wilhelm, J., Richter, K., Höfer, U., and Huber, R. (2021). Tunable non-integer high-harmonic generation in a topological insulator. *Nature*, 593, pp. 385.
26. Smirnova, D., Leykam, D., Chong, Y., and Kivshar, Y. (2020). Nonlinear topological photonics. *Appl. Phys. Rev.*, 7, pp. 021306.
27. Lu, L., Joannopoulos, J., and Soljačić, M. (2014). Topological photonics. *Nat. Photonics*, 8, pp. 821.
28. Jisha, C. P., Nolte, S., and Alberucci, A. (2021). Geometric phase in optics: from wavefront manipulation to waveguiding. *Laser Photonics Rev.*, 15, pp. 2100003.
29. Kravets, V. G., Schedin, F., Jalil, R., Britnell, L., Gorbachev, R. V., Ansell, D., Thackray, B., Novoselov, K. S., Geim, A. K., Kabashin, A. V., and Grigorenko, A. N. (2013). Singular phase nano-optics in plasmonic metamaterials for label-free single-molecule detection. *Nat. Mater.*, 12, pp. 304.
30. Hasman, E., Kleiner, V., Biener, G., and Niv, A. (2003). Polarization dependent focusing lens by use of quantized Pancharatnam–Berry phase diffractive optics. *Appl. Phys. Lett.*, 82, pp. 328.
31. Nassar, H., Yousefzadeh, B., Fleury, R., Ruzzene, M., Alù, A., Daraio, C., Norris, A. N., Huang, G., and Haberman, M. R. (2020). Nonreciprocity in acoustic and elastic materials. *Nat. Rev. Mater.*, 5, pp. 667.
32. Muhammad, M. and Lim, C. W. (2022). From photonic crystals to seismic metamaterials: A review via phononic crystals and acoustic metamaterials. *Arch. Comput. Methods Eng.*, 29, pp. 1137.
33. Xue, H., Yang, Y., and Zhang, B. (2022). Topological acoustics. *Nat. Rev. Mater.*, 7, pp. 974.

34. Qi, M., Wang, D., Cao, P.-C., Zhu, X.-F., Qiu, C.-W., Chen, H., and Li, Y. (2022). Geometric phase and localized heat diffusion. *Adv. Mater.*, 34, pp. 2202241.
35. Wang, Z., Wang, L., Chen, J., Wang, C., and Ren, J. (2022). Geometric heat pump: Controlling thermal transport with time-dependent modulations. *Front. Phys.*, 17, pp. 1.
36. Yan, H.-R., Cao, P.-C., Wang, Y.-X., Zhu, X.-F., and Li, Y. (2024). Geometric heat pumping under continuous modulation in thermal diffusion. arXiv preprint arXiv:2406.19100., https://arxiv.org/abs/2406.19100.
37. Wang, Z., Chen, J., and Ren, J. (2022). Geometric heat pump and no-go restrictions of nonreciprocity in modulated thermal diffusion. *Phys. Rev. E*, 106, pp. L032102.
38. Ordonez-Miranda, J., Anufriev, R., Nomura, M., and Volz, S. (2022). Net heat current at zero mean temperature gradient. *Phys. Rev. B*, 106, pp. L100102.
39. Li, Y. X., Dang, Y. D., Zhang, S., Li, X. R., Chen, T. L., Choudhury, P. K., Jin, Y., Xu, J. B., Abdallah, P. B., Ju, B. F., and Ma, Y. G. (2024). Observation of heat pumping effect by radiative shuttling. *Nat. Commun.*, 15, pp. 5465.
40. Ju, R., Cao, P.-C., Wang, D., Qi, M., Xu, L., Yang, S., Qiu, C.-W., Chen, H., and Li, Y. (2024). Nonreciprocal heat circulation metadevices. *Adv. Mater.*, 36, pp. 2309835.
41. Cao, P.-C., Ju, R., Wang, D., Qi, M., Liu, Y.-K., Peng, Y.-G., Chen, H., Zhu, X.-F., and Li, Y. (2024). Observation of parity-time symmetry in diffusive systems. *Sci. Adv.*, 10, pp. eadn1746.
42. Liu, Y.-K., Cao, P.-C., Qi, M., Huang, Q.-K.-L., Gao, F., Peng, Y.-G., Li, Y., and Zhu, X.-F. (2024). Observation of non-Hermitian skin effect in thermal diffusion. *Sci. Bull.*, 69, pp. 1228–1236.
43. Liu, Z., Cao, P.-C., Xu, L., Xu, G., Li, Y., and Huang, J. (2024). Higher-order topological in-bulk corner state in pure diffusion systems. *Phys. Rev. Lett.*, 132, pp. 176302.
44. Ju, R., Xu, G., Xu, L., Qi, M., Wang, D., Cao, P.-C., Xi, R., Shou, Y., Chen, H., Qiu, C.-W., and Li, Y. (2023). Convective thermal metamaterials: exploring high-efficiency, directional, and wave-like heat transfer. *Adv. Mater.*, 35, pp. 2209123.
45. Yang, F., Zhang, Z., Xu, L., Liu, Z., Jin, P., Zhuang, P., Lei, M., Liu, J., Jiang, J., Ouyang, X., Marchesoni, F., and Huang, J. (2024). Controlling mass and energy diffusion with metamaterials. *Rev. Mod. Phys.*, 96, pp. 015002.

Part III

Transient Heat Transfer

Chapter 6

Effective Hamiltonian and Thermal Modes

6.1. Introduction

This chapter uses the effective Hamiltonian method to study the decaying thermal modes in transient heat transfer. We assign physical meanings to the parameters in the waveform solution derived from Fourier's law and describe the evolution dynamics of the thermal diffusion systems with eigenmodes. We explain the temperature field modulation mechanism in toy model units, especially the amplitude attenuation due to pure diffusion and the phase variation due to pure convection. We construct the effective Hamiltonian to characterize the mode coupling process in different units and study the physical symmetry and non-Hermitian physics in thermal diffusion with convection and coupling modulations. We show that the non-Hermitian potentials in thermal diffusion can be generated in two ways: by introducing fluctuation and by directly contacting different sizes/materials. We then investigate the anti-parity-time (APT) and parity-time (PT) symmetry, respectively, uncovering the coupling mechanism in thermal modes by extending the inherently imaginary coupling to the real coupling. We highlight the importance of phase modulation in thermal regulation and provide a paradigm for exploring the counterintuitive non-Hermitian diffusive effects.

After discussing the effective non-Hermitian Hamiltonian in thermal diffusion, we put the singularity of the Riemann surface, named the exceptional point, on the table. Based on it, we introduce the topology in effective non-Hermitian Hamiltonian in terms of Weyl exceptional rings.

6.2. Thermal Unit Cell

In physics and engineering, a "mode" refers to a distinct pattern of oscillation or vibration exhibited by a system. This concept is prevalent across various disciplines, including acoustics, optics, electromagnetics, and mechanics [1, 2]. Modes characterize the behavior of a system in response to external stimuli or internal interactions [3]. In the context of oscillatory systems, such as vibrating strings, acoustic cavities, or electromagnetic resonators, modes represent specific configurations of energy distribution within the system [4]. Each mode corresponds to a motion or field distribution pattern of frequency, spatial structure, and energy content [5].

The study of modes is essential for understanding the behavior and properties of complex systems, as it provides insights into how energy propagates and interacts within them [6]. Mode analysis is widely used in signal processing, structural dynamics, photonics, and microwave engineering to optimize system performance, design efficient devices, and predict system behavior under different operating conditions [7, 8].

In addition to its applications in oscillatory systems, the concept of modes finds resonance in the realm of thermal diffusion processes [9]. Thermal diffusion, or heat conduction, describes the spontaneous transfer of heat energy from higher-temperature to lower-temperature regions within a material medium [10]. Like oscillatory systems, thermal diffusion systems exhibit distinct patterns of behavior that can be characterized by modes [11]. In thermal diffusion, modes represent specific configurations of temperature distribution within a medium [12]. Each mode corresponds to a unique spatial arrangement of temperature gradients characterized by thermal conductivity, diffusivity, and overall heat transfer properties [13]. These modes encapsulate the fundamental dynamics of thermal diffusion, shedding light on how heat energy propagates and interacts within a given system [14]. Studying thermal diffusion modes is essential for understanding heat transfer phenomena in diverse fields such as thermal engineering, materials science, and geophysics [15]. From the cooling of electronic devices to the thermal management of spacecraft, modes provide valuable insights into the mechanisms driving heat conduction processes and the factors influencing their behavior [16].

Thus, by extending the concept of modes to thermal diffusion systems, we shall gain a deeper understanding of the underlying principles governing heat transfer and thermal dynamics, enhancing our ability to address critical issues related to energy efficiency, sustainability, and thermal comfort [17].

6.3. Eigenmode of a Single Structure

6.3.1. *Pure Diffusion Unit*

In the domain of pure thermal diffusion, the flow of heat is elegantly described by Fourier's law:

$$\boldsymbol{q} = -\kappa \nabla T \tag{6.1}$$

where $\boldsymbol{q}$ represents the heat flux vector, κ denotes the thermal conductivity, and ∇T symbolizes the temperature gradient.

The heat equation governs the dynamic evolution of temperature over space and time,

$$\frac{\partial T}{\partial t} = D\nabla^2 T \tag{6.2}$$

where $\partial T/\partial t$ denotes the temporal rate of change of temperature, D represents the thermal diffusivity, and $\nabla^2 T$ signifies the Laplacian operator applied to temperature.

Eigenmodes emerge from this mathematical landscape, the characteristic pattern of temperature distribution within the system. With the separation of variables, the temperature field can be decomposed by the product of position-dependent parts $\phi_k(\boldsymbol{r})$ and time-harmonic part $\exp(-i\omega_k t)$, where ω_k is the frequency of this thermal mode,

$$T(\boldsymbol{r}, t) = \sum_{\boldsymbol{k}} \phi_k(\boldsymbol{r}) \exp(-i\omega_k t). \tag{6.3}$$

By substituting it into Eq. (6.2), each eigenmode $\phi_k(\boldsymbol{r})$, with its unique spatial signature, satisfies the eigenvalue problem,

$$D\nabla^2\phi_k(\boldsymbol{r}) = \lambda_k\phi_k(\boldsymbol{r}) \tag{6.4}$$

where $\lambda_k = -i\omega_k$ represents the eigenvalue, determining the mode's influence on the temperature field. From Eq. (6.4), a little thought shows that each eigenvalue is a positive real number, indicating the frequency in pure diffusion thermal mode is a positive pure imaginary number, $\omega_k = ia$, where $a \in R^+$. This nature gives the field in pure diffusion the dissipative characteristic, and it will not exhibit periodic oscillatory behaviors but decays uniformly over time, preventing us from exploring the following coupling between different thermal modes.

6.3.2. *Convection-induced Phase Change*

Thermal convection is a fundamental process in which heat transfer occurs by moving fluids (liquids or gases) driven by temperature differences. This phenomenon arises from the tendency of fluids to expand when heated and contract when cooled, resulting in density variations within the fluid [13, 15]. When subjected to a temperature gradient (difference in temperature across space), the warmer, less dense fluid rises, while the cooler, denser fluid sinks [18]. Mathematically, this buoyancy-driven flow is described by the *buoyancy force equation*,

$$\mathbf{F}_{buoyancy} = \rho\mathbf{g}\beta(T - T_0) \tag{6.5}$$

where $\mathbf{F}_{buoyancy}$ is the buoyancy force, ρ is the fluid density, $\mathbf{g}$ is the gravitational acceleration, β is the volumetric thermal expansion coefficient, T is the temperature of the fluid, and T_0 is the reference temperature [19]. This buoyancy-driven flow sets up continuous circulation patterns known as convection or thermal cells. Mathematically, these flow patterns are described by the *Navier-Stokes equations* for fluid motion,

$$\rho\left(\frac{\partial\mathbf{u}}{\partial t} + (\mathbf{u}\cdot\nabla)\mathbf{u}\right) = -\nabla p + \mu\nabla^2\mathbf{u} + \rho\mathbf{g} + \mathbf{F}_{buoyancy} \tag{6.6}$$

where **u** is the fluid velocity, p is the pressure, and μ is the dynamic viscosity [20]. These convection cells vary in size and shape depending on the system's geometry, the magnitude of the temperature gradient, and the properties of the fluid. They range from small-scale turbulent eddies to large-scale circulation patterns observed in natural systems like the atmosphere and oceans [21].

Convection enhances heat transfer rates compared to pure conduction by promoting the mixing and transport of heat within the fluid. This efficient heat exchange is described by the *conservation of energy (heat transfer) equation*,

$$\rho c_p \left(\frac{\partial T}{\partial t} + (\mathbf{u} \cdot \nabla) T \right) = \nabla \cdot (\kappa \nabla T) + Q \qquad (6.7)$$

where c_p is the specific heat capacity at constant pressure, κ is the thermal conductivity, and Q is any volumetric heat source or sink [11]. It is crucial in applications where rapid heat dissipation or redistribution is necessary, such as cooling systems or thermal management [22].

Without loss of generality, a 1D convective thermal diffusion model can help understand convection's role in phase evolution and change. The onsite convection-driven thermal mode can be analyzed from the eigenvalues and eigenstates [23]. In the convective single thermal unit with homogenous media, according to the heat transfer equation, the temperature field can be expressed into

$$\rho c_p \frac{\partial T}{\partial t} = \kappa \frac{\partial^2 T}{\partial x^2} - \rho c_p v \frac{\partial T}{\partial x} \qquad (6.8)$$

where v is the linear velocity of the convective thermal unit. By substituting the wave-like form of temperature field $T = T_0 + Ae^{i(kx-\omega t)}$ into Eq. (6.8), where T_0 is the reference temperature and A is the modulated amplitude of temperature, the eigenvalue problem is derived as

$$\omega A = (-ik^2 D + kv) A \qquad (6.9)$$

where $D = \kappa/\rho C_p$ is the diffusivity of the thermal unit. From Eq. (6.9), the dispersion can be obtained as the modulated amplitude

A will be eliminated,

$$\omega = -ik^2D + kv. \tag{6.10}$$

We can find the imaginary part of ω is $-k^2D$, indicating the decay rate of the temperature field, which can be recorded by $A(t)$. $k = n\frac{2\pi}{L}$ (n is a positive integer, and L is the length of the layer) is the wavenumber of the temperature field in the layer, which is fixed as $k = \frac{2\pi}{2\pi R} = \frac{1}{R}$ (R is the radius of the ring structure) because only the first-order thermal mode can survive in the ring structure. Therefore, the real part of eigenfrequency evolves into $\mathrm{Re}(\omega) = kv = \Omega$, suggesting the temporal phase evolution of temperature fields in the ring. The real part is zero when the channel is purely dissipative and becomes kv when the convection is added. The original steady spatial phase of the temperature field will oscillate with time in the direction of convection. The thermal modes and the evolution of the temperature field in a single convective thermal unit cell are illustrated in Fig. 6.1. The figure shows an exponentially decaying

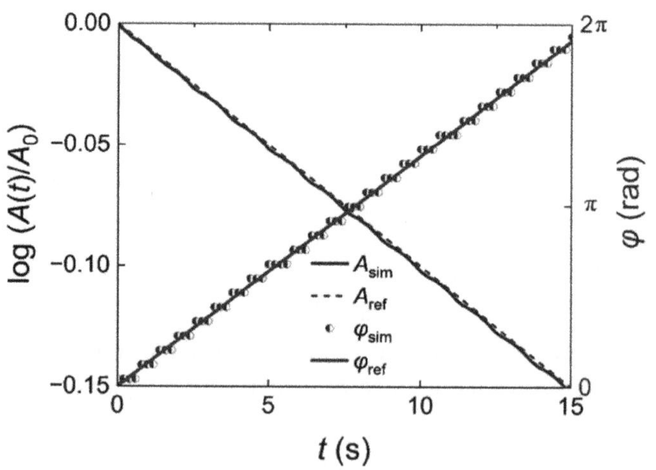

Fig. 6.1. Convection-driven thermal phase fluctuation in a single thermal unit cell, adapted from Ref. [26]. It shows the temperature field evolution with the rotating velocity $\Omega = kv$, where the linear phase evolution is at the speed of Ω.

amplitude alongside a linear phase evolution, providing a detailed visualization of how the temperature field evolves within the channel. The decay in amplitude reflects the dissipative nature of the thermal process, while the linear phase shift underscores the consistent progression of thermal waves within the convective environment.

6.4. Mode Coupling and Effective Hamiltonian

A mode represents a distinct pattern of oscillation or vibration the system exhibits. However, these modes do not exist in isolation in many real-world scenarios. Instead, they can interact, leading to a phenomenon known as mode coupling. Mode coupling occurs when different modes of oscillation or vibration within a system interact or influence each other. This interaction can arise from various mechanisms, including nonlinear effects, geometric and material properties, external perturbations, and coupling between different physical domains.

One of the primary consequences of mode coupling is the transfer of energy between different modes. This energy exchange can redistribute energy within the system, leading to changes in its dynamic behavior. Mode coupling can also result in mixing different modes, creating new vibrational patterns, or transitioning between existing modes. Understanding mode coupling is essential in various fields of science and engineering. In structural dynamics, for example, mode coupling plays a crucial role in analyzing vibrations in complex structures, affecting their stability, response to external forces, and susceptibility to fatigue failure. In optics and photonics, mode coupling is exploited in optomechanical systems for applications such as sensing, cavity opto-mechanics, and phonon-photon coupling. Overall, mode coupling is a phenomenon that adds richness and complexity to the behavior of dynamic systems. By studying and understanding mode coupling effects, researchers and engineers can gain insights into the underlying dynamics of these systems, develop predictive models, and design innovative solutions for a wide range of practical applications.

6.4.1. *Convection Model — Anti-parity-time Symmetry*

The temperature distribution $T(x,t)$ in a 1D background with velocity v satisfies the convective-diffusion equation,

$$\frac{\partial T}{\partial t} = D\frac{\partial^2 T}{\partial x^2} - v\frac{\partial T}{\partial x} \tag{6.11}$$

where $D = \kappa/\rho c$ is the diffusivity, κ is the thermal conductivity, and ρ and c are the density and heat capacity of the media, respectively. The material properties are all assumed to be temperature-independent for simplicity. This equation is dissipative due to the first-order time derivative. However, the advection term introduces a wave-like behavior given that, for $D = 0$, Eq. (6.11) reduces to a first-order wave (advection) equation. The eigenfrequencies of this system can be obtained by substituting $T(x,t) = A\exp(ikx - i\omega t)$, which results in the dispersion relation in Eq. (6.10). For an arbitrary v, the system is dissipative with a complex eigenvalue, while the advection term gives rise to real eigenfrequencies with a linear dispersion.

Things will be interesting when the two convective thermal units are in contact with each other in the y direction. Consider one thermal unit with velocity v_1 and the other thermal unit with velocity v_2, whose illustrated structure is shown in Fig. 6.2. A new spatial dimension y is introduced here, and the model has to be extended into two dimensions, with one dimension indicating convective background and the other indicating coupling direction. For the sake of comprehensiveness, we give the thermal units sickness b_1 and b_2 in the coupling dimension, and the contact layer (coupling layer) sickness d.

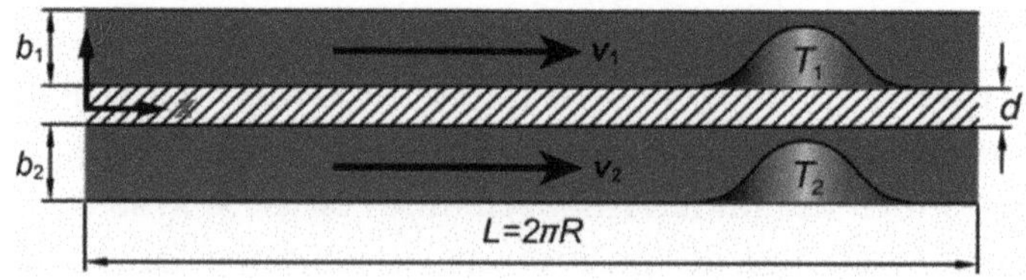

Fig. 6.2. Structure of two coupled thermal units. Adapted from Ref. [23].

The fields of these three parts satisfy the following equation respectively:

$$\frac{\partial T}{\partial t}=\begin{cases} D_1\left(\dfrac{\partial^2 T_1}{\partial x^2}+\dfrac{\partial^2 T_1}{\partial y^2}\right)-v_1\dfrac{\partial T_1}{\partial x}, & \dfrac{d}{2}\leq y\leq\dfrac{d}{2}+b_1 \\ D_c\left(\dfrac{\partial^2 T_c}{\partial x^2}+\dfrac{\partial^2 T_c}{\partial y^2}\right), & -\dfrac{d}{2}\leq y\leq\dfrac{d}{2} \\ D_2\left(\dfrac{\partial^2 T_2}{\partial x^2}+\dfrac{\partial^2 T_2}{\partial y^2}\right)-v_2\dfrac{\partial T_2}{\partial x}, & -b_2-\dfrac{d}{2}\leq y\leq-\dfrac{d}{2} \end{cases} \tag{6.12}$$

where D_1, D_2, and D_c are the diffusivity of thermal units and coupling layer, respectively. T_c is the temperature field in the coupling layer. Like what one deals with in the coupled mode theory, Eq. (6.12) is supposed to reduce into a two-variation equation to dive into more information about the coupling effect. The matching conditions, where one is the continuity of temperature field in the interface and the other is the continuity of heat flux in the interface, are given as follows,

$$\begin{cases} T_1=T_c, \kappa_1\dfrac{\partial T_1}{\partial y}=\kappa_c\dfrac{\partial T_c}{\partial y}, & y=\dfrac{d}{2} \\ T_2=T_c, \kappa_2\dfrac{\partial T_2}{\partial y}=\kappa_c\dfrac{\partial T_c}{\partial y}, & y=-\dfrac{d}{2} \end{cases} \tag{6.13}$$

where κ_c is the thermal conductivity of the coupling layer. Equation (6.13) tells us key information: the coupling layer plays a role in heat exchange between two thermal units. Therefore, the boundary condition could be treated as a heat source term h_{c_1} and h_{c_2} acting on thermal units, causing Eq. (6.12) to express,

$$\frac{\partial T}{\partial t}=\begin{cases} D_1\dfrac{\partial^2 T_1}{\partial x^2}-v_1\dfrac{\partial T_1}{\partial x}+\dfrac{h_{c_1}}{\rho_1 c_1}, & \dfrac{d}{2}\leq y\leq\dfrac{d}{2}+b_1 \\ D_2\dfrac{\partial^2 T_2}{\partial x^2}-v_2\dfrac{\partial T_2}{\partial x}+\dfrac{h_{c_2}}{\rho_2 c_2}, & -b_2-\dfrac{d}{2}\leq y\leq-\dfrac{d}{2} \end{cases}. \tag{6.14}$$

The temperature variance in the y direction $\partial^2 T/\partial y^2$ has been neglected in the system. The heat flux q comes from the boundary where $y=\pm d/2$. When width $b_{1,2}$ is small, the heat source could be assumed to be uniformly distributed along the y direction: $h_{c1,c2} = q_{c1,c2}$,

$$q_{c1} = -\kappa_1 \frac{\partial T_1}{\partial y} = -\kappa_c \frac{\partial T_c}{\partial y}, \quad y = \frac{d}{2} \tag{6.15}$$

$$q_{c2} = \kappa_2 \frac{\partial T_2}{\partial y} = \kappa_c \frac{\partial T_c}{\partial y}, \quad y = -\frac{d}{2}. \tag{6.16}$$

If T_c is linear with y, we have $\partial T_c/\partial y = \left(T_1|_{y=\frac{d}{2}} - T_2|_{y=-\frac{d}{2}}\right)/d$. Equation (6.14) turns out to be

$$\frac{\partial T}{\partial t} = \begin{cases} D_1 \dfrac{\partial^2 T_1}{\partial x^2} - v_1 \dfrac{\partial T_1}{\partial x} + h_1(T_2 - T_1), & \dfrac{d}{2} \le y \le \dfrac{d}{2} + b_1 \\ D_2 \dfrac{\partial^2 T_2}{\partial x^2} - v_2 \dfrac{\partial T_2}{\partial x} + h_2(T_1 - T_2), & -b_2 - \dfrac{d}{2} \le y \le -\dfrac{d}{2} \end{cases} \tag{6.17}$$

where $h_{1,2} = \kappa_c(\rho_{1,2}c_{1,2})^{-1}(b_{1,2}d)^{-1}$ is defined as thermal coupling and assumes little difference between T_1 (T_2) at the $y = d/2$ ($y = -d/2$) boundary and inside the thermal unit along the y direction, as width $b_{1,2}$ is small.

Using plane wave solutions $T_{1,2}(x,t) = A_{1,2}\exp(ikx - i\omega t)$ and substituting it into (6.17), we have

$$\omega A_1 = -ik^2 D_1 A_1 + kv_1 A_1 + ih_1(A_2 - A_1) \tag{6.18}$$

$$\omega A_2 = -ik^2 D_2 A_2 + kv_2 A_2 + ih_2(A_1 - A_2). \tag{6.19}$$

Equations (6.18) and (6.19) could be reformed into matrix form if we define the amplitude column $(A_1, A_2)^{\mathrm{T}}$,

$$\omega \begin{pmatrix} A_1 \\ A_2 \end{pmatrix} = \begin{pmatrix} -ik^2 D_1 - ih_1 + kv_1 & ih_1 \\ ih_2 & -ik^2 D_2 - ih_2 + kv_2 \end{pmatrix} \begin{pmatrix} A_1 \\ A_2 \end{pmatrix}. \tag{6.20}$$

It is just the eigenproblem of the matrix H_{eff},

$$H_{\mathit{eff}} = \begin{pmatrix} -ik^2 D_1 - ih_1 + kv_1 & ih_1 \\ ih_2 & -ik^2 D_2 - ih_2 + kv_2 \end{pmatrix}. \tag{6.21}$$

The matrix H_{eff} is named as the effective Hamiltonian of the coupled thermal units, which is non-Hermitian because of the dissipative nature of the diffusive system. The off-diagonal element ih_1 and ih_2 characterized the coupling between these two thermal units. The relation $h_{1,2} = \kappa_c(\rho_{1,2}c_{1,2})^{-1}(b_{1,2}d)^{-1}$ indicates the purely imaginary coupling $ih_{1,2}$. With an effective Hamiltonian, we can analyze the system's properties like one did in quantum mechanics.

For some conditions, such as the parameters of thermal units being identical, this effective Hamiltonian will exhibit intriguing properties. Under this circumstance, $ih_1 = ih_2$, implying equal coupling. As easily seen, if the background velocities of the thermal unit are opposite, $v_1 = -v_2 = v$, H_{eff} will be anti-parity-time (APT)-symmetric,

$$H_{\mathit{eff}} = \begin{pmatrix} -ik^2D - ih + kv & ih \\ ih & -ik^2D - ih - kv \end{pmatrix} \tag{6.22}$$

$$\{H_{\mathit{eff}}, \mathcal{PT}\} = 0 \tag{6.23}$$

where $\mathcal{P} = (0,1;1,0)^{\mathrm{T}}$ is the parity operator which swaps the two thermal units, and $\mathcal{T}$ is the time reversal operator representing complex conjugation. The eigenvalues of this APT-symmetric Hamiltonian, which are also the system's eigenfrequencies in Eq. (6.20), could be calculated,

$$\omega_{\pm} = -i(k^2D + h) \pm \sqrt{k^2v^2 - h^2}. \tag{6.24}$$

Now, we analyze the properties by Eq. (6.24). In the last section, we offer the conclusion that convection is the origin of phase in heat transfer, as eigenfrequency will have nonzero real part $\mathrm{Re}(\omega) = kv$ in a single convective thermal unit. However, things could be more interesting for coupled convective thermal units. In the uncoupled limit with $h = 0$, the coupled system exhibits only a single phase, in which the eigenvalues always have a nonzero real part at $v \neq 0$. When $h \neq 0$, this system exhibits two different phases as v varies.

For small v where $k^2v^2 < h^2$, $\omega_{\pm}$ are purely imaginary with eigenvectors $\boldsymbol{u}_{\pm} = (A_1^{\pm}, A_2^{\pm})^{\mathrm{T}} = \left(\frac{kv}{h} \pm i\sqrt{1-(\frac{kv}{h})^2}, i\right)^{\mathrm{T}}$. The fields in these two thermal units only differ by a phase because $|A_1^{\pm}| = |A_2^{\pm}|$. This scenario can be labeled as an exact APT regime because, under a proper normalization, the eigenvectors preserve the symmetry of the Hamiltonian as $\mathcal{PT}\boldsymbol{u}_{\pm} = -\boldsymbol{u}_{\pm}$. Here, the eigenvectors can be rewritten as $\boldsymbol{u}_{+} = i\left(e^{-\frac{i\phi}{2}}, e^{\frac{i\phi}{2}}\right)^{\mathrm{T}}$, $\boldsymbol{u}_{-} = i\left(e^{\frac{i(\pi-\phi)}{2}}, e^{-\frac{i(\pi-\phi)}{2}}\right)^{\mathrm{T}}$ with $\phi = \sin^{-1}\left(\frac{kv}{h}\right)$. The eigenvectors can be gauged real by combining the forward and backward solutions $(T_1, T_2)^{\mathrm{T}} = e^{-i\omega t}e^{i(\pi-\phi)/2}[A\boldsymbol{u}(k)e^{ikx} + B\boldsymbol{u}(-k)e^{-ikx}]$. When $A = B = 1$, two standing-wave solutions are

$$(T_1, T_2)^{\mathrm{T}} = 2e^{-i\omega t}(\cos kx, \pm\cos(kx \pm \phi))^{\mathrm{T}}. \quad (6.25)$$

When increasing convection intensity to reach the condition $k^2v^2 = h^2$, where the critical convective velocity is $v_{EP} = h/|k|$, the two eigenvectors will coalesce to the same as $\boldsymbol{u}_{+} = \boldsymbol{u}_{-} = i\left(e^{-\frac{i\pi}{4}}, e^{\frac{i\pi}{4}}\right)^{\mathrm{T}}$. At this specific value, both eigenvalues and eigenvectors degenerate, and we call this unusual degeneracy the exceptional point.

If the convection exceeds the exceptional point and causes $k^2v^2 > h^2$, the eigenvectors exhibit a finite real part according to Eq. (6.24), $\mathrm{Re}(\omega_{\pm}) = \pm\sqrt{k^2v^2 - h^2}$. The corresponding eigenvectors become $\boldsymbol{u}_{\pm} = i\left(e^{-\frac{i\pi}{4}}e^{\pm\frac{\psi}{2}}, e^{\frac{i\pi}{4}}e^{\mp\frac{\psi}{2}}\right)^{\mathrm{T}}$, with $\psi = sgn(k)\cosh^{-1}\left(\frac{|k|v}{h}\right)$. Opposite to the small convection case, the difference arises between eigenvector intensities in the two thermal units: while one eigenvector is primarily concentrated in one thermal unit, the other is located in the other. Additionally, the two eigenvectors exhibit a constant $\pi/2$ phase difference between the thermal units. It is worth noting that the APT symmetry of the eigenvectors is broken because the APT operation interchanges the eigenvectors: $\mathcal{PT}\boldsymbol{u}_{\pm} = -\boldsymbol{u}_{\mp}$. As a result, the eigenvector of H_{eff} is no longer an eigenvector after PT operation, indicating that APT symmetry is broken. A significant consequence of this symmetry-breaking phase is that the eigenvectors can no longer be gauged to be real due to the rescaling coefficient e^{ψ}. If we continue to use the same gauge, where $A = B = 1$, which produces a real solution for T_1, the eigenstate becomes

$$(T_1, T_2)^{\mathrm{T}} = e^{-i\omega t}\left(2\cos kx, e^{-\psi}e^{i\left(kx+\frac{\pi}{2}\right)} + e^{\psi}e^{-i\left(kx+\frac{\pi}{2}\right)}\right)^{\mathrm{T}}. \quad (6.26)$$

From Eq. (6.26), we find that the opposite waves have reciprocal amplitudes in thermal unit 2. The opposite waves in both channels

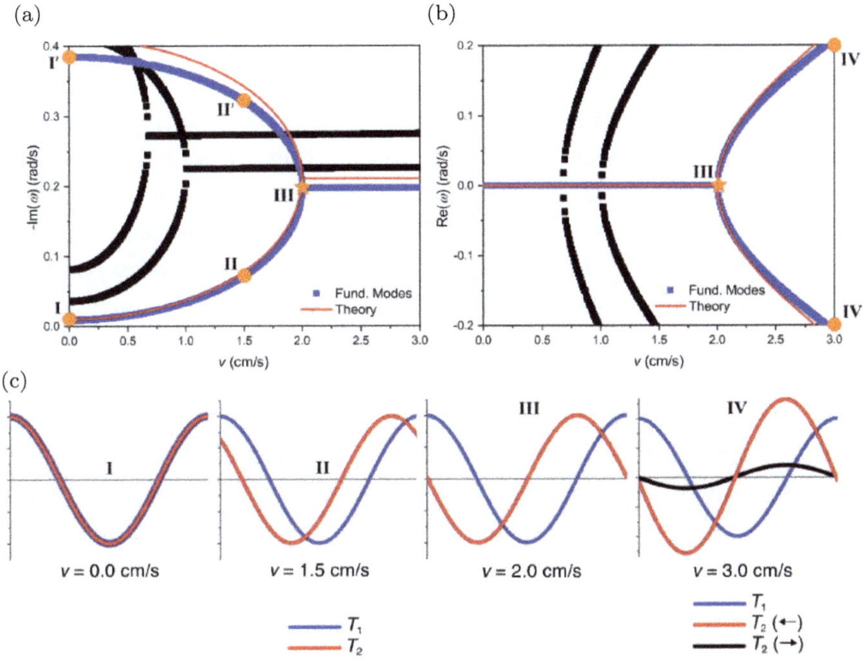

Fig. 6.3. Eigenvalues and eigenstates of the APT-symmetric system. (a) Imaginary part (decay rate). (b) Real part (eigenfrequency). The spectrum of the fundamental wave is depicted in blue, while the red lines represent the theoretical solutions to Eq. (6.22). (c) Eigenstates at different positions within the spectrum (orange markers in a and b), with T_1 aligned to the same cosine function. Adapted from Ref. [23].

cannot be simultaneously balanced. With a nonzero real part of the eigenfrequency $\text{Re}(\omega)$, a standing-wave solution does not exist, leading to a movement of the temperature profile in at least one channel. The eigenvalues and eigenstates are presented in Fig. 6.3, from Ref. [23].

Heat convection offers a distinct advantage in studying APT symmetry due to its unique characteristics. Specifically, the real part $\text{Re}(\omega_\pm)$ and the imaginary part $\text{Im}(\omega_\pm)$ of the eigenfrequencies are independently influenced by two factors: the diffusive term, which is associated with the diffusion coefficient D, and the convective term, which is related to the background velocity v. This distinction allows researchers to independently control the symmetric (non-Hermitian) and anti-symmetric (Hermitian) aspects of the Hamiltonian by adjusting D and v accordingly.

An important feature of this system is that the two thermal units exhibit opposite convection directions, leading to velocities v with opposite signs. This phenomenon is reminiscent of certain concepts in relativistic quantum mechanics, where positive and negative energy solutions coexist [27]. In such a context, the negative energy solution is sometimes interpreted as moving backward in time, a concept that parallels the behavior observed in our system. This analogy highlights the deep connections between different areas of physics and enriches our understanding of the underlying principles.

Although moving media can induce frequency shifts — either positive or negative — in propagating waves [28, 29], these shifts are not directly useful in this context because the system already starts with a nonzero frequency. Moreover, the effects of moving background on heat conduction have been extensively studied [30], primarily focusing on the spatiotemporal modulation of density and thermal conductivity. These studies often aim to induce nonreciprocity, a feature allowing for asymmetric wave propagation. PT symmetry breaking for electromagnetic waves in moving media in the relativistic regime has also been studied [31]. However, the mechanism involved [32, 33] differs from the one we discuss.

6.4.2. *Convection Model — Parity-time Symmetry*

In the previous section, we discussed the coupling phenomenon that occurs when two thermal units come into contact. By solving the governing equations and applying the matching conditions, we can calculate the strength of this coupling. Newton's law of cooling describes the underlying physics of the coupling between static-contacting thermal units, where the key coefficient h represents the heat exchange coefficient in Newton's law of cooling.

When convection is introduced into the thermal units, the eigenfrequencies gain a real component, leading to phase effects. Introducing this phase in the heat transfer system provides a new degree of freedom in controlling heat transfer. This phase change originates from convection, making convection crucial in analyzing the intrinsic properties of thermal units and the coupling effects between them. Therefore, introducing convection into the coupling mechanism may significantly impact the coefficient.

In the diffusive coupling case, the coupling between two thermal units mainly manifests as imaginary coupling, meaning the heat transfer process primarily involves energy dissipation without significant phase changes. However, when convection is added to the coupling, the temperature fields between two coupled thermal units will be influenced by both diffusion and convection. This convection effect can lead to the appearance of a real component in the eigenfrequencies of the thermal units, potentially resulting in real coupling.

Real coupling indicates the presence of energy transfer and storage in the system, which may lead to phenomena similar to those observed in wave systems, such as resonance and phase shifts. Therefore, studying the impact of adding convection on the coupling is crucial for understanding and controlling the interactions between thermal units, particularly in complex heat transfer systems. In the following, we will explore whether the introduction of convection leads to a transition from imaginary to real coupling and how this transition affects system behavior.

Before introducing how to achieve real coupling between thermal units, let us first review how imaginary coupling between thermal units is constructed. In the previous section, we introduced a static intermediate layer between two thermal units to bring them into contact, and using the boundary conditions, we derived the governing equations for the coupling of the two thermal units. The coupling term in these equations essentially corresponds to Newton's law of cooling, and by calculating the effective Hamiltonian, we demonstrated that this coupling is purely imaginary.

Since static thermal contact is widely prevalent, most couplings in heat transfer systems are imaginary. According to previous studies, we realized that the convection could carry real potential. The idea of convection-induced real potential could be illustrated when we look at the on-site term in the effective Hamiltonian given by Eq. (6.21). The on-site term gives

$$H_{\mathit{eff}}^{11,22} = -ik^2 D_{1,2} - ih_{1,2} + kv_{1,2}. \tag{6.27}$$

By separating real and imaginary terms, the effective Hamiltonian in Eq. (6.22) could also be rewritten as

$$H_{\mathit{eff}} = ih\hat{\sigma}_x + kv\hat{\sigma}_z - iS\hat{I} \tag{6.28}$$

where ih is the coupling strength, $\hat{\sigma}_x$ and $\hat{\sigma}_z$ are the Pauli matrices and $-iS = -i(k^2 D + h)$. We see that only the convective term contributes to real values.

The real potential induced by thermal convection in heat transfer systems inspired us to consider whether we could modify this coupling term through thermal contact with thermal convection.

As we did in the last section, we consider two temperatures, T_1 and T_2, coupled through a convective middle layer at T_3. The convective layer rotates with an angular velocity Ω. According to Fourier's law, the three-level thermal coupling equations are

$$\begin{aligned}
\rho_1 c_{p1}\frac{\partial T_1}{\partial t} &= \kappa_1 \frac{\partial^2 T_1}{\partial x^2} + h_1(T_3 - T_1), \\
\rho_2 c_{p2}\frac{\partial T_2}{\partial t} &= \kappa_2 \frac{\partial^2 T_2}{\partial x^2} + h_2(T_3 - T_2), \\
\rho_3 c_{p3}\frac{\partial T_3}{\partial t} &= \kappa_3 \frac{\partial^2 T_3}{\partial x^2} - \rho_3 c_{p3} v \frac{\partial T_3}{\partial x} + h_3(T_1 + T_2 - 2T_3),
\end{aligned} \tag{6.29}$$

where $\rho_{1,2,3}$, $c_{p1,2,3}$, $T_{1,2,3}$, and $\kappa_{1,2,3}$ are the mass densities, specific heat capacity at constant pressure, temperature fields, and thermal conductivities, respectively. $h_{1,2,3}$ are the heat exchange coefficients for thermal unit 1, thermal unit 2, and convective layer, respectively. The term v represents the linear velocity of T_3.

To introduce nonzero real potentials, we consider convective ring structures. Given the periodic distribution of the temperature eigenfield in the ring, the temperature profile in each channel can be assumed to exhibit a waveform described by $T = T_0 + Ae^{i(kx-\omega t)}$, where T_0 is the reference temperature, A is the amplitude of temperature modulation, k is the wavenumber, and ω is the eigenfrequency. In the straight channel case, k equals $\frac{2\pi}{L}$ (where L is the length) under periodic boundary conditions. In the ring structure, k equals $\frac{1}{R}$ (where R is the radius) as only the first-order thermal mode can survive.

We assume the channel thicknesses are thin enough to regard the temperature field in the vertical dimension as uniform. Based on the continuity of the temperature field and heat flux at the vertical boundaries, we derive the imaginary coupling strengths as $ih_{1,2,3} = i\frac{\kappa_i}{\rho_{1,2,3}c_{p1,2,3}bd}$. Here, the coupling strengths can be regarded as ih_0 for simplicity, and the reference temperature T_0 is set to be zero. Due to strong convection, which significantly enhances thermal conduction, the temperature field T_3 in the convective layer attenuates much faster than T_1 and T_2. This property means the convective coupling layer T_3 will quickly reach a steady state compared with thermal units T_1 and T_2.

To illustrate this, we rewrite the coupled governing equation for the convective layer,

$$\frac{\partial T_3}{\partial t} = D_3\frac{\partial^2 T_3}{\partial x^2} - v\frac{\partial T_3}{\partial x} + h_0(T_1 + T_2 - 2T_3). \tag{6.30}$$

We continue using the wave-like form for the temperature field to analyze the temperature behavior. They are expressed into: $T_{1,2,3}(x,t) = T_0 + A_{1,2,3}(x,t)e^{-i\omega_{1,2,3}t}$. $D_3 = \frac{\kappa_3}{\rho_3 c_{p3}}$ is the thermal diffusivity. The temperature amplitude of the strong convective coupling layer T_3 evolving with time can be expressed as

$$\frac{\partial A_3}{\partial t} = \frac{\partial T_3}{\partial t}e^{i\omega_3 t} + i\omega_3 A_3 \tag{6.31}$$

which gives rise to

$$\frac{\partial A_3}{\partial t} = -(k^2 D_3 + 2h_0 + ikv)A_3 + h_0[A_1 e^{i(\omega_3-\omega_1)t} + A_2 e^{i(\omega_3-\omega_2)t}] \tag{6.32}$$

for which $A_3(t)$ could be solved to be

$$A_3(t) = h_0 \int_0^t [A_1(\tau)e^{i(\omega_3-\omega_1)\tau} + A_2(\tau)e^{i(\omega_3-\omega_2)\tau}] \times e^{-(k^2D_3+2h_0+ikv)(t-\tau)}d\tau. \tag{6.33}$$

Since A_3 attenuates and becomes steady rapidly, we denote this relaxation time as τ_0, and $A_3(t) = h_0 \int_{t-\tau_0}^{t} [A_1(\tau')e^{i(\omega_3-\omega_1)\tau'} + A_2(\tau)e^{i(\omega_3-\omega_2)\tau'}]e^{-(k^2D_3+2h_0+ikv)(t-\tau')}d\tau'$. A_3 is generated as

$$A_3(t) \approx \frac{h_0}{k^2D_3+2h_0+ikv}A_1(t)e^{i(\omega_3-\omega_1)t} + \frac{h_0}{k^2D_3+2h_0+ikv}A_2(t)e^{i(\omega_3-\omega_2)t}. \tag{6.34}$$

In this way, the field of convective coupling layer T_3 could be replaced by the combination of thermal units' field $T_{1,2}$,

$$T_3(t) \approx \frac{h_0}{k^2D_3+2h_0+ikv}[T_1(t)+T_2(t)] \tag{6.35}$$

where k^2D_3 denotes the decay rate in the decoupled convective coupling layer. By substituting T_3 into the three-level thermal coupling equations, the three-level thermal system can be simplified to a two-level system. In this reduced model, the imaginary coupling between the temperature fields becomes complex and is given by

$$h = \frac{h_0^2}{[kv - i(k^2D_3+2h_0)]}. \tag{6.36}$$

This relation tells us that as the strength of convection varies, the coupling will change from imaginary to complex and finally to real. We mark this in Fig. 6.4. When the strength of convection becomes dominant, specifically when $|kv| \gg k^2D_3+2h_0$, this complex coupling simplifies to a predominantly real value, approximately

$$h = \frac{h_0^2}{kv}. \tag{6.37}$$

The thermal phase fluctuation is effectively transferred to the real coupling of the temperature fields, illustrating how the convective flow's strength influences the system's dynamic behavior.

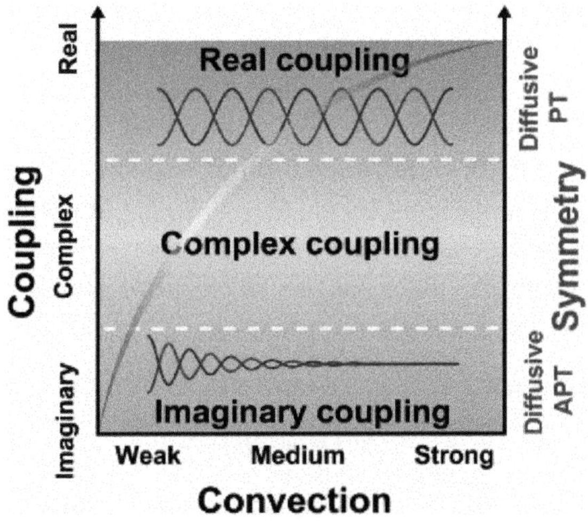

Fig. 6.4. Schematic of convection strength-dependent thermal coupling that evolves from imaginary to real, from Ref. [26]. When the convection strength increases, the APT and PT symmetry in diffusive systems can be constructed via the imaginary and real coupling, respectively.

In this dual-level system, the simplified coupling indicates a significant interaction between the temperature fields of T_1 and T_2, mediated through the effects of the convective coupling layer originally at T_3. The term $h_0^2/kv = h_0^2/\Omega$ highlights the pivotal role of convection in modulating the thermal dynamics of the system. This transformation from a three-level to a dual-level model underscores the importance of convection in thermal systems, demonstrating that even complex interactions can be reduced to more manageable forms while retaining their essential physical characteristics.

Therefore, this reduction not only simplifies the mathematical treatment of the system but also provides deeper insights into how phase fluctuations in the coupling layer translate into tangible changes in the coupling strength between the two thermal units. This understanding is crucial for designing and analyzing thermal systems where precise control over temperature fields is required, such as in advanced heat management applications and thermal engineering designs.

Based on real coupling, diffusive PT symmetry can thus be constructed by introducing different decay rates to the coupled temperature fields, similar to the case of passive PT-symmetric wave systems [35–37]. Before that, giving a universal form of the effective Hamiltonian under thermal real coupling is better.

In the last part, we have taken $h_{1,2} = h_0$ for simplicity. However, h_1 and h_2 can be not the same when the thermal conductivities or materials are different. Back to the third-order Hamiltonian, it can be deduced with asymmetric coupling as

$$H = \begin{pmatrix} -i(k^2D_1 + h_1) & ih_1 & 0 \\ ih_3 & -i(k^2D_3 + 2h_3) + kv & ih_3 \\ 0 & ih_2 & -i(k^2D_2 + h_2) \end{pmatrix}. \tag{6.38}$$

The second-order effective Hamiltonian can be obtained when we replace the convective background temperature T_3 by $\frac{h_3(T_1+T_2)}{ikv}$,

$$H_{\mathit{eff}} = \begin{pmatrix} -i(k^2D_1 + h_1) + h_1h_3/(kv) & \dfrac{h_1h_3}{kv} \\ \dfrac{h_2h_3}{kv} & -i(k^2D_2 + h_2) + h_2h_3/(kv) \end{pmatrix}. \tag{6.39}$$

Eigenvalues are solved as

$$\omega_\pm = -i\left(k^2\frac{D_1 + D_2}{2} + \frac{h_1 + h_2}{2}\right) \pm \sqrt{h_1h_2\left(\frac{h_3}{kv}\right)^2 - \left(k^2\frac{D_1 - D_2}{2} + \frac{h_1 - h_2}{2}\right)^2} \tag{6.40}$$

and eigenstates are

$$u_\pm = \left[-\left(i\frac{v}{w} \pm \sqrt{1 - \left(\frac{v}{w}\right)^2}\right), 1\right]^{\mathrm{T}} \tag{6.41}$$

with $v = \frac{\sqrt{h_1h_2}h_3}{kv}$ and $w = k^2\frac{D_1-D_2}{2} + \frac{h_1-h_2}{2}$.

From the expression of eigenvalues and eigenstates, one could easily find that when the parameters are set to satisfy $v = w$, both eigenvalues and eigenstates coalesce, indicating the exceptional point (EP).

Specifically, in the case of symmetric coupling, where $h_{1,2} = h_0$, the second-order effective Hamiltonian becomes

$$H_{\textit{eff}} = \begin{pmatrix} -i(k^2 D_1 + h_0) + h_0^2/(kv) & \dfrac{h_0^2}{kv} \\ \dfrac{h_0^2}{kv} & -i(k^2 D_2 + h_0) + h_0^2/(kv) \end{pmatrix}. \tag{6.42}$$

Using Pauli matrices, this dual-level $H_{\textit{eff}}$ with real coupling is derived as

$$H_{\textit{eff}} = h\hat{\sigma}_x + i\delta\hat{\sigma}_z + (h - iS_0)\hat{I} \tag{6.43}$$

where $\hat{\sigma}_x$ and $\hat{\sigma}_z$ are the Pauli matrices, $h = h_0^2/\Omega$ is the real coupling strength, $\delta = \frac{1}{2}\frac{k^2}{\rho C_p}(\kappa_1 - \kappa_2) + h_0$ and $S_0 = \frac{1}{2}\frac{k^2}{\rho C_p}(\kappa_1 + \kappa_2) + h_0$ are the detuning and average decay rates, respectively. It should be noted that the imaginary coupling plays a significant role in the decay rates associated with thermal diffusion. As a result, the decay rate detuning is influenced by the asymmetric imaginary couplings of temperature fields. In this discussion, we will focus on the case of symmetric coupling. For this scenario, the eigenfrequency of the effective Hamiltonian can be expressed as

$$\tilde{\omega}_{1,2} = \pm\sqrt{h^2 - \delta^2} \tag{6.44}$$

where $\tilde{\omega}_{1,2} = \omega_{1,2} - (h - iS_0)$ with $h - iS_0$ being a reduced term. The eigenfrequency of the convective coupling layer is given by $\tilde{\omega}_3 = \Omega$. In diffusive systems, the definition of eigenfrequency is similar to that in wave systems with gain and/or loss, where the imaginary part represents the amplitude decay or increase, and the real part denotes the phase evolution.

From Eq. (6.44), it is evident that the eigenfrequencies degenerate at $h = \delta$, where the real coupling strength equals the decay rate detuning. Before reaching this EP, where $h > \delta$, the convection is relatively weak, and $\tilde{\omega}_{1,2}$ remains real. As the convection strength increases and surpasses the EP, such that $h < \delta$, $\tilde{\omega}_{1,2}$ becomes imaginary. This transition indicates a phase shift in the system's dynamics, where the coupling strength is now insufficient to counterbalance the decay rate detuning. The EP is the self-bifurcation singularity point of both eigenstates and eigenfrequencies of Riemann surfaces in non-Hermitian parameter space.

The non-orthogonal eigenvectors, which represent the eigenstates of the temperature fields, are given by

$$\Psi_{1,2} = (1, \pm e^{\pm i\theta})^{\mathrm{T}} \tag{6.45}$$

for the diffusive PT symmetry phase, where $\theta = \mathrm{asin}\frac{\delta}{h}$. In this phase, the eigenstates $\Psi_{1,2}$ return to themselves after a PT operation, signifying that PT symmetry is preserved.

In the diffusive PT symmetry-broken phase, the eigenvectors are

$$\Psi_{1,2} = (1, ie^{\pm \varnothing})^{\mathrm{T}} \tag{6.46}$$

with $\varnothing = \mathrm{acosh}\frac{\delta}{h}$. In this case, $\mathcal{PT}\Psi_{1,2} \neq \Psi_{1,2}$, indicating that PT symmetry is broken.

In the PT symmetry phase, the coupled temperature fields maintain a steady phase lag θ and have the same amplitude, ensuring stable oscillatory behavior. However, in the PT symmetry-broken phase, the temperature fields exhibit an unchanging $\pi/2$ phase lag and have unbalanced amplitudes, reflecting the loss of symmetry and the onset of instability in the system.

In diffusive APT-symmetric systems, the phase fluctuation kv is directly imposed on the thermal unit's temperature fields, making the phase oscillation quite pronounced, especially in the broken phase. Conversely, in diffusive PT-symmetric systems, the phase fluctuation kv is introduced indirectly through the relatively weak real coupling. This indirect introduction means that the impact of kv on the temperature fields is mitigated, resulting in significantly suppressed phase oscillations in both PT-symmetric and PT symmetry-broken phases. This suppression is advantageous as it allows for more precise control over the spatial phases of the temperature fields, ensuring stable and predictable thermal behavior even when phase fluctuations are present. This characteristic is particularly beneficial for applications requiring exact thermal management and stability.

6.5. Effective Non-Hermitian Topology in Thermal Diffusion

Topology, a branch of mathematics dealing with properties that remain invariant under continuous deformations, has become a significant area of interest in condensed matter physics. Topological insulators, for instance, are materials that conduct electricity on their surface but act as insulators in their interior. These materials have opened new avenues for understanding quantum phenomena and have potential applications in next-generation electronic devices.

Non-Hermitian topology extends these concepts to systems where the Hamiltonian (the operator corresponding to the system's total energy) is not Hermitian. In traditional quantum mechanics, Hermitian operators ensure real eigenvalues corresponding to measurable physical quantities. However, non-Hermitian operators, which can have complex eigenvalues, have been found to describe systems with gain and loss, such as optical and acoustic systems. The study of non-Hermitian topology has revealed phenomena like EPs, where eigenvalues and eigenvectors coalesce, leading to unique physical properties.

Applying non-Hermitian topology to thermal systems is a novel and exciting development. Thermal diffusion systems, traditionally described by the heat equation, can be enriched by incorporating synthetic dimensions and non-Hermitian terms. This approach allows for new topological phenomena, such as Weyl exceptional rings. These are closed loops in momentum space where eigenvalues coalesce and eigenvectors degenerate, leading to robust, topologically protected states [35, 36].

This section explores the theoretical framework of effective Hamiltonians in thermal diffusion systems. By introducing synthetic dimensions through spatiotemporal modulation, we derive an effective Hamiltonian that encapsulates the topological properties of the system. This approach not only broadens our understanding of thermal diffusion but also highlights the interdisciplinary nature of modern physics, where concepts from topology and non-Hermitian physics intersect to reveal new physical phenomena.

6.5.1. *Derivation of the Effective Hamiltonian*

Synthetic dimensions are introduced by adding spatiotemporally modulated advections to the system. These advections create additional degrees of freedom, allowing us to explore more complex behavior. The modified heat equation with the advection is given by

$$\frac{\partial T}{\partial t} = D\nabla^2 T + \sum_i v_i \frac{\partial T}{\partial x_i} \tag{6.47}$$

where $v_i(x,t)$ are the velocity components of the modulated advections. These terms introduce a non-Hermitian component to the system, as the advections can represent gain and loss mechanisms. Assuming plane-wave solutions for the temperature fields, $T(x,t) = Aexp(i(kx - \omega t))$, we can derive the effective Hamiltonian for the system. Consider the propagative directions of heat flux is $\pm x$. The direction of the periodic heat flux is simultaneously influenced by the four advective projections of spinor fields in the orthogonal $x-y$ and $x-z$ spaces. Therefore, we can further divide the systematic thermal process into two subparts: (1) Heat transfer in the $x-y$ space and (2) Heat transfer in the $x-z$ space. The horizontal (T_{hor}) and vertical (T_{ver}) components are

$$T_{hor}(x,t) = A_{1,hor}e^{i(k_{hor}x+\omega_{hor}t-Dk_{hor}^2t)} + A_{2,hor}e^{i(k_{hor}x-\omega_{hor}t-Dk_{hor}^2t)} \tag{6.48}$$

where the first term stands for $T_{1,hor}$ and the second term stands for $T_{2,hor}$, and

$$T_{ver}(x,t) = A_{1,ver}e^{i(k_{ver}x+\omega_{ver}t-Dk_{ver}^2t)} + A_{2,ver}e^{i(k_{ver}x-\omega_{ver}t-Dk_{ver}^2t)} \tag{6.49}$$

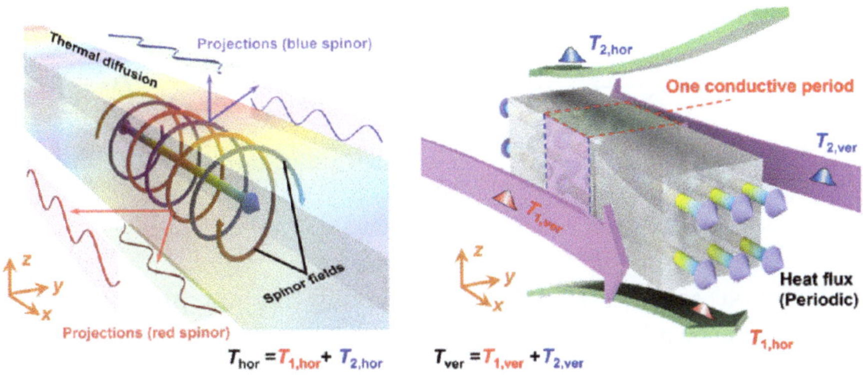

(a) (b)

Fig. 6.5. (a) A schematic illustration of helicities in periodic thermal diffusion around an existing heat flux from Ref. [34]. The light green and light purple regions represent the orthogonal surfaces that contain the projections of the red and blue advective spinor fields, respectively. The red and blue curves on these surfaces indicate the orthogonal projections of the corresponding spinor fields. (b) A model underlying the formation of a Weyl exceptional ring (WER) in a non-Hermitian diffusive system coupled with four counter advections on orthogonal surfaces. The rainbow-colored periodic heat flux within the central medium represents the initial thermal distributions.

where the first term stands for $T_{1,ver}$ and the second term stands for $T_{2,ver}$. The relative quantities are presented in Fig. 6.5.

Considering these hybrid thermal effects, the coupled thermal system leads to the following thermal processes.

In the $x-y$ space, the thermal process satisfies

$$
\begin{aligned}
c\frac{\partial T_{hor}}{\partial x} &= \kappa\frac{\partial^2 T_{1,hor}}{\partial x^2} + \rho c v_{1,hor}\frac{\partial T_{1,hor}}{\partial x} + \frac{h_{1,hor}}{b_{1,ver}}(T_{2,hor} - T_{1,hor}) \\
&\quad + \kappa\frac{\partial^2 T_{2,hor}}{\partial x^2} + \rho c v_{2,hor}\frac{\partial T_{2,hor}}{\partial x} + \frac{h_{2,hor}}{b_{2,hor}}(T_{1,hor} - T_{2,hor}).
\end{aligned} \tag{6.50}
$$

The first row describes the parallel thermal process in the $+x$ direction, while the second row describes the parallel thermal process in the $-x$ direction.

In the $x-z$ space, the thermal process satisfies,

$$\rho c\frac{\partial T_{ver}}{\partial x}=\kappa\frac{\partial^2 T_{1,ver}}{\partial x^2}+\rho c v_{1,ver}\frac{\partial T_{1,ver}}{\partial x}+\frac{h_{1,ver}}{b_{1,ver}}(T_{2,ver}-T_{1,ver})$$
$$+\kappa\frac{\partial^2 T_{2,ver}}{\partial x^2}+\rho c v_{2,ver}\frac{\partial T_{2,ver}}{\partial x}+\frac{h_{2,ver}}{b_{2,ver}}(T_{1,ver}-T_{2,ver}). \quad (6.51)$$

The first row describes the orthogonal thermal process in the $+x$ direction, while the second row describes the orthogonal thermal process in the $-x$ direction.

Taking the wave-like form of T_{hor} and T_{ver} into the governing equations describing thermal processes, we can obtain the effective Hamiltonian of the coupled thermal system,

$$H=(im_{hor}-m_{ver})\sigma_x-k_{ver}v_{ver}\sigma_y+k_{hor}v_{hor}\sigma_z$$
$$-i(Dk_{hor}^2+iDk_{ver}^2+m_{hor}+im_{ver})I_{2\times 2} \quad (6.52)$$

where σ_i are the Pauli matrices, k_{hor} and k_{ver} are the wavenumbers associated with the respective planes and $I_{2\times 2}$ is the 2×2 identity matrix. The real space vectors can be indicated as $d_x=-m_{ver}$, $d_y=-k_{ver}v_{ver}$, and $d_z=k_{hor}v_{hor}$.

The eigenvalues of this Hamiltonian are complex, reflecting the non-Hermitian nature of the system. The eigenvectors, denoted as $\psi_\pm$, can be found by solving the eigenvalue equation $H\psi_\pm=E_\pm\psi_\pm$. The corresponding eigenvalues $E_\pm$ are

$$E_\pm=-i(Dk_{hor}^2+iDk_{ver}^2+m_{hor}+im_{ver})$$
$$\pm\sqrt{(k_{hor}v_{hor})^2-(k_{ver}v_{ver})^2+(m_{hor}-im_{ver})^2}. \quad (6.53)$$

The term $(Dk_{hor}^2+iDk_{ver}^2+m_{hor}+im_{ver})$ only leads to the global shift of the eigenvalues, and can be neglected for further analyzing the eigenvectors. The complex nature of the eigenvalues indicates the presence of gain and loss mechanisms in the system, leading to phenomena such as EPs and non-Hermitian topology.

It is important to note that the eigenvalues of the current Hamiltonian are complex, extending beyond the simple PT/APT symmetry typically observed in similar systems. This characteristic distinguishes it significantly from the 2D non-Hermitian Hamiltonian used for thermal diffusion, which exhibits APT symmetry in heat transfer. The complexity of the eigenvalues in the current Hamiltonian introduces a layer of intricacy that warrants deeper exploration.

Weyl Exceptional Rings (WERs) are a typical complex nature of the eigenvalues. In 3D non-Hermitian systems, WERs can form. These are closed loops in momentum space where pairs of eigenvalues coalesce, and their corresponding eigenvectors become degenerate. WERs are the non-Hermitian counterparts of Weyl points in Hermitian systems, which are isolated points in momentum space with linear dispersion relations.

The formation of WERs can be understood by examining the eigenvalue spectrum of the effective Hamiltonian derived earlier. Eigenvalues $E_{\pm}$ given by Eq. (6.53) tell us that, at the EPs, the discriminant of the square root term becomes zero, leading to the coalescence of E_{+} and E_{-}. By tuning the system's parameters, such as the advection velocities and thermal exchange terms, these EPs trace out closed loops in momentum space, forming WERs.

The eigenvalue given by Eq. (6.53) results in a WER characterized by a constant non-Hermitian horizontal heat exchange (m_{hor}), as depicted in Fig. 6.6a and 6.6b. Additionally, there is a nearly flat real band within the WER (Fig. 6.6a), corresponding to the regions between the EP pair with low advective velocities in the complex plane of $i \times m_{hor} - d_z$ as shown in Fig. 6.6c and 6.6d.

WERs exhibit robust topological properties protected by the non-Hermitian nature of the system. These rings can host topologically protected edge states, leading to unidirectional transport phenomena immune to certain disorder types. Heat can be directed along specific paths without being scattered, which has potential applications in thermal management and energy harvesting. Incorporating non-Hermitian topology into thermal diffusion systems reveals a rich landscape of phenomena epitomized by WERs. These structures underscore the profound impact of non-Hermitian physics on classical systems, bridging the gap between quantum mechanics and classical thermal processes.

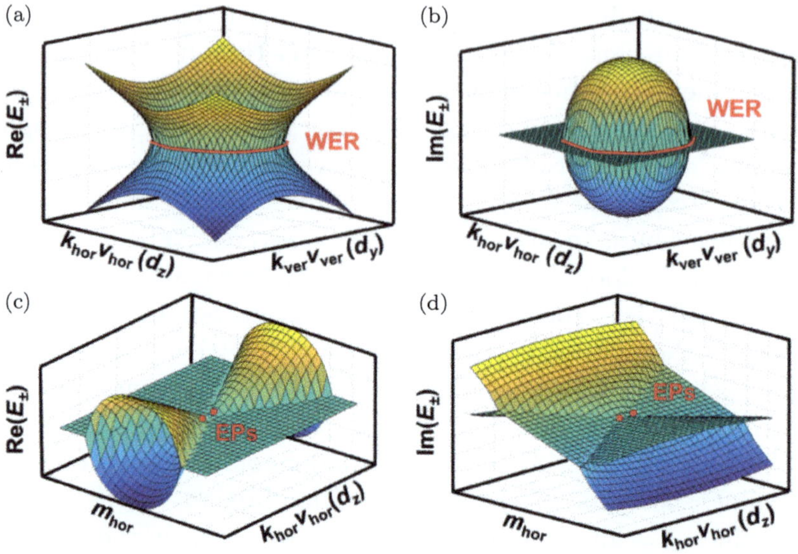

Fig. 6.6. (a, b) The band structures in the $d_y - d_z$ plane under the condition $m_{ver} = 0$, showing the presence of a WER marked in red. (c, d) The band structures in the $i \times m_{hor} - d_z$ plane reveal two EPs for $k_{ver}v_{ver} = 0$ (with m_{ver} also vanishing due to the absence of orthogonal advection). These two EPs correspond to points within the WER shown in A and B, and the $i \times m_{hor} - d_z$ plane intersects the $d_y - d_z$ plane at two locations, passing through the EP pair [34].

6.5.2. *Non-Hermitian Topology*

To thoroughly investigate the topological transitions of this non-Hermitian Weyl Hamiltonian, it is essential first to determine the topological charge, which can be achieved by calculating the Berry connection and Berry curvature. The Berry connection, which provides information about the geometric phase acquired over a closed path in parameter space, and the Berry curvature, which serves as the field strength of the Berry connection, are pivotal in understanding the topological nature of the system.

Calculating these quantities will reveal the presence of topological invariants that characterize the system's phase transitions. These invariants are crucial for identifying and classifying the topological phases of the Hamiltonian, and they provide insight into the robustness of the system's edge states and their response to perturbations.

Understanding these topological properties is fundamental for harnessing the potential applications of non-Hermitian systems in areas such as thermal management, energy transfer, and even quantum computing.

We use the effective Hamiltonian in Eq. (6.52) to analyze. To establish the periodic condition on the central strip with length L, we define the effective wavelengths as L for horizontal and vertical advections. Consequently, the effective wavenumbers are characterized as $k_{hor} = k_{ver} = 1/L$. Equation (6.52) describes a non-Hermitian Hamiltonian possessing distinct left and right eigenvectors. The following functions can represent these eigenvectors,

$$\langle \psi_{\pm}^{L}(k)|H(k) = E_{\pm}(k)\langle \psi_{\pm}^{L}(k)| \tag{6.54}$$

$$H(k)|\psi_{\pm}(k)\rangle = E_{\pm}(k)|\psi_{\pm}(k)\rangle. \tag{6.55}$$

By solving these two equations, the eigenvectors are

$$\langle \psi_{\pm}^{L}(\boldsymbol{k})| = \left(\frac{1}{N_{\pm}}\right)\begin{pmatrix} i(m_{hor} + k_{ver}v_{ver}) - m_{ver}, \\ -i(\pm M) - k_{hor}v_{hor} \end{pmatrix} \tag{6.56}$$

$$|\psi_{\pm}^{R}(k)\rangle = \left(\frac{1}{N_{\pm}}\right)\begin{pmatrix} i(m_{hor} - k_{ver}v_{ver}) - m_{ver}, \\ -i(\pm M) - k_{hor}v_{hor} \end{pmatrix} \tag{6.57}$$

with

$$M = \sqrt{(m_{hor})^2 - (k_{hor}v_{hor})^2 - (k_{ver}v_{ver})^2 - (m_{ver})^2 + 2im_{hor} \cdot m_{ver}}. \tag{6.58}$$

And $N_{\pm}$ is the normalization factor,

$$(N_{\pm})^2 = 2E_{\pm}(E_{\pm} - k_{hor}v_{hor}). \tag{6.59}$$

Then, we could adopt the Berry connection calculated by the left and right eigenvectors as a representation,

$$A_{\pm}^{LR}(\mathbf{k}) = i \cdot \langle \psi_{\pm}^{L}(\mathbf{k})|\nabla_{\mathbf{k}}|\psi_{\pm}^{R}(\mathbf{k})\rangle. \tag{6.60}$$

Based on the above Berry connection, we can obtain the Berry curvature,

$$\Omega_{\pm}^{LR}(\mathbf{k}) = \nabla_{\mathbf{k}} \times A_{\pm}^{LR}(\mathbf{k}). \tag{6.61}$$

Then the topological charge can be subsequently calculated,

$$\gamma_{\pm} = \frac{1}{2\pi} \cdot \oint_S \Omega_{\pm}^{LR}(\mathbf{k})dS \tag{6.62}$$

where S denotes the integration surface. To determine the integration surface conveniently, we carry out a spatial transformation to choose a sphere integration surface with a radius of

$$m_R = \sqrt{(m_{ver})^2 + (k_{hor}v_{hor})^2 - (k_{ver}v_{ver})^2}. \tag{6.63}$$

The spatial components of the real space can be defined as

$$\begin{aligned} d_x &= -m_{ver} = -m_R \sin(\theta)\cos(\phi) \\ d_y &= -k_{ver}v_{ver} = -m_R \sin(\theta)\sin(\phi) \\ d_z &= k_{hor}v_{hor} = m_R \cos(\theta). \end{aligned} \tag{6.64}$$

It is noted that the integration surface determines the topological charge. Here, we consider two typical cases:

Scheme 1: the selected integration surface encircles the entire WER $|m_{hor}| < |m_R|$.

Scheme 2: the selected integration surface does not enclose the WER (within or outside the WER).

For scheme 1, with delicate calculation given by [34], the topological charge is

$$\gamma_{\pm} = \frac{1}{2\pi} \cdot \oint_S \Omega_{\pm}^{LR}(\mathbf{k})dS = \frac{1}{2\pi} \cdot \oint_S (\Omega_R^{LR}, \Omega_\vartheta^{LR}, \Omega_\phi^{LR})dS. \tag{6.65}$$

It can be indicated that the Berry curvatures of Ω_R^{LR} and Ω_ϕ^{LR} lead to zero components; hence, the integration results in

$$\begin{aligned} \gamma_{\pm} &= \pm\frac{1}{2\pi} \cdot \int_0^{\pi} d\theta \int_0^{2\pi} d\phi \cdot \Omega_\theta^{LR} \\ &= \pm\frac{1}{2\pi} \cdot \int_0^{\pi} -\pi \cdot \sin(\theta)d\theta = \mp 1. \end{aligned} \tag{6.66}$$

The result shows that the topological charge is a quantized integer when the integration surface encircles the entire WER in this thermal system.

We can examine the topological charge using two integration surfaces for Scheme 2, which does not involve encircling any WER. The first type is completely outside the WER for arbitrary values of $|m_R|$, while the second type is within the WER under the condition $|m_{hor}| > |m_R|$. This approach allows us to investigate the behavior and properties of the topological charge in a more detailed manner, accounting for the specific conditions and characteristics of the integration surfaces relative to the WER.

Using these two types of integration surfaces, we can comprehensively understand the topological charge in different scenarios, whether the surfaces are inside or outside the WER. This analysis is crucial for revealing the nature of topological transitions and the stability of the system's eigenstates under various conditions.

For the first type, the integration surface is selected within one separative band (upper or lower), distinct from the former scheme that encircles the entire WER around the degenerate bands. Therefore, a complete path ranging from 0 to 2π should be followed in the θ direction, which results in a zero topological charge.

The chosen surface comprises branch points within the WER for the second case. Referring to the strategy outlined in [37], we can rewrite the eigenvectors by following a single band in the current system,

$$\begin{aligned}\langle\psi_1^L(\mathbf{k})| &= \langle\psi_+^L(\mathbf{k})|\Theta(\mathrm{Re}(d_x)) + \langle\psi_-^L(\mathbf{k})|\Theta(-\mathrm{Re}(d_x)),\\ \langle\psi_2^L(\mathbf{k})| &= \langle\psi_-^L(\mathbf{k})|\Theta(\mathrm{Re}(d_x)) + \langle\psi_+^L(\mathbf{k})|\Theta(-\mathrm{Re}(d_x)).\end{aligned} \quad (6.67)$$

$$\begin{aligned}|\psi_1^R(\mathbf{k})\rangle &= \Theta(\mathrm{Re}(d_x))|\psi_+^R(\mathbf{k})\rangle + \Theta(-\mathrm{Re}(d_x))|\psi_-^R(\mathbf{k})\rangle,\\ \langle\psi_2^L(\mathbf{k})| &= \Theta(\mathrm{Re}(d_x))|\psi_-^R(\mathbf{k})\rangle + \Theta(-\mathrm{Re}(d_x))|\psi_+^R(\mathbf{k})\rangle\end{aligned} \quad (6.68)$$

where Θ denotes the Heaviside function. With the conditional eigenvectors, the Berry connection and curvature can be calculated. Under the condition of $|m_{hor}| > |m_R|$, only a zero topological charge can be subsequently observed.

In general, the topological charges remain zero when the integration surface S does not encircle the WER for the proposed system, regardless of whether the selected integration surface is within or outside the WER. The topological properties of the system are preserved, and the topological charge does not change as long as the integration surface avoids encircling the WER.

References

1. Auld, B. A. (1990). *Acoustic Fields and Waves in Solids*, 2nd Ed. Krieger Publishing.
2. Jackson, J. D. (1999). *Classical Electrodynamics*, 3rd Ed. John Wiley & Sons.
3. Morse, P. M. and Ingard, K. U. (1968). *Theoretical Acoustics.* Princeton University Press.
4. Kinsler, L. E., Frey, A. R., Coppens, A. B., and Sanders, J. V. (2000). *Fundamentals of Acoustics*, 4th Ed. John Wiley & Sons.
5. Ramo, S., Whinnery, J. R., and Van Duzer, T. (1984). *Fields and Waves in Communication Electronics*, 2nd Ed. John Wiley & Sons.
6. Zienkiewicz, O. C. and Taylor, R. L. (2005). *The Finite Element Method for Solid and Structural Mechanics*, 6th Ed. Butterworth-Heinemann.
7. Hargrave, P. A. and Cooke, D. C. (1997). *Microwave Engineering: Concepts and Fundamentals.* McGraw-Hill.
8. Yariv, A. (1989). *Quantum Electronics*, 3rd Ed. John Wiley & Sons.
9. Carslaw, H. S. and Jaeger, J. C. (1959). *Conduction of Heat in Solids*, 2nd Ed. Oxford University Press.
10. Özisik, M. N. (1993). *Heat Conduction*, 2nd Ed. John Wiley & Sons.
11. Kaviany, M. (2002). *Principles of Heat Transfer in Porous Media*, 2nd Ed. Springer.
12. Cengel, Y. A. (2003). *Heat Transfer: A Practical Approach*, 2nd Ed. McGraw-Hill.
13. Incropera, F. P. and DeWitt, D. P. (2002). *Fundamentals of Heat and Mass Transfer*, 5th Ed. John Wiley & Sons.
14. Crank, J. (1975). *The Mathematics of Diffusion*, 2nd Ed. Oxford University Press.
15. Bejan, A. (2013). *Convection Heat Transfer*, 4th Ed. John Wiley & Sons.
16. Kakac, S. and Yener, Y. (1993). *Heat Conduction.* Hemisphere Publishing Corporation.

17. Kaviany, M. (1995). *Principles of Convective Heat Transfer*. Springer, USA.
18. Kundu, P. K., Cohen, I. M., and Dowling, D. R. (2015). *Fluid Mechanics*, 6th Ed. Academic Press.
19. Holman, J. P. (2010). *Heat Transfer*, 10th Ed. McGraw-Hill.
20. Batchelor, G. K. (1967). *An Introduction to Fluid Dynamics*. Cambridge University Press.
21. Spiegel, E. A. (1971). Convection in stars, *Annu. Rev. Astron. Astrophys.*, 9, pp. 323–352.
22. Cengel, Y. A. and Ghajar, A. J. (2011). *Heat and Mass Transfer: Fundamentals and Applications*, 4th Ed. McGraw-Hill.
23. Li, Y., Peng, Y.-G., Han, L., Miri, M.-A., Li, W., Xiao, M., Zhu, X.-F., Zhao, J., Alù, A., Fan, S., and Qiu, C.-W. (2019). Anti-parity-time symmetry in diffusive systems. *Science*, 364, pp. 170–173.
24. Ewins, D. J. (2000). *Modal Testing: Theory, Practice and Application*, 2nd Ed. Research Studies Press.
25. Aspelmeyer, M., Kippenberg, T. J., and Marquardt, F. (2014). Cavity optomechanics. *Rev. Mod. Phys.*, 86, pp. 1391–1452.
26. Cao, P.-C., Ju, R., Wang, D., Qi, M., Liu, Y.-K., Peng, Y.-G., Chen, H., Zhu, X.-F., and Li, Y. (2024). Observation of parity-time symmetry in diffusive systems. *Sci. Adv.* 10, pp. eadn1746.
27. Feynman, R. P. (1949). The theory of positrons. *Phys. Rev.*, 76, pp. 749–759.
28. Fleury, R., Sounas, D. L., Sieck, C. F., Haberman, M. R., and Alù, A. (2014). Sound isolation and giant linear nonreciprocity in a compact acoustic circulator. *Science*, 343, pp. 516–519.
29. Huang, R., Miranowicz, A., Liao, J.-Q., Nori, F., and Jing, H. (2018). Nonreciprocal photon blockade. *Phys. Rev. Lett.*, 121, pp. 153601.
30. Torrent, D., Poncelet, O., and Batsale, J.-C. (2018). Nonreciprocal thermal material by spatiotemporal modulation, *Phys. Rev. Lett.*, 120, pp. 125501.
31. Silveirinha, M. G. (2014). Spontaneous parity-time-symmetry breaking in moving media. *Phys. Rev. A*, 90, pp. 013842.
32. Silveirinha, M. G. (2014). Optical instabilities and spontaneous light emission by polarizable moving matter. *Phys. Rev. X*, 4, pp. 031013.
33. Morgado, T. A. and Silveirinha, M. G. (2017). Negative Landau damping in bilayer graphene. *Phys. Rev. Lett.*, 119, pp. 133901.
34. Xu, G., Li, W., Zhou, X., Li, H., Li, Y., Fan, S., Zhang, S., Christodoulides, D. N., and Qiu, C.-W. (2022). Observation of Weyl exceptional rings in thermal diffusion. *Proc. Natl. Acad. Sci. U.S.A.*, 119, pp. e2110018119.

35. Wu, Y. and Chen, X. (2020). Weyl exceptional rings and topological phases in non-Hermitian systems. *Phys. Rev. Lett.*, 124, pp. 016803.
36. Yao, S. and Wang, Z. (2018). Edge states and topological invariants of non-Hermitian systems. *Phys. Rev. Lett.*, 121, pp. 086803.
37. Cerjan, A., Huang, S., Wang, M., Chen, K. P., Chong, Y., and Rechtsman, M. C. (2019). Experimental realization of a Weyl exceptional ring. *Nat. Photonics*, 13, pp. 623–628.

Chapter 7

Thermal Lattices

7.1. Introduction

Thermal lattices are structures that efficiently conduct or dissipate heat, which can be used in various thermal management applications, such as electronics, power systems, and heat sinks. A thermal lattice typically comprises interconnected nodes or struts with a geometric pattern. These nodes are often made of thermally conductive materials such as metals and ceramics. The thermal lattice facilitates conduction, convection, and radiation heat transfer by providing a large surface area. Thermal lattices have various applications, including electronics cooling, aerospace, automotive, and power generation. They can improve heat dissipation in electronic devices, enhance the efficiency of thermal energy storage systems, and provide thermal stability to structures exposed to high temperatures.

Lattice structures have been widely studied in electronic systems and condensed matter physics for their unique electronic properties and physical phenomena. For example, the unique electronic properties of lattice structures such as kagome lattices or Su-Schrieffer-Heeger (SSH) lattices and emergent phenomena such as topological phases or exotic magnetic behaviors are studied in condensed matter physics. Previously, "thermal lattices" referred specifically to lattice structures for thermal management and heat dissipation in engineering and have nothing to do with electronic systems and condensed matter physics. With the equivalent Hamiltonian describing and predicting the temperature field in heat transfer, we can also use similar methods to maximize heat dissipation and discover interesting phenomena, meeting the needs of thermal lattice-optimized heat dissipation and thermal management. This property opens a new gate to the design and development of thermal lattices.

7.2. Theoretical Basis

7.2.1. *Bloch's Theorem*

Bloch's Theorem: This principle is foundational in describing how particles (like electrons or phonons) move through a periodic potential, which is common in solid-state physics. The theorem states that the wave functions can be described as a product of a plane wave and a periodic function.

Bloch's theorem underlies the concept that in a periodic potential, the wave function can be expressed as $\varphi(x) = e^{ikx}u_{k(x)}$, where $k(x)$ is a periodic function with the same periodicity as the lattice, and k is the wave vector. Though the analogy principle is mainly applied to quantum systems, it can also describe wave-like phenomena in other cases, such as thermal diffusion in periodic structures due to lattice symmetry.

Take the 1D SSH model as an example [1]. A schematic model consisting of interconnected spheres and rods is introduced, featuring uniform density (ρ), heat capacity (c_p), and thermal conductivity (κ), as depicted in Fig. 7.1a. This setup has a conceptual similarity to the SSH model. In this model, lengths occupied by the two spheres of radius R in one unit are b_1 and b_3 ($=b_1$). The lengths of the intra-cell and inter-cell rods with radius R_0 are b_2 and b_4. A crucial parameter

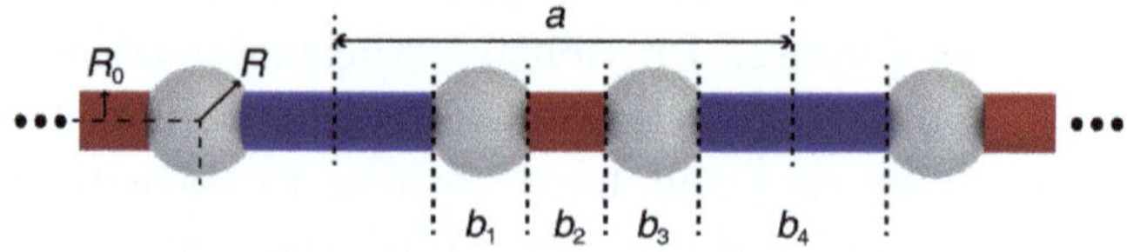

Fig. 7.1. Heat diffusion in a thermal lattice. (a) A 3D model of spheres and rods. The inter-cell and intra-cell rods are blue and red, respectively. Adapted from Ref. [2].

$c = (b_4 - b_2)/2$ represents the length difference, while a signifies the unit length. The design allows for the adjustment of rod lengths to effectively manage heat transfer efficiencies both within and between the units.

We define the middle of the intra-cell rod ($x = 0$) as the center of a unit. The 1D governing equation of the temperature field $T(x, t)$ is

$$\rho c_p(x) \frac{\partial T}{\partial t} = \frac{\partial}{\partial x} \left(\kappa(x) \frac{\partial T}{\partial x} \right) \tag{7.1}$$

where T is temperature, x is the direction of the chain model, and t is time. $\rho c_p(x) = \rho c_{pm}$ and $\kappa(x) = \kappa_m$ are piecewise functions, with $m = 1$ for the sphere parts and $m = 2$ for the rod parts.

7.2.2. *Tight-Binding Model*

The tight-binding model is used in solid-state physics to calculate electronic band structures. It relies on an approximate set of wave functions based on superpositions of wave functions for isolated atoms located at each atomic site. The tight-binding model provides an initial framework for understanding discrete interactions within the system before recognizing its limitations and moving to a continuum model approach. However, it is specifically stated that the tight-binding model is inadequate for rigorously describing the system due to the specific characteristics of heat diffusion and its non-local nature.

The SSH model illustrates the limitations of applying quantum mechanical models directly to thermal systems. The SSH model provides a framework with $H = \sum(t_1 c_{n,A}^{\dagger} c_{n,B} + t_2 c_{n,B}^{\dagger} c_{n+1,A} + \text{h.c.})$, where t_1 and t_2 are hopping terms between sites (A and B) within and between unit cells. It falls short in accurately modeling nonelectronic systems like heat diffusion due to the nonlocality and dissipative nature of heat.

Using the SSH model to discretize the thermal lattice, we mark the left and right spheres in a unit as A and B, respectively. T_{An} and T_{Bn} are the average temperatures of spheres A and B. According to Fourier's law in heat conduction, the rate of internal energy change of each sphere is equal to the heat flow in and out,

$$\rho c_p V_{\mathrm{sphere}} \frac{\partial T_{A,Bn}}{\partial t} = \Phi_{\mathrm{left}} - \Phi_{\mathrm{right}} \tag{7.2}$$

where Φ_{left} is the heat flow into the sphere from its left boundary, and Φ_{right} is the heat flow out of the sphere from its right boundary.

Assuming that the temperature distribution on the rod is linear, the heat flow can be considered proportional to the temperature difference

$$\Phi = -\pi R_0^2 \kappa \frac{dT}{dx} \approx -\pi R_0^2 \kappa \frac{T_{\mathrm{right}} - T_{\mathrm{Left}}}{b_{2,4}} \tag{7.3}$$

where T_{right} and T_{left} are the temperatures at the right and left ends of the rod, respectively. We have $T_{\mathrm{right}} \approx T_{Bn}$ and $T_{\mathrm{left}} \approx T_{An}$ for intra-cell, and $T_{\mathrm{right}} \approx T_{An}$ and $T_{\mathrm{left}} \approx T_{Bn}$ for inter-cell. A discretized version of Eq. (7.3) is

$$\begin{cases} \rho c V_{\mathrm{sphere}} \dfrac{\partial T_{An}}{\partial t} = \pi R_0^2 \kappa \left(\dfrac{T_{Bn-1} - T_{An}}{b_4} + \dfrac{T_{Bn} - T_{An}}{b_2} \right) \\ \rho c V_{\mathrm{sphere}} \dfrac{\partial T_{Bn}}{\partial t} = \pi R_0^2 \kappa \left(\dfrac{T_{An} - T_{Bn}}{b_2} + \dfrac{T_{An+1} - T_{Bn}}{b_4} \right) \end{cases} \tag{7.4}$$

with $V_{sphere} \approx 4\pi R^3/3$. The inter-cell coupling coefficient τ_1 and the intra-cell coupling coefficient τ_2 are

$$\begin{cases} \tau_1 = \dfrac{3{R_0}^2}{4R^3 b_2} D \\ \tau_2 = \dfrac{3{R_0}^2}{4R^3 b_4} D \end{cases} \tag{7.5}$$

where $D = \kappa/\rho c_p$ is thermal diffusivity.

The Hamiltonian for this diffusive SSH model is

$$H_{SSH}(k) = -i \begin{pmatrix} \tau_1 + \tau_2 & -\tau_1 - \tau_2 e^{-ika} \\ -\tau_1 - \tau_2 e^{ika} & \tau_1 + \tau_2 \end{pmatrix} \tag{7.6}$$

where k is wavenumber. Setting $\Delta = \tau_1/\tau_2$, the spectrum of the system is

$$\omega(k) = -i\tau_2\left[(1+\Delta) \pm \sqrt{(\Delta+\cos k)^2+\sin^2 k}\right]. \tag{7.7}$$

The discrete Hamiltonian matrix of the SSH model predicts the eigenvalues of the system. Yet, the SSH model is only approximately valid under certain conditions because the actual temperature field is nonlinear in the sphere-and-rod model.

The eigenvalues of the discrete Hamiltonian matrix from the SSH model do not agree with the simulation results in Fig. 7.2.

Therefore, the direct application of the quantum mechanical model to thermal systems has limitations. Due to the nonlocality and dissipation of heat, it cannot accurately model nonelectronic systems such as thermal diffusion. In contrast, the continuum model is universal, accurate, and rigorous.

7.2.3. *Continuum Model*

The continuum model avoids the limitations of the tight-binding approximation and allows for the analytical determination of

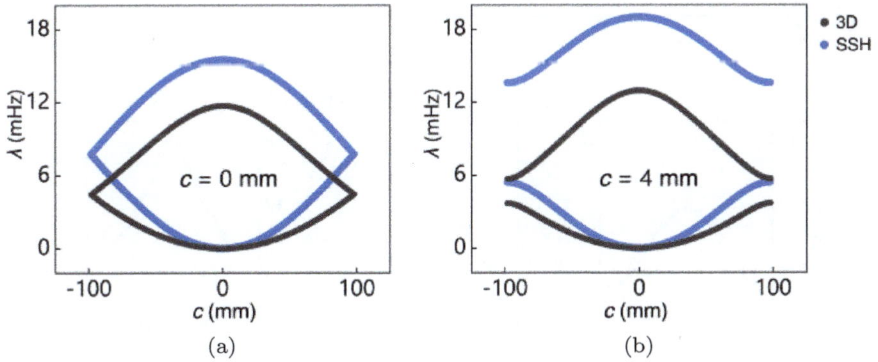

Fig. 7.2. SSH model. Band structures of the system with (a) $c = 0\,\text{mm}$ and (b) $c = 4\,\text{mm}$. Blue and black lines represent the SSH prediction and 3D simulation solutions. Adapted from Ref. [2].

band structures and geometric phases in heat diffusion systems. This method focuses on solving for eigenvalues in a manner that respects the nonlocality of thermal fields.

In a continuum model, the temperature $T(x,t)$ in a 1D thermal lattice is determined by the eigenfunction $\boldsymbol{H}u = \omega u$, where $\boldsymbol{H}$ is the effective Hamiltonian, and u is the temperature profile of the model. The Hamiltonian is anti-Hermitian due to the dissipative nature of heat transfer, leading to purely imaginary eigenvalues $\omega = -i\lambda$.

We consider a 1D equivalent model to analyze the model rigorously, as shown in Fig. 7.3a. The region of the sphere parts is modeled as effective heat capacity c_{p1}, density ρ_1, and thermal conductivity κ_1. The rod parts have effective heat capacity $c_{p2} = c_{p3D}$, density $\rho_2 = \rho_{3D}$, and thermal conductivity $\kappa_2 = \kappa_{3D}$.

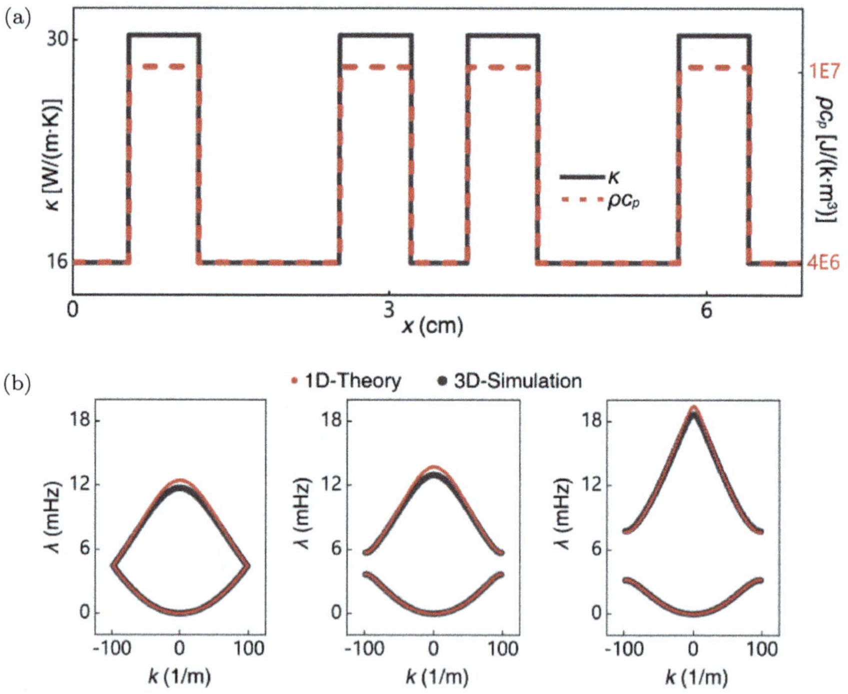

Fig. 7.3. Heat diffusion in thermal lattices. (a) The distributions of κ and ρc_p in a 1D equivalent model. (b) The band structures of the system. Adapted from Ref. [2].

Assuming that the solution to Eq. (7.1) is $T(x,t) = T_0 + u(x) e^{-i\omega t}$, where T_0 is a constant, it becomes an eigenvalue problem

$$\omega u(x) = \frac{i}{\rho c_p(x)} \frac{\partial}{\partial x}\left(\kappa(x)\frac{\partial u}{\partial x}\right) = \boldsymbol{H} u(x) \tag{7.8}$$

where $\boldsymbol{H}$ is the effective Hamiltonian. We define the inner product of two states u and v as

$$\langle v|u\rangle = \int_{-a/2}^{a/2} \rho c_p(x) v(x)^* u(x) dx. \tag{7.9}$$

Since the effective Hamiltonian is anti-Hermitian, $\omega = -i\lambda$ is a pure imaginary number, where λ is the decay rate. We assume that $u(x)$ has been normalized for $\langle u|u\rangle = 1$.

With the plane-wave expansion, $u(x)$ can be written as a piecewise function in the nth unit $(-a/2 \leq x - na \leq a/2)$

$$u(x) = X^{(j)}{}_n e^{ik_m(x-na)} + Y^{(j)}{}_n e^{-ik_m(x-na)}, \quad \zeta_j < x - na < \zeta_{j+1} \tag{7.10}$$

where $k_m = (\lambda \rho_m c_{pm}/\kappa_m)^{1/2}$ is obtained from the dispersion relations. ζ_j represents the coordinate of the interface.

7.2.4. *Transfer Matrix Method*

This method solves the transmission properties and band structure of a system by considering boundary conditions and interface matching. The matrix encapsulates how heat diffuses through different sections of the modeled structure, predicting behavior at interfaces and allowing for the identification of band gaps and edge states [3].

The transfer matrix M connects the states across different units of the lattice, effectively capturing how variations in structural parameters like the lengths of rods influence the thermal properties. The general form of the transfer matrix equation is

$$M\begin{bmatrix} u_n \\ u_{n+1} \end{bmatrix} = \begin{bmatrix} u_{n+1} \\ u_{n+2} \end{bmatrix} \tag{7.11}$$

where u_n denotes the temperature or thermal state at the nth unit. This matrix is crucial for determining the propagation of heat and identifying the presence of localized states or bandgaps.

Application in 1D Heat Diffusion: It is particularly noted for enabling the calculations that link geometric phases with physical phenomena such as edge states. The transfer matrix method addresses boundary conditions in a 1D heat diffusion system. The heat flow from one unit to the next is calculated using the following expression for the transfer matrix,

$$M = \begin{bmatrix} M_{11} & M_{12} \\ M_{21} & M_{22} \end{bmatrix} \tag{7.12}$$

where M_{11}, M_{12}, M_{21}, and M_{22} are calculated based on the thermal conductivity and heat capacity parameters at the interface.

According to the matching conditions, we can obtain the transfer matrix $\boldsymbol{M}$ from unit $n-1$ to unit n,

$$\boldsymbol{M}\begin{pmatrix} X_n^{(-2)} \\ Y_n^{(-2)} \end{pmatrix} = \begin{pmatrix} X_{n+1}^{(-2)} \\ Y_{n+1}^{(-2)} \end{pmatrix} = e^{ika}\begin{pmatrix} X_n^{(-2)} \\ Y_n^{(-2)} \end{pmatrix}. \tag{7.13}$$

The second equation uses the Bloch theorem $u(x) = \psi(x)e^{ikx}$, where $\psi(x)$ is periodic in a unit cell, and k is the wavenumber.

Equation (7.13) is an eigenvalue problem for $\boldsymbol{M}$, whose eigenvalue is

$$\Lambda_{\pm} = e^{ika} = \mathrm{Re}(M_{11}) \pm i\sqrt{1 - \mathrm{Re}(M_{11})^2} \tag{7.14}$$

which can be simplified as

$$\cos ka = \mathrm{Re}(M_{11}). \tag{7.15}$$

The eigenvector is

$$\begin{pmatrix} X_{\pm}^{(-2)} \\ Y_{\pm}^{(-2)} \end{pmatrix} \propto \begin{pmatrix} -M_{12} \\ M_{11} - \Lambda_{\pm} \end{pmatrix} = \begin{pmatrix} -M_{12} \\ i[\mp\sqrt{1 - \mathrm{Re}(M_{11})^2} + \mathrm{Im}(M_{11})] \end{pmatrix}. \tag{7.16}$$

When the two rods have the same lengths ($b_2 = b_4$, $c = 0$), we have

$$\cos ka = 2(\cosh\gamma \sin\alpha \sin\beta - \cos\alpha \cos\beta)^2 - 1. \tag{7.17}$$

Since the right-hand-side ≥ -1, the bandgap is closed at $k = \pm\pi/a$,

$$\cot\alpha \cot\beta = \cosh\gamma. \tag{7.18}$$

When the two rods have different lengths ($b_2 \neq b_4$, $c \neq 0$), a bandgap is opened because the right-hand side of Eq. (7.17) could be smaller than -1. Thus, Eq. (7.17) has no real solution for k. The upper and lower edges of the bandgap are reached at $k = \pm\pi/a$,

$$\sin\beta(\cosh\gamma - \cot\alpha \cot\beta) = \pm \sinh\gamma \sin\delta. \tag{7.19}$$

The band structure can be obtained by solving λ from Eq. (7.15). The results are shown as red dots in Fig. 7.3b. When $c = 0$, the two bands degenerate at $k = \pm\pi/a$. When $|c| > 0$, the right-hand side of Eq. (7.15) satisfies $\mathrm{Re}(M_{11}) < -1$, resulting in a bandgap opened at $k = \pm\pi/a$.

7.2.5. *Geometric Phase*

The topological properties of a non-Hermitian system often differ from those of a Hermitian system. The non-Hermitian skin effect modifies the bulk-boundary correspondence. The Zak phase is a crucial parameter describing topological properties in 1D systems. The calculation of the Zak phase [4, 5] is based on the symmetry and periodicity of the wave function,

$$\mathrm{Z} = \int_{-\pi/a}^{\pi/a} \langle \psi(x,k) | i\partial_k | \psi(x,k) \rangle \mathrm{d}k. \tag{7.20}$$

The Zak phase in systems with inversion symmetry is typically an integer multiple of π, closely related to the topological nontriviality of the system. For inversion-symmetric systems with $x = 0$ at the inversion center, $\mathrm{Z} = 0$ when $u(x, \pm\pi/a)$ and $u(x, 0)$ have the same parity under spatial inversion $x \to -x$, whereas $\mathrm{Z} = \pi$ when they have different parities.

In our example, $u(x, k)$ is determined by Eq. (7.10), where the coefficients $X^{(-2)}$ and $Y^{(-2)}$ are solved as the eigenvector of the transfer matrix $\boldsymbol{M}$ from Eq. (7.13). Based on the expressions of entries of $\boldsymbol{M}$, it follows that the eigenvector has the form

$$\begin{pmatrix} X^{(-2)} \\ Y^{(-2)} \end{pmatrix} \propto \begin{pmatrix} e^{ik_2a/2}U_1 \\ e^{-ik_2a/2}U_2 \end{pmatrix} \tag{7.21}$$

where $U_1 = \mathrm{Im}[M_{12}\mathrm{exp}(-ik_2a)]$ and $U_2 = \mathrm{sin}ka - \mathrm{Im}(M_{11})$ are two real numbers. We define a state vector $\boldsymbol{U} = (U_1, U_2)^T$ to illustrate the properties of $u(x, k)$.

Based on the transfer matrix, we can show that the Zak phase is determined by whether U_1 changes its sign in a band. We also verify analytically and numerically that the Zak phases calculated from the lower and upper bands are the same. Therefore, the Zak phase Z can be used as a topological invariant. In the lower band, $\mathrm{Z} = \pi$ when U_1 changes its sign, and $\mathrm{Z} = 0$ when the sign of U_1 remains unchanged. Considering that $U_1 > 0$ in the vicinity of $\lambda = 0$, we have

$$\exp(i\mathrm{Z}) = \mathrm{sgn}(\underline{U}_1) \tag{7.22}$$

where $\underline{U}_1$ is the value of U_1 at the upper edge of the lower band. The Zak phase reveals two distinct topological regimes: $\mathrm{Z} = \pi$ for $c < 0$ and $\mathrm{Z} = 0$ for $c > 0$. Figure 7.4 shows the trajectories of $\boldsymbol{U}$ in the lower and upper bands at $c = -8$ mm and 8 mm. All our theoretical predictions are confirmed, especially the close relationship between the sign of U_1 and the Zak phase.

These formulas and descriptions offer a deep understanding of the theoretical basis for the heat topological models and methods proposed in the article, as well as how to validate the correctness of these theories experimentally. The generalization of this approach could have significant implications for the development of new thermal management technologies.

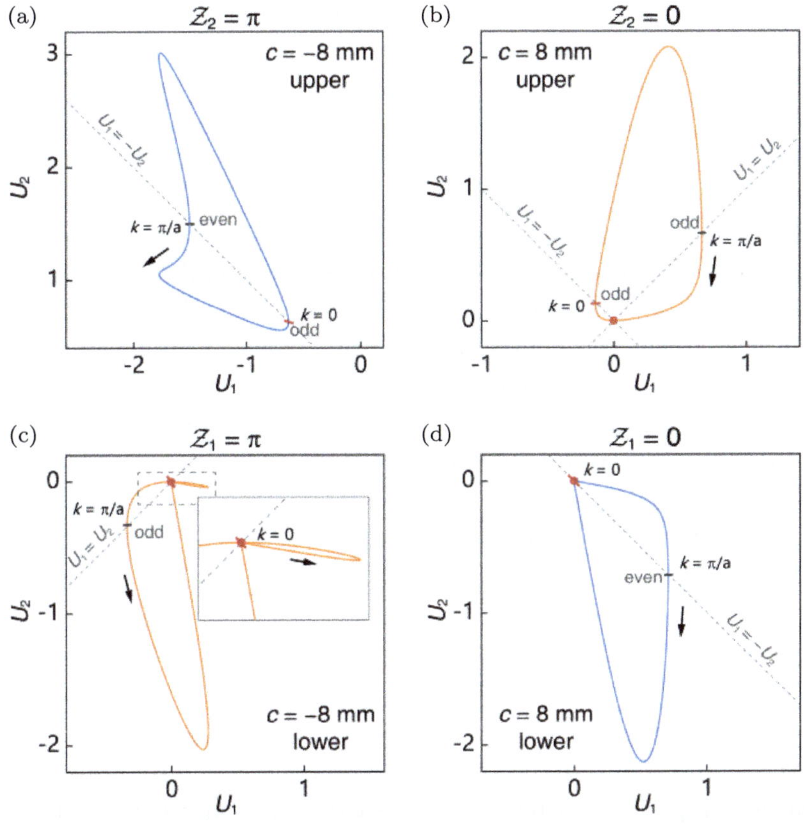

Fig. 7.4. Trajectories of the state vector. The trajectories of U as k are increased from $-\pi/a$ to π/a in the upper (a,b) and lower (c,d) bands for $c = -8\,\text{mm}$ (a,c) and $8\,\text{mm}$ (b,d). The orange-colored trajectories in (b,c) contain sign changes of U_1. The points where $k = \pm\pi/a(k = 0)$ are marked by short grey (red) bars. Red dots mark the $(0, 0)$ points. The grey dashed lines represent symmetric states ($U_1 = U_2$) or anti-symmetric ($U_1 = -U_2$) under spatial inversion around the center of the inter-cell rod. The parity (even or odd) of the state is marked at $k = 0$ and $\pm\pi/a$. Adapted from Ref. [2].

7.3. Topological Effects

An effective Hamiltonian can describe and predict the transient temperature field evolution [6, 7]. Due to the dissipative nature of heat transfer, the effective Hamiltonian is non-Hermitian. Non-Hermitian wave systems show unusual symmetric and topological properties [8, 9], like unidirectional transparency [10], coherent perfect absorption [11], single-mode lasing [12], and skin effect [13]. Thus, thermal systems were also developed to study the topology of heat transfer [14–18].

The thermal lattice structure can be customized to optimize specific thermal properties using topological principles, which can be divided into three categories: pure diffusion, thermal convection, and asymmetric coupling. Different novel topological phenomena exist in different models, such as local thermal diffusion, thermal diodes, and non-Hermitian skin effects. It provides a new perspective and future for the design and development of thermal lattices.

7.3.1. *Pure Diffusion Model*

We reveal the general topology of pure heat diffusion. The effective Hamiltonian for thermal diffusion is non-Hermitian within an appropriate Hilbert space. Its eigenvalues are purely imaginary, indicating the decay rates of each characteristic mode. Due to the absence of driving forces at imaginary frequencies, resonant phenomena do not occur in diffusion systems. This starkly contrasts with wave-based scenarios (Fig. 7.5a), where selective excitation and examination of

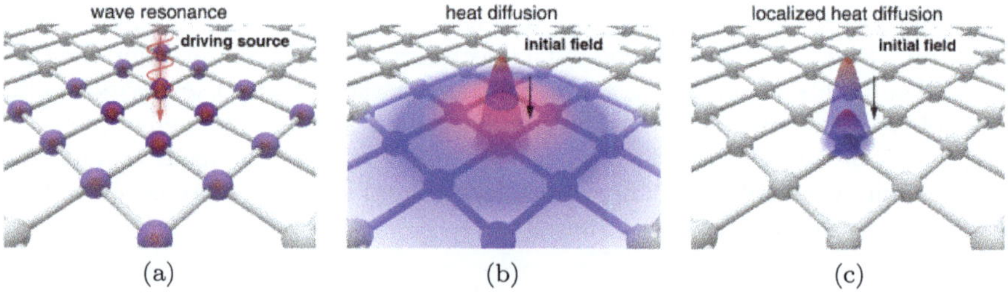

Fig. 7.5. Wave resonance and heat diffusion in lattice structures. (a) A wave mode localized on the units can be excited by a driving source at the resonant frequency. (b) Heat diffusion mode cannot be selectively excited. The field evolution is determined by the initial field and composed of extended modes. (c) A localized heat diffusion should contain just one localized mode that does not spread from its initial position. Adapted from Ref. [2].

diffusion modes are nearly impossible. Generally, transient thermal diffusion, as shown in Fig. 7.5b, involves the gradual spread and attenuation of localized heat spots. Such fields are constituted by many modes, among which those with lower decay rates and extensive ranges are observed predominantly. Since observable fields mainly comprise extended modes, confining diffusion modes within thermal materials akin to wave resonators is challenging. Therefore, field localization in thermal diffusion is rare, besides trivial instances in insulating components. Localized thermal diffusion, as shown in Fig. 7.5c, aids in the dissipating hot spot without affecting their adjacent areas; however, its realization and the underlying physical mechanisms remain elusive.

The topological edge state is a localized mode originating from the nontrivial band topology of a lattice structure. This bulk-boundary correspondence is a remarkable finding in various wave systems [19], which might be the awaited mechanism to localize a diffusion mode.

The continuum model and transfer matrix theory give band structures and geometric phases. The bulk-boundary correspondence [20, 21] in wave systems implies a topological edge state for nontrivial bulk bands. This work provides a comprehensive and deterministic relationship between geometric phases and edge states, thus being the first formal proof of the correspondence between body boundaries in thermal diffusion. The topological edge state was experimentally observed as local thermal diffusion. Different topological phases controlled by boundary conditions were predicted and demonstrated. A common problem in general heat transfer systems is the heat dissipation of hot spots. Thus, it is necessary to have a fast decay rate of hot spot temperature without heating the surrounding environment. The topological edge states satisfy the conditions well due to their local distribution of exponential decay. Therefore, they are expected to contribute to the effective heat dissipation of hot spots without affecting their neighborhoods.

The bulk-boundary correspondence links edge states with topological invariants, but it has not been demonstrated rigidly for heat diffusion. We provide rigorous proof based on our continuum model. Under the parameters in Fig. 7.3a, we take $N + 1 = 20$ units to form a chain and apply constant temperature boundary conditions $T_0 = 293.15\,\mathrm{K}$ to both ends.

We consider two cases of boundaries to construct a finite-length chain. For Case 1, we take the center of the inter-cell rod as the boundary of the chain, as shown in Fig. 7.6a. For Case 2, the rods at both ends are extended by $b_5/2$.

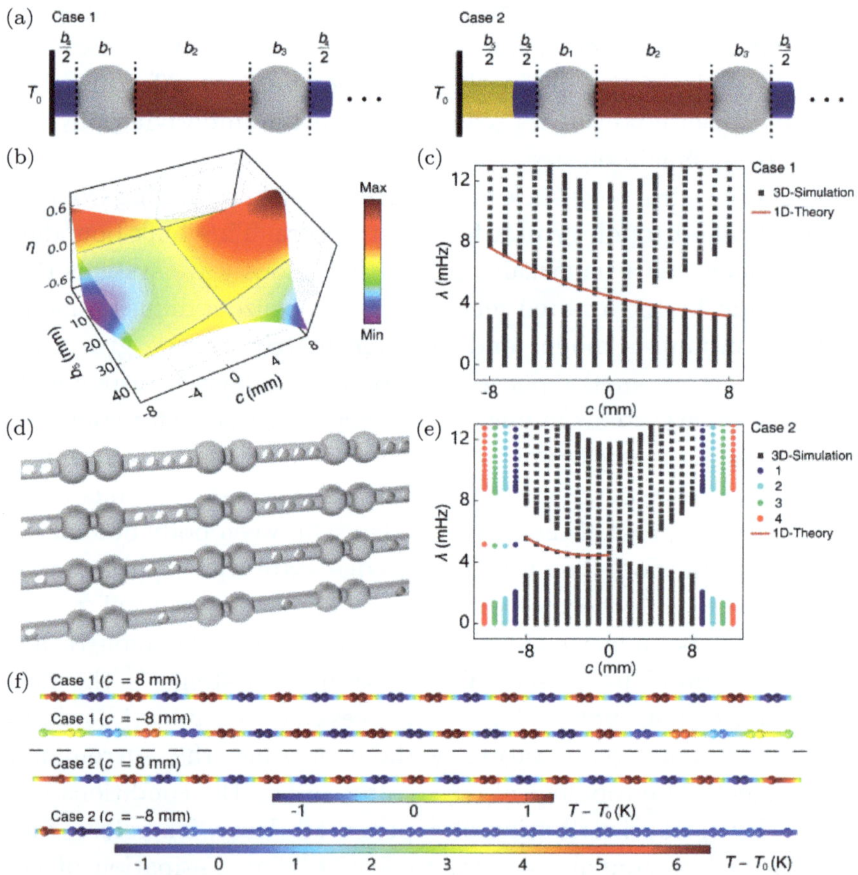

Fig. 7.6. Influence of boundary conditions and optimization. (a) Different boundary selections. (b) 3D diagram of η. (c) Eigenvalue distribution for Case 1. (d) Optimized models with 1–4 holes on the rods. (e) Eigenvalue distribution for Case 2. The colored scatters are the results for the models with holes on the inter-cell or intra-cell rods. (f) Temperature distributions of states in or at the edge of the bandgap. Adapted from Ref. [2].

As shown in Fig. 7.6c, no edge state exists in Case 1. For Case 2, we show that $[\text{Re}(M_{11})^2 - 1]^{1/2} = -U_1 \sin k_2 b_5$, which supports edge states when a solution exists or $U_1 \sin k_2 b_5 < 0$. Figure 7.6e shows the solutions for $b_5 = b_1 + b_4$, which agree with the numerical results.

It is convenient to introduce the indicator of edge state as $\eta = U_1 \sin k_2 b_5$. We note that U_1 should be nonzero in the bandgap, since $|M_{12}|^2 = \text{Re}(M_{11})^2 + \text{Im}(M_{11})^2 - 1 > 0$. Therefore, the sign of U_1 does not change inside the bandgap and is determined by its sign at the upper edge of the lower band $\text{sgn}(\underline{U}_1)$. Using Eq. (7.22), we have

$$\text{sgn}(\eta) = \exp(iZ)\text{sgn}(\sin k_2 b_5). \tag{7.23}$$

The system is thus proved to possess bulk-boundary correspondence, which depends on the extension length b_5. The dependence of η on b_5 and c is plotted in Fig. 7.6b. Edge state exists in regimes below the zero plane.

The dependence of bulk-boundary correspondence on boundary conditions is a distinctive aspect of 1D topological systems [22, 23]. A common approach to addressing these scenarios involves incorporating an edge potential to define the boundary conditions more precisely.

Expanding the bandgap interval is necessary to enhance the model's performance in the edge state. This expansion can be achieved by diminishing the intra-cell coupling strength and effectively introducing holes into the intra-cell rods when $c = -8\,\text{mm}$, as illustrated in Fig. 7.6d. The eigenvalue map, shown in Fig. 7.6e, indicates that the bandgap widens progressively with increased holes. Figure 7.6f displays the temperature fields for two scenarios with $c = \pm 8\,\text{mm}$. In Case 1, where there is no edge state, the temperature distribution across the model is periodic. In Case 2, the edge state emerges, resulting in a temperature field that is highly concentrated at the boundary when $c = -8\,\text{mm}$. These edge states hold the potential for developing an effective heat dissipation system. By strategically positioning the hot spot at a location corresponding to the boundary, rapid cooling of the hot spot can occur without impacting the surrounding areas.

Finite-element simulations complement the theoretical predictions, verifying the presence of edge states and the accuracy of the band structure calculations. The experimental setup measures the thermal response of the system under varying conditions, aligning closely with the theoretical predictions.

As presented in Fig. 7.7a, models I, II, and III have four holes with $c = -8$ mm, no holes with $c = 0\,$mm, and four holes with

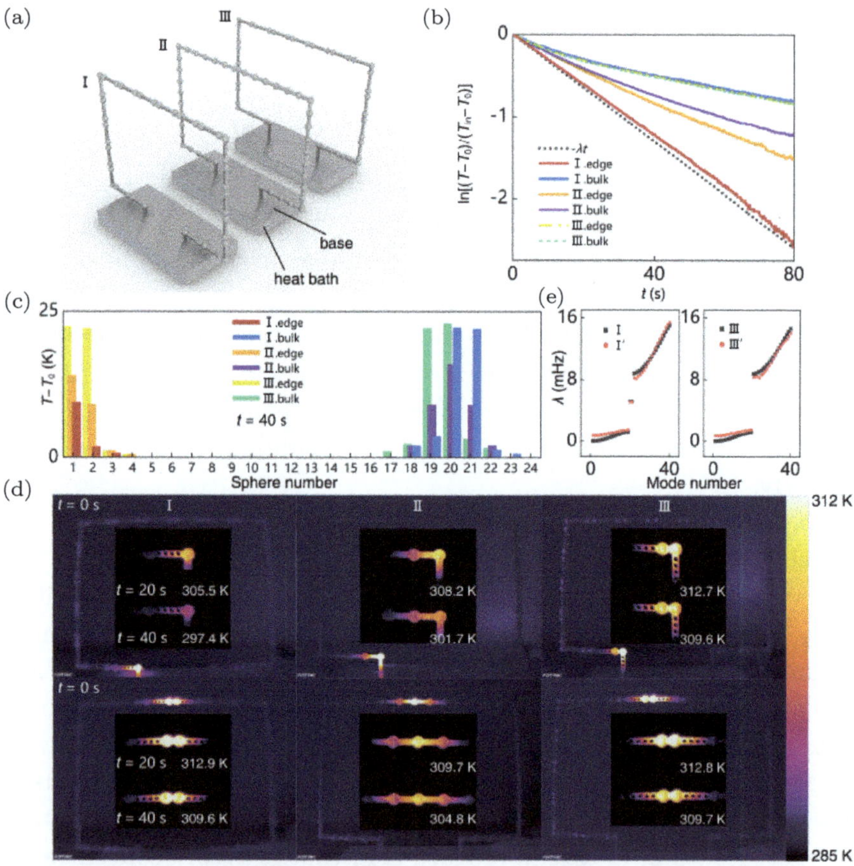

Fig. 7.7. Experimental setup and results. (a) Physical models of I, II, III. (b) Maximum temperature decay rates. (c) Temperature distributions of the models. The abscissa denotes the sphere number (1–40). (d) Thermal images. The values represent the highest temperature. (e) Measured (red) and simulated (black) eigenvalues of models I and III. Adapted from Ref. [2].

$c = 8\,\text{mm}$. According to theoretical predictions, the eigenstate of the system should decay exponentially with time, and its decay rate λ corresponds to the imaginary part of the eigenvalue.

The experimentally observed decay rates confirm the nontrivial topological characteristics predicted by the model, showcasing the potential of topological principles in designing systems with controlled heat diffusion properties.

The entire spectrum can be examined using a series of experimental measurements. We define a column vector as $\hat{T}(t) = (T_1, T_2, \ldots, T_{40})^{\mathrm{T}}$ to represent the temperature field (the room temperature is subtracted), with $T_n(t)$ referring to the temperature of the nth sphere at the time t. We perform 40 series of experiments. For the nth experiment, an initial temperature T_{in} is applied to the nth sphere, so we have $\hat{T}(t) = e^{-i\hat{\boldsymbol{H}}t}\hat{T}_n(0)$, where $\hat{\boldsymbol{H}}$ is a 40×40 matrix discretized from the Hamiltonian $\boldsymbol{H}$ in Eq. (7.8). Thus, the eigenvalues of $\exp(-i\hat{\boldsymbol{H}}t)$ should approximate $\exp(-\lambda t)$. As shown in Fig. 7.7e, the experimental eigenvalues align with the numerical predictions. This direct observation demonstrates that model I exhibits edge states, whereas model III does not, highlighting distinct topological phases.

While the sphere/rod configuration is a compelling example, the underlying principles can be extended to various other systems. The robust thermal topological mode boundary states identified in our study are versatile and can be adapted to diverse applications. Drawing from the insights gained from the 1D line model, we can reshape the sphere-rod model into different geometries. For instance, as depicted in Fig. 7.8, we adapted Case 2 with $c = -8\,\text{mm}$ into a LEGO-based model. This structure demonstrates that if the effective parameters align with our theoretical framework, the edge state of the thermal topology remains stable across various structural forms.

This work overcomes the above challenges and performs systematic studies on the topology of heat diffusion. We analytically obtain the band structure and geometric phase based on a continuum model without using the tight-binding approximation. A deterministic parameter is found to link the geometric phase with the edge state,

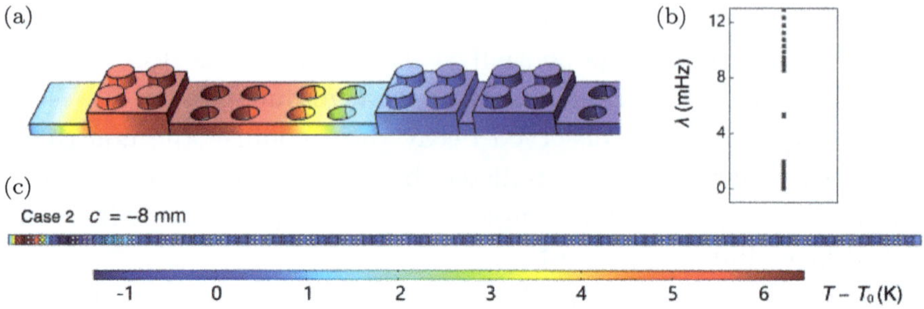

Fig. 7.8. The LEGO model. (a) Close-up of the edge of the LEGO model. (b) Eigenvalue of the LEGO model. (c) Temperature distribution of the edge state. Adapted from Ref. [2].

thereby proving the bulk-boundary correspondence for heat diffusion. It has been experimentally demonstrated that the topological edge state is a local thermal diffusion, and its dependence on boundary conditions has been verified. The approach is general, rigorous, and able to accurately reveal rich knowledge about the system. The findings established a solid foundation to explore the topology in novel thermal management applications. The localized heat diffusion may be further investigated to facilitate the design of novel thermal metamaterials and devices for efficient heat dissipation and thermal management.

We have explored the unique scenario of anti-Hermitian systems characterized by purely imaginary eigenvalues. Now, let us delve into the broader realm of non-Hermitian systems, where eigenvalues can be complex, encompassing both real and imaginary components. This transition enables us to examine the intriguing dynamics and phenomena that emerge when dissipative and conservative elements influence the system, providing a richer tapestry of physical behaviors and theoretical implications.

7.3.2. *Convection Model*

Exploring topological insulators [6, 24–27] has significantly advanced our understanding of new quantum states of matter. In solid-state physics, topological insulators have insulating interiors and conductive surfaces or edges, where conduction happens without dissipation. We already know that analogous phenomena have been investigated in thermal diffusion, where the conduction of heat, rather than electric current, is considered in terms of topological insulating phases. This study delves into how topological concepts influence and control thermal convection patterns in fluidic systems by designing specific lattice structures imposed by spatiotemporal advection.

Topological modes, typically associated with quantum systems and electronic materials, have found a surprising counterpart in thermal diffusion processes through spatiotemporally modulated advection. These advections, applied over a fluidic surface, create an effective periodic propagating potential akin to those seen in electronic topological insulators but within a purely thermal context. It was found that the Hamiltonian associated with classical thermal exchange conforms to a skew-Hermitian [28] relationship, suggesting the potential for creating topological insulating phases within thermal diffusion systems. However, the inherent lack of periodic potential and dynamic coherence in thermal transport presents a significant challenge. Replicating the periodic potential and coherent interference is crucial for a topological insulating phase in a purely dissipative diffusion system, similar to those in photonics and acoustics. This adaptation could rejuvenate Bloch's theory, paving the way for diffusive topological transport akin to its photonic and acoustic analogs [12, 29–38].

We discuss diffusion-based topological states enabled by spatiotemporally modulated advection layered over a fluidic surface, thereby simulating a periodic propagating potential within an effective thermal lattice. Observations include edge and bulk states within purely nontrivial and trivial lattices. The diffusive system exhibits interface states manifesting inhomogeneous thermal properties on the fluidic surface at interfaces between these lattice types. This work delves into the innovative study of diffusive topological transport in spatiotemporal thermal lattices, focusing on the implementation and theoretical underpinnings of topological phenomena. Experiments found that thermal topological modes with edge and interface states and traditional body states exhibit steady-state and deviating temperature distributions, respectively. These findings provide a feasible method for creating effective periodic lattices on the surface of advection fluids, bringing more opportunities and applications for topological thermal diffusion. The scheme of diffusion topology mode proposed in this work can further inspire the manipulation of general diffusion fields [39–43].

Topological insulators use a Hermitian Hamiltonian in quantum systems, but inherently dissipative and non-Hermitian thermal systems require a different approach. The Hamiltonian associated with classical thermal transport is a skew-Hermitian relation, suggesting engineering topological states in a system without the traditional periodic potential and dynamic coherence prerequisites. $\boldsymbol{H} = -\boldsymbol{H}^\dagger$ is the foundational adjustment in transitioning from quantum mechanical systems to thermal systems. While the matrix elements are complex, their non-Hermitian nature adequately represents the dissipative, irreversible thermal diffusion processes.

Consider a planar fluid surface with periodic advections in a simulated space, as depicted in Fig. 7.9a, to explore non-Hermitian topological modes in thermal diffusion. Due to the inherently dissipative nature of thermal diffusion, the system exhibits non-Hermiticity. For coherent resonance across this fluid surface, spatiotemporal modulation is applied to designated advective regions, highlighted with specific colors in Fig. 7.9a. These regions are organized into four modulated units that form an effective lattice. This four-unit lattice structure facilitates the creation of effective dimers and a four-band system through strategic decoupling of units influenced by the advections. This setup allows for examining diverse band properties, topological transitions, and midgap edge states across various parameters, offering a simpler arrangement than lattices with fewer units. A 1D chain is formed on the fluid surface by arranging these lattices periodically, enhancing the study of heat transfer processes between adjacent supercells via their interfaces in this virtual space,

$$\rho c \frac{\partial T_n}{\partial t} = \kappa \frac{\partial^2 T_n}{\partial x_p^2} \pm \rho c v_n \frac{\partial T_n}{\partial x_p} + \frac{h_n}{a_n}(T_{n+1} - T_n) \tag{7.24}$$

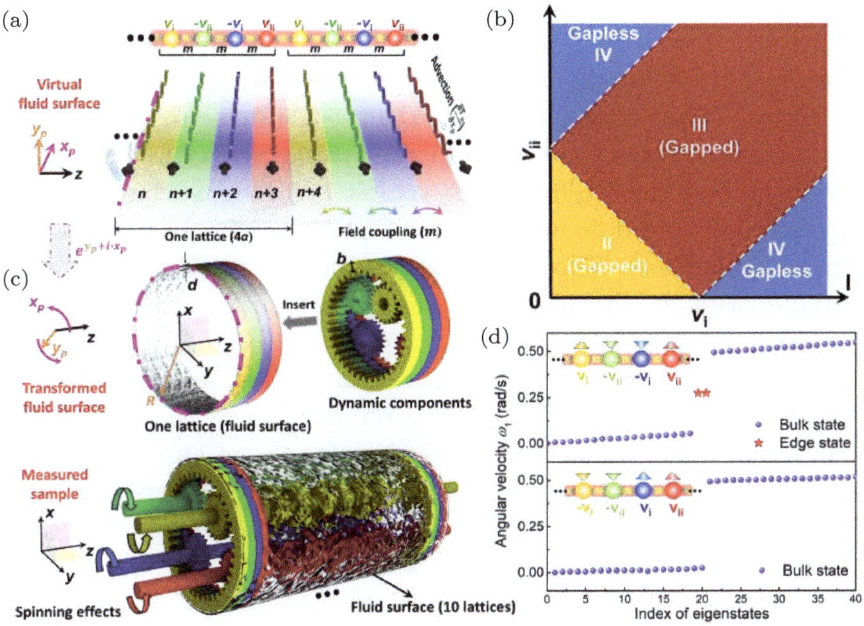

Fig. 7.9. Comprehensive overview of topological thermal diffusion. (a) A schematic setup where periodic advections are applied to a planar fluid surface in a virtual space, creating an effective "lattice" of four units, contributing to the topological structure. (b) The phase diagram identifies different phases based on the modulation of advections. (c) The practical model in a transformed spatial configuration, derived from the schematic via conformal mapping, with specific units demarcated by purple dash-dot lines, representing boundaries. (d) The effects on nontrivial and trivial lattice configurations in a gapped phase. Adapted from Ref. [16].

where ρ, c, and κ are the density, specific heat, and thermal conductivity of the planer fluid surface. T_n is the temperature of a particular unit of one lattice shown in Fig. 7.9a, while the subscript n indicates the unit index counting from 1. a_n and h_n are the width and convective heat transfer coefficient of each unit. Equation (7.24) is the general thermal energy equation describing heat transfer and exchange between two neighboring units with a joint interface.

Figrue 7.9 shows the initial setup of the thermal lattice, highlighting the arrangement of nodes and the application of initial temperature gradients. It sets the stage for understanding how boundary conditions and initial states influence heat flow evolution within the lattice. Introducing spatiotemporally modulated advection over a fluidic surface mimics a periodic potential essential for inducing topological behavior analogous to that seen in electronic systems. The modulation aims to create an effective lattice within the fluid, allowing the thermal diffusion to exhibit band-like structures. The interfaces between different advective configurations manipulate the edge and bulk states. These configurations lead to distinct topological characteristics at the interfaces, governed by the continuity and changes in advective parameters.

Figrue 7.10 captures the temperature profiles and distribution across the fluid surface, highlighting the distinct thermal behaviors corresponding to different topological states. Edge and bulk states manifest as variations in temperature distribution, underscoring the influence of topological configurations on thermal characteristics.

The thermal behaviors of edge and bulk states during different phases are depicted in Fig. 7.10. Temperature profiles, captured at various moments within the nontrivial lattice configuration, show robust and stable temperatures throughout the process. Yet, some deviations are noted near the system's boundaries (Fig. 7.10a,c). By analyzing the maximum temperature locations, they align with the first and last lattices at the boundaries, indicating a nonzero winding number. The system's temperature distribution remains unchanged across most units, except near boundaries where slight movements occur.

Like electrical resistance in topological circuits, thermal resistance varies significantly at the boundaries compared to other areas, indicating the presence of edge states (Fig. 7.10e,g). This effect is demonstrated by the larger resistance values at the boundaries, which affect the direction and rate of heat flow. Unlike the nontrivial configurations, the trivial lattice setups exhibit more pronounced temperature shifts from initial conditions, particularly near the boundaries, suggesting a stable bulk state without significant edge effects (Fig. 7.10b–g).

This study presents a novel method of exploring thermal topological modes for the first time by layering periodic advections onto

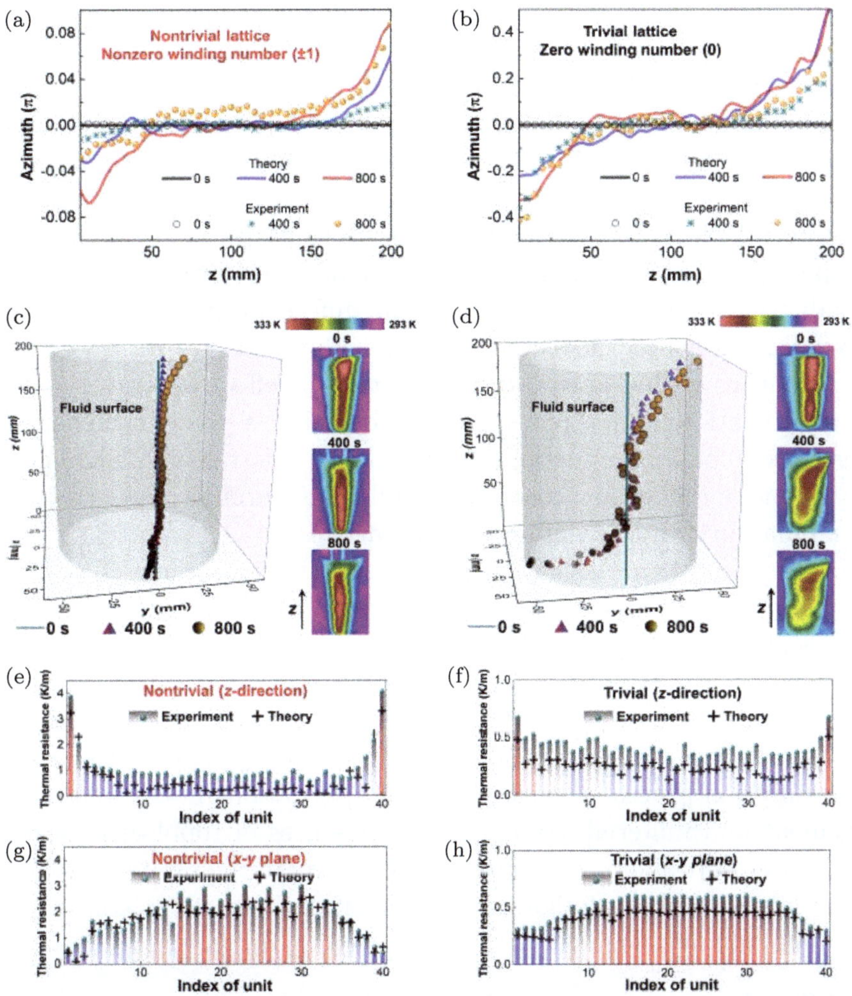

Fig. 7.10. Thermal behaviors in topologically nontrivial and trivial lattices during a gapped phase. (a,b) Azimuths of maximum temperature ($T_{\max}$) across various units on the fluid surface at specific times. (c,d) Visual representations of $T_{\max}$ locations and corresponding thermal images. In nontrivial lattices, the temperature profile remains confined along the x and y directions, but it expands significantly in the trivial lattice configurations. (e,f) Total effective thermal resistance along the z-direction for each lattice type. (g,h) Effective thermal resistances in the x-y plane for both lattice types, highlighting the differences in thermal response between nontrivial and trivial configurations. Adapted from Ref. [16].

a fluid surface. These modulated advections manipulate momentum space in a dissipative system like thermal diffusion, creating non-trivial and trivial lattice structures. Adjusting these advections in a gapped phase identifies various thermal topological states, including edge, interface, and bulk. This approach advances beyond traditional topological physics, which relies on oscillatory fields, providing a distinctive platform to investigate non-Hermitian topological modes in purely diffusive systems. A key technique involves generating a conduction-advection framework within different diffusive fields using actual or simulated advective elements. Suppose natural advection is absent, as with electrical conductivity. One effective strategy is to employ spatiotemporal parameters to formulate a practical advective term from basic conduction. Additionally, this research highlights a unique method for controlling diffusion through the strategic combination of nontrivial and trivial lattices.

The theoretical insights provided in this document showcase a remarkable synthesis of concepts from quantum topology and classical thermal dynamics. The study enriches the understanding of thermal transport and opens up new avenues for controlling and exploiting heat in novel technological applications. The findings of this study pave the way for new methods to manipulate heat transfer and diffusion processes, with potential applications in thermal management and material science. The robustness of topological states against defects and continuous variations in material properties offers a promising tool for designing more efficient thermal devices.

7.3.3. *Asymmetric Coupling Model*

Overall, the thermal lattice exploits unique topological properties in pure diffusion, convection, and asymmetric coupling modes to provide an efficient method of thermal management to ensure optimal thermal performance in various applications.

7.3.3.1. *Diffusive Skin Effect and Topological Heat Funneling*

Non-Hermitian Hamiltonians are essential for describing systems that interact with the energy of their surrounding environment.

It is worth noting that systems that maintain parity-time symmetry at outliers can support phenomena such as invisibility cloak [44], single-mode laser [45], and wireless power transmission [46, 47]. In addition, the modification of the system to break the symmetry of time reversal has led to the emergence of new equipment, such as gyroscopes, isolators, and circulators. In addition, the discovered non-Hermitian skin effect (with states located at the boundary) challenges traditional body boundary relationships [12, 13, 21, 48–53] and provides new possibilities for controlling unidirectional energy flow. This effect has been explored on multiple platforms, including optical lattices [54], mechanical devices [55], quantum networks [56], and electrical systems [57].

This work delves into integrating the skin effect, traditionally observed in non-Hermitian wave systems, into thermally diffusive systems. By employing asymmetrically coupled metamaterial channels, researchers demonstrate the feasibility of directing and controlling heat flow in a manner previously thought to be limited to non-dissipative systems. Additionally, the study introduces the phenomenon of topological heat funneling, where the temperature field is not only directed but also robustly maintained against perturbations and defects.

In diffusive systems, the non-Hermitian Hamiltonians that describe asymmetric couplings are imaginary due to their inherent dissipative characteristics, as illustrated in Fig. 7.11a. The temperature changes during energy transfer are not symmetrical across different components, leading to the diffusive skin effect. This effect enables eigenstates across all bulk bands to localize at the edges under open boundary conditions, a divergence from Hermitian systems where only edge modes localize at boundaries. Unlike wave systems, where asymmetric couplings are created through temporal adjustments or engineered gain/loss, in diffusive systems, such adjustments are straightforwardly achieved using direct contact with thermal metamaterials. Additionally, we introduce a method to develop a periodic non-Hermitian Hamiltonian from an aperiodic setup, underpinning the demonstrated skin effect-induced heat funneling, where temperature fields reliably concentrate at a

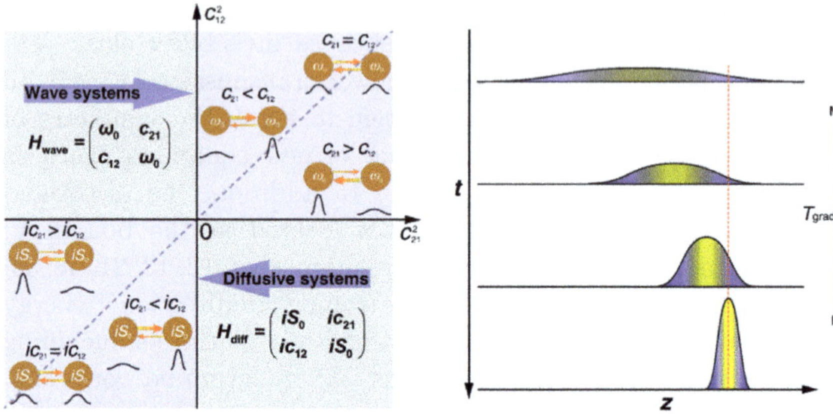

(a) (b)

Fig. 7.11. (a) Asymmetric coupling of fields in wave and thermal diffusion systems. Compared to the Hamiltonian of the wave system, the elements in the Hamiltonian of the thermal diffusion system all carry an imaginary i to represent the dissipation characteristics of the heat conduction process. (b) Conceptual diagram of the thermal funnel. The temperature field rapidly converges to a specified location and reaches a steady state. Adapted from Ref. [13].

predetermined area, shown in Fig. 7.11b, regardless of starting conditions and with high defect tolerance. These stable temperature gradients could advance thermoelectric energy production and harvesting. Furthermore, the pronounced sensitivity of the diffusive skin effect to boundary changes facilitates the development of topological thermal sensing [6].

To understand the asymmetric coupling diffusive systems intuitively, we start from a double-ring toy model experimentally demonstrated [58]. As shown in Fig. 7.12a, the two rings are vertically coupled in the z-direction through an interlayer. According to Fourier's law in heat conduction, the coupling equations can be written as [7, 58, 59]

$$\begin{aligned}\frac{\partial T_1}{\partial t} &= D_1\frac{\partial^2 T_1}{\partial x^2} + h_{21}(T_2 - T_1),\\ \frac{\partial T_2}{\partial t} &= D_2\frac{\partial^2 T_2}{\partial x^2} + h_{12}(T_1 - T_2),\end{aligned} \tag{7.25}$$

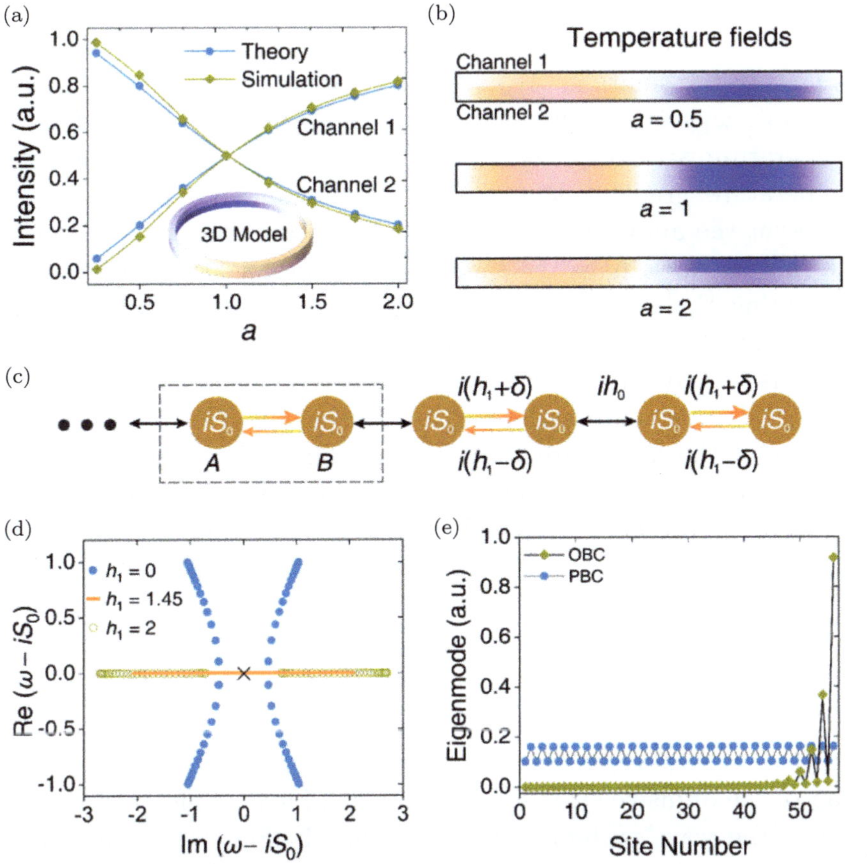

Fig. 7.12. Asymmetric coupling model and diffusive skin effect. (a,b) Temperature field intensities and distributions of two directly coupled channels versus the asymmetric factor a. (c) The diffusive tight-binding SSH model for OBC with asymmetric intracell coupling $i(h_1 \pm \delta)$ and symmetric intercell coupling ih_0. iS_0 is the unified decay rate on each site. (d) Spectra in the complex plane at different coupling parameters h_1. (e) Amplitude distributions of eigenstates for OBC and PBC. Adapted from Ref. [13].

where T_1 (T_2) is the temperature field in the two channels and $D_1 = \frac{\kappa_1}{\rho_1 C_1} \left(D_2 = \frac{\kappa_2}{\rho_2 C_2} \right)$ is thermal diffusivity with κ_1 (κ_2), ρ_1 (ρ_2), and C_1 (C_2) being the thermal conductivity, mass density, and heat capacity, respectively. x is the position along the channel. h_{12} and h_{21} are the heat exchange rates that introduce asymmetry into the system.

Since the temperature field in the ring structure is spatially periodic, Eq. (7.25) has a wave-like solution of $T_{1,2} = A_{1,2}e^{i(\beta x - \omega_{1,2}t)} + T_0$ [58]. In the solution, $A_{1,2}$ is the amplitude, β is the propagation constant, $\omega_{1,2}$ is the complex frequency, and T_0 is the reference temperature set to be zero for simplicity. Here, we employ the temperature gradient and position of maximum temperature to represent the amplitude and phase of the wave-like solution. As all the ring channels have the same size, the propagating constants of circulating "heat wave" are $\beta = m\frac{2\pi}{2\pi R} = \frac{m}{R}$, where m is the mode order and R is the radius of the ring. Usually, only the first-order mode is considered since it is very stable and can be selectively excited. Temperature variations in each channel can be considered uniform if the layers are thin enough. Thus, the heat transfer in these channels possesses coherent properties. Substituting the wave-like solution into Eq. (7.25) and considering the boundary continuity condition, we can deduce an effective Hamiltonian about $(A_1, A_2)^T$ [7, 58, 59]:

$$H_0 = \begin{pmatrix} iS_1 & ih_{21} \\ ih_{12} & iS_2 \end{pmatrix}, \tag{7.26}$$

where $S_1 = -(\beta^2 D_1 + h_{21})$ and $S_2 = -(\beta^2 D_2 + h_{12})$. The Hamiltonian of the diffusive system is imaginary, which is different from the one in the wave system (real), as shown in Fig. 7.11a. Moreover, when the material parameters $\rho_1 C_1 \neq \rho_2 C_2$, the ring coupling becomes asymmetric with $ih_{21} \neq ih_{12}$. The main diagonal elements in Eq. (7.25) can be unified into iS_0 by adjusting the diffusivities $D_{1,2}$ along with $h_{21,12}$, where $ih_{21,12}$ are replaced by $ic_{21,12}$ in Fig. 7.11a. The amplitudes in channels are asymmetric in the steady state. They are always time-dependent with the factor of $e^{-i\omega_\pm t}$ due to the system being dissipative. Therefore, our structure does not require dynamic modulation or gain-loss engineering to achieve asymmetric couplings.

Using non-Bloch band theory and taking the densities of directly coupled channels varying with a ratio of a^2, Fig. 7.12a,b displays the asymmetric coupling temperature fields with different a. We can connect the asymmetric coupling unit cells with symmetric couplings to implement the non-Hermitian diffusive SSH model [50], as shown in Fig. 7.12c. Each unit cell consists of sublattices (A + B), where the asymmetric intracell couplings and symmetric intercell coupling are $i(h_1 \pm \delta)$ and ih_0, respectively. For the open boundary condition (OBC), the Hamiltonian is [21, 50]

$$H_{\text{OBC}} = iS_0 \sum_l [\hat{A}_l^\dagger \hat{A}_l + \hat{B}_l^\dagger \hat{B}_l] + \sum_l [i(h_1 + \delta)\hat{B}_l^\dagger \hat{A}_l + i(h_1 - \delta)\hat{A}_l^\dagger \hat{B}_l + ih_0(\hat{B}_l^\dagger \hat{A}_{l+1} + \hat{A}_l^\dagger \hat{B}_{l-1})] \tag{7.27}$$

where $\hat{A}_l^\dagger$ and $\hat{B}_l^\dagger$ are the creation operators of A and B sites in the lth period. For the periodic boundary condition (PBC), the Bloch-mode Hamiltonian in momentum space is expressed into

$$H_{\text{PBC}}(K) = iS_0\hat{I} + i(h_1 + h_0 \cos K)\hat{\sigma}_x + i(h_0 \sin K - i\delta)\hat{\sigma}_y. \tag{7.28}$$

The corresponding amplitudes of eigenstates for PBC and OBC at the exceptional point (EP) are shown in Fig. 7.12e.

Figure 7.13a displays a heat funneling setup featuring two mirrored SSH chains intersecting at a designated interface, with five

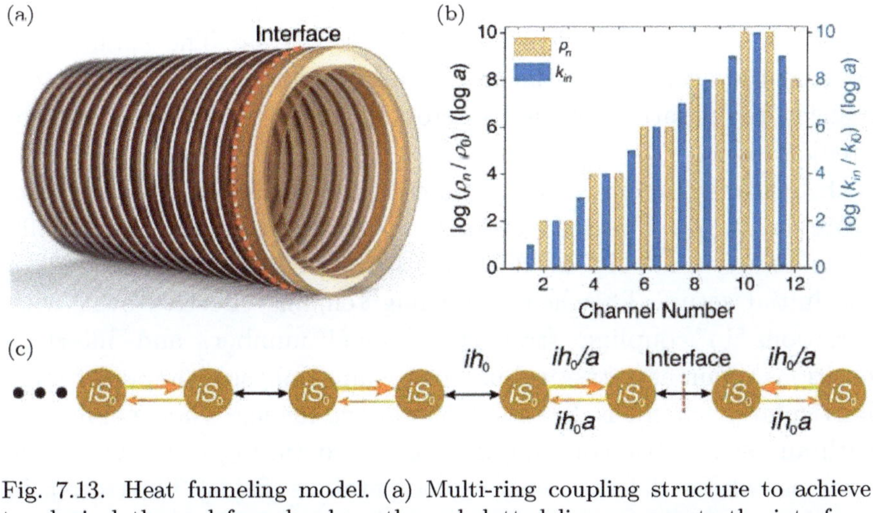

Fig. 7.13. Heat funneling model. (a) Multi-ring coupling structure to achieve topological thermal funnel, where the red dotted line represents the interface. (b) Distribution of ring density and inter-ring coupling layer thermal conductivity. (c) Equivalent tight-binding model, where the asymmetric coupling coefficients are ih_0a and ih_0/a. Adapted from Ref. [13].

periods on the left chain and one on the right. The system uses thermal metamaterials for the ring channels and coupling interlayers, whose properties such as thermal conductivity, mass density, and heat capacity are varied according to the channel number. The propagation constant of temperature fields is defined, and coupling coefficients between channels are specifically arranged to maintain translation symmetry in the non-Hermitian diffusive SSH model. This configuration is in Fig. 7.13b.

Achieving the required gradient parameters in natural materials is challenging but can be addressed using composite metamaterials according to the effective medium theory. The desired effective material parameters are achieved by adjusting the doping rates of materials with highly contrasting properties. The asymmetric and symmetric couplings within the cells are adjusted to form a uniform eigenfrequency across the system, as demonstrated in the semi-infinite model (Fig. 7.12c) and further explored in Fig. 7.13c.

The progression of temperature fields is examined in a symmetrically coupled structure with random stimulation to validate the topological heat funneling concept. The temperature naturally focuses toward the center channels, as the eigenmode with the lowest decay rate is activated, as shown in Fig. 7.14a. The simulations confirm the accuracy of the tight-binding model, aligning the normalized temperature gradients across channels with the theoretical predictions.

In contrast, under asymmetric coupling conditions, the temperature fields are directed towards a specific interface, specifically channels 10 and 11, as depicted in Fig. 7.14b, irrespective of the initial setup. This heat funneling remains effective even with variations in coupling strength, channel number, and interface positions, demonstrating robustness against physical imperfections. Particularly, channels 2 and 3 exhibit resilience when configured with an asymmetric coupling defect, differentiating from symmetric systems where defects localize temperature fields, illustrated in Fig. 7.14c,d.

Further studies, shown in Fig. 7.14e,f, underscore the consistency of the heat funneling effect with theoretical models, as previously

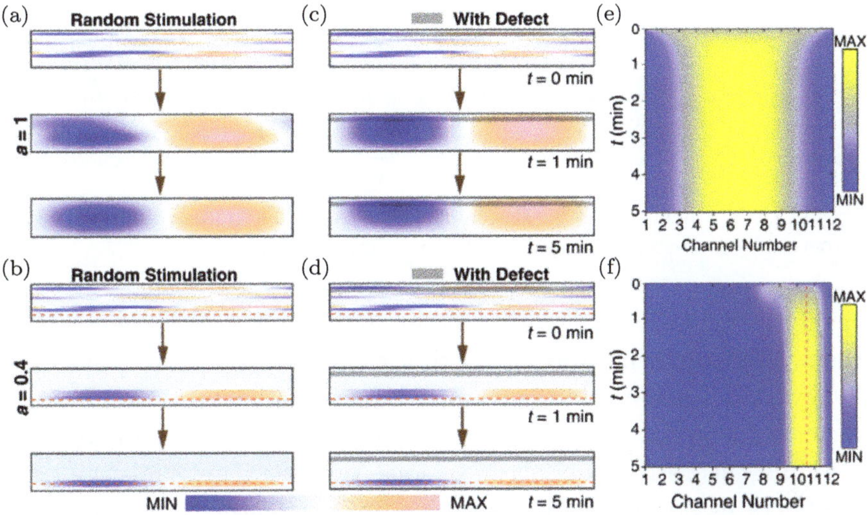

Fig. 7.14. Heat funneling effect. (a,b) Temperature field evolution with time under symmetric coupling and asymmetric coupling. (c,d) The influence of coupling defects (gray shading) on the evolution of the temperature field. Compared to symmetric coupling, the thermal funnel effect induced by asymmetric coupling is more robust to coupling defects. Adapted from Ref. [13].

outlined in Fig. 7.11b. These observations are grounded in a gapless system with consistent intracell and intercell couplings, extending to scenarios where intercell couplings exceed asymmetric intracell couplings, focusing heat in channel 12 due to uniform heat flow directions.

This research explores the mechanisms behind the diffusive skin effect and topological heat funneling, demonstrating these phenomena through directly contacted thermal metamaterials, unlike wave systems. It details how a periodically driven Hamiltonian can be established within an aperiodic structure, showing how diffusive systems uniquely facilitate the study of topological physics and non-Hermitian dynamics. The findings suggest that this approach provides a novel platform for investigating advanced effects like higher-order diffusive skin effects, topological heat flow transfer, and efficient thermoelectric generation. Furthermore, it opens up possibilities for practical applications in heat harvesting and thermal

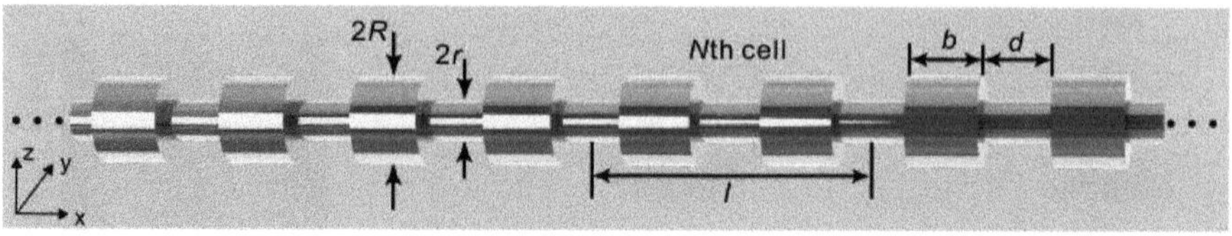

Fig. 7.15. A schematic of a 1D cavity chain for the diffusive NHSE. The geometrical parameters are marked. Adapted from Ref. [60].

sensing, previously challenging areas in thermal management. This study highlights the special attributes of non-Hermitian systems in new applications and suggests broad implications for energy efficiency, material science, and thermal management technologies.

While theoretical models have long predicted unusual topological effects in non-Hermitian systems, experimental verification has remained elusive, especially in diffusive systems. The next study bridges this gap by demonstrating the non-Hermitian skin effect (NHSE) experimentally in a system characterized by asymmetric temperature field couplings facilitated through precise material parameter modulation.

The experimental setup consists of a 1D lattice of thermally conductive rods interconnected by thermal coupling sticks (Fig. 7.15a), forming a lattice structured to facilitate NHSE. The lattice exhibits asymmetric intra-cell coupling coefficients, which are essential for inducing NHSE while maintaining the uniformity of inter-cell coupling coefficients. Thermal properties such as conductivity and specific heat capacities are strategically varied across the lattice to control the temperature field dynamically, aiming for concentration at a target boundary. The entire setup is placed in a vacuum chamber to minimize external thermal interference, ensuring the integrity of the experiment.

A significant hurdle is establishing specific heat capacity and thermal conductivity to experimentally demonstrate the diffusive NHSE. This challenge was addressed using thermal metamaterials to engineer effective coupling coefficients. The experimental setup, depicted in Fig. 7.16a, utilized an aluminum alloy (AlSi10Mg) sample of eight rods linked with coupling sticks configured to

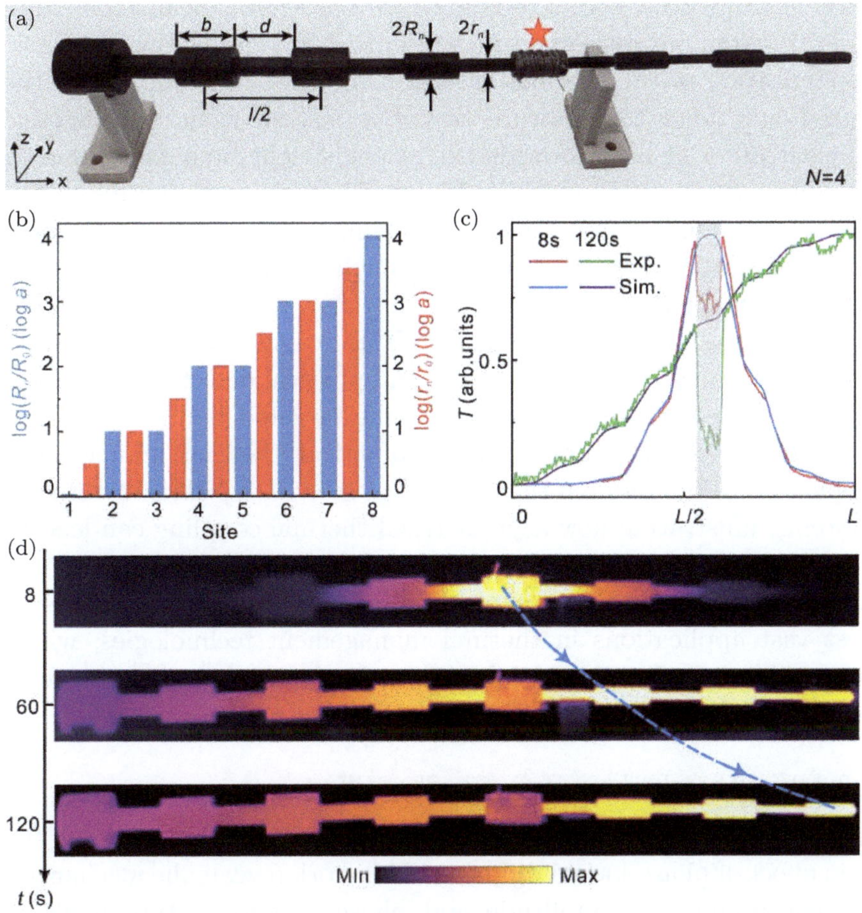

Fig. 7.16. Experimental observation of diffusive NHSE. (a) Photograph of the 3D-printed samples. The red asterisk shows the location of the initial heat excitation. (b) Variation of site rods and coupling sticks at different sites. (c) Experimentally measured and numerically calculated temperature fields. (d) Thermography of the heated sample at different times. Adapted from Ref. [60].

facilitate left-to-right and right-to-left thermal couplings. Precise structural parameters, including rod and stick dimensions set to optimize 3D metal printing capabilities, were crucial for maintaining the translation symmetry essential for observing the NHSE. The experiment was conducted in a vacuum chamber to minimize external thermal influences, with a resistance wire providing localized heating to initiate temperature variations. Figure 7.16b details the geometric configuration of the thermal lattice, while Fig. 7.16c illustrates the initial and final temperature distributions, showing the effective concentration of heat towards the lattice's right boundary, thereby visually confirming the NHSE. Figure 7.16d presents thermographs at various times, visually tracing the temperature field's progression towards the designated boundary, further affirming the NHSE's presence in the experimental setup.

The experiment highlighted the NHSE's robustness, where despite various initial excitation conditions and physical defects, the temperature fields consistently localized at the targeted boundary, underscoring the NHSE's potential for practical thermal management applications. It validated the theoretical models of NHSE in diffusive systems, illustrating how asymmetrical thermal coupling can lead to controlled and predictable thermal behavior, promising significant advancements in thermal regulation technologies. The findings suggest vast applications in thermal management technologies, where controlling heat flow is crucial, such as in electronic cooling systems or energy-efficient building designs.

7.3.3.2. *Phase-Locking Diffusive Skin Effect*

Combining convection and asymmetric coupling further yields the skin effect of phase-locked diffusion. This work reveals the mechanism for controlling the amplitude and phase of temperature fields in thermal coupled systems and has the potential for non-Hermitian topology in thermal diffusion.

There are two methods to achieve non-Hermitian wave systems. One introduces gain and/or loss materials, and the other uses asymmetric coupling [6, 9, 62–64].

In inherently dissipative or non-Hermitian thermal systems, wave oscillations can be effectively simulated through counteracting thermal convection [6, 12, 23–28]. For instance, anti-parity-time symmetry and the corresponding phase transitions at EPs are achieved by employing two interconnected, oppositely directed coupled-ring channels [7]. Before the EP, temperature fields within these anti-parity-time symmetric systems become dynamically localized, maintaining a consistent phase difference at the point of maximum temperature and a balanced distribution of temperature gradients across both channels. Although heat and temperature differ fundamentally, asymmetric thermal coupling can be achieved by connecting channels made from different materials, which does not violate the principles of reciprocal heat conduction [16]. Adjusting the mass densities within these channels allows the amplitude distribution to be easily controlled. Thus, the non-Hermitian SSH model incorporating asymmetric coupling was developed, confirming the existence of the diffusive skin effect [13, 52, 65, 66]. Unlike systems with symmetric coupling [17, 49], where bulk and edge modes are distinct, in asymmetrically coupled systems, all modes localize at the boundaries when open boundary conditions are applied, a phenomenon known as the diffusive skin effect.

The asymmetric coupled dual channel model is fundamental in which the temperature field exhibits unbalanced amplitude and phase locking before reaching the anomaly point (EP). They extended this concept to 1D tightly bound non-Hermitian lattices and studied the phase locking and diffusion skin effects caused by convection and asymmetric coupling, respectively. Finally, the skin effect of phase-locked diffusion was demonstrated by combining complex thermal convection and asymmetric coupling. This work demonstrates the unique temperature field modulation mechanism in coupled diffusion systems, signifying the promising potential of thermal topology.

As depicted in Fig. 7.17, two channels, each with distinct mass densities, are set to move in opposite directions at specified velocities of $\mp v$. These channels, identified by their lengths L and widths b, exhibit temperature fields $T_{1,2}$ in the upper and lower channels, respectively. These fields are dynamically localized, showing a consistent phase difference and varied amplitudes. The interaction of these temperature fields can be analyzed using diffusive coupled mode theory, suggesting that the temperature distribution in the ring-like structure of the channels mimics a wave-like behavior $T = Ae^{i(\beta x-\omega t)} + T_0$ [23]. A symbolizes the amplitude, indicating the temperature gradient across the channel, β is the wave propagation

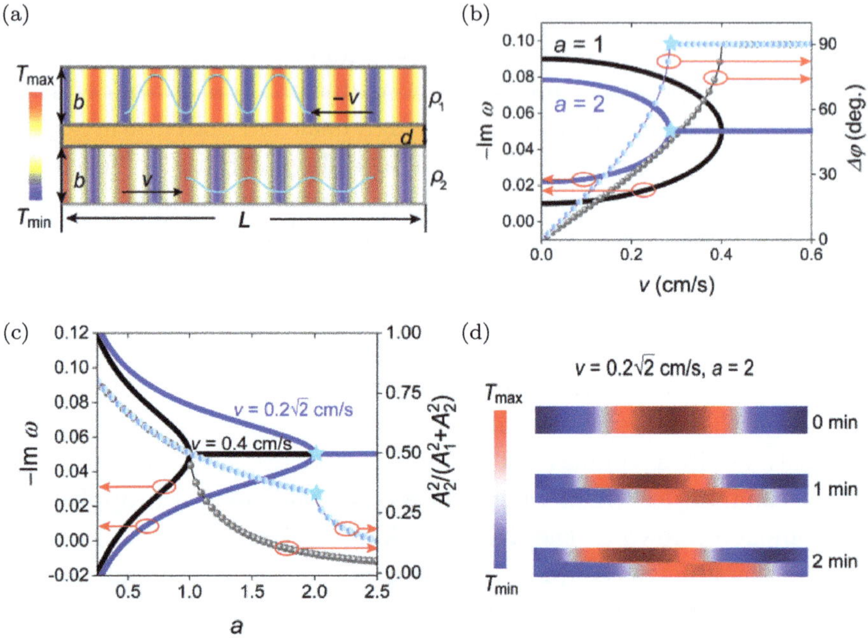

Fig. 7.17. (a) Schematic of the double-channel system with opposite convection v ($-v$) and asymmetric coupling. The mass-density ratio determines the asymmetric coupling $a = \rho_2/\rho_1$. The temperature field is asymmetrically distributed and phase locked in channels. (b) Decay rate ($-\text{Im}\omega$) and the phase difference ($\Delta\varphi$) as a function of the convection velocity. (c) Decay rate and intensity distribution *vs.* the asymmetric coupling. (d) Temperature field evolution at the EP in the coupled channels. Adapted from Ref. [14].

constant, ω represents the decay rate, and T_0 is the baseline temperature. For simplicity, this discussion considers only the first-order mode and sets T_0 to zero. Following Fourier's law of heat conduction, the coupled mode equation is derived to explore the dynamics of this thermal system [23],

$$\rho_{1,2}C_{1,2}\frac{\partial T_{1,2}}{\partial t} = \kappa_{1,2}\frac{\partial^2 T_{1,2}}{\partial x^2} \pm \rho_{1,2}C_{1,2}v\frac{\partial T_{1,2}}{\partial x} + h_{s1,s2} \tag{7.29}$$

where $C_{1,2}$ and $\kappa_{1,2}$ are the heat capacities and thermal conductivities in the upper and lower channels, respectively, while $T_{1,2}(x,t)$ denotes the temperature fields within these channels. Equation (7.29) specifies t as the time variable and x as the axis of convection. Given the narrow width of the thermal channels, $T_{1,2}(x,t)$ is assumed to be uniform along the axis of coupling. The effective Hamiltonian for the system, incorporating non-Hermitian elements, is expressed as

$$\boldsymbol{H} = \begin{bmatrix} -i(\beta^2 D_1 + h_1) + \beta v & ih_1 \\ ih_2 & -i(\beta^2 D_2 + h_2) - \beta v \end{bmatrix} \tag{7.30}$$

where $D_{1,2} = \frac{\kappa_{1,2}}{\rho_{1,2}C_{1,2}}$ are diffusivities and $h_{1,2}$ are heat exchange rates. Note that $PT\boldsymbol{H} = -\boldsymbol{H}PT$ is satisfied only at $h_1 = h_2$. Thus, the Hamiltonian is not anti-parity-time symmetric in a general case,

$$\begin{aligned} u_\pm &= (\pm\sqrt{h_1/h_2}e^{\pm i\varphi}, 1)^T, \quad (v = v_{\mathrm{EP}} \sin\varphi \le v_{\mathrm{EP}}), \\ u_\pm &= (\pm\sqrt{h_1/h_2}e^{\pm\phi}, i)^T, \quad (v = v_{\mathrm{EP}} \cosh\phi \ge v_{\mathrm{EP}}). \end{aligned} \tag{7.31}$$

EP locates at $v_{\mathrm{EP}} = \frac{\sqrt{h_1 h_2}}{\beta}$, where eigenvalues and eigenvectors are both degenerated. In the state before EP ($v \le v_{\mathrm{EP}}$), the eigenvalues are imaginary, and the coupling dominates the diffusion process. In this case, phase difference $\Delta\varphi$ is only related to the convection velocity v, and the amplitude keeps at $\sqrt{h_1/h_2} = \sqrt{\rho_2/\rho_1} = \sqrt{a}$. In the state after the EP ($v \ge v_{\mathrm{EP}}$), eigenvalues begin to have real parts. $\Delta\varphi$ holds at $\pi/2$ and $\sqrt{h_1/h_2}$ varies with the convection.

Figure 7.17b–d shows a simulation example. EPs exist for symmetric and asymmetric coupling cases. On condition of eigenstate excitation, the phase difference between the two coupled ring channels changes from 0 to $\pi/2$ at $v = 0 \to v_{\mathrm{EP}}$ and keeps $\pi/2$ at $v > v_{\mathrm{EP}}$. Here, the EPs are located at $v_{\mathrm{EP}} = 0.4\,\mathrm{cm/s}$, $0.2\sqrt{2}\,\mathrm{cm/s}$, with $a = 1, 2$, respectively. The intensity distributions of temperature fields (A_1^2 and A_2^2) in channels 1 and 2 are presented. Interestingly, the intensity distributions of temperature fields in the upper and lower channels can be modulated from near 0/1 to almost 1/0 in succession, balanced at 0.5/0.5 with the symmetric couplings of $a = 1$. For each case, the EP connects the regions of different intensity change rates concerning the variation of mass density ratio. Figure 7.17d shows the transient temperature field evolution in the channels when $v = v_{\mathrm{EP}} = 0.2\sqrt{2}\,\mathrm{cm/s}$ and $a = 2$. The result shows that the temperature field reaches a steady state finally, with the phase lag fixed at $\pi/2$ and the field amplitude asymmetric.

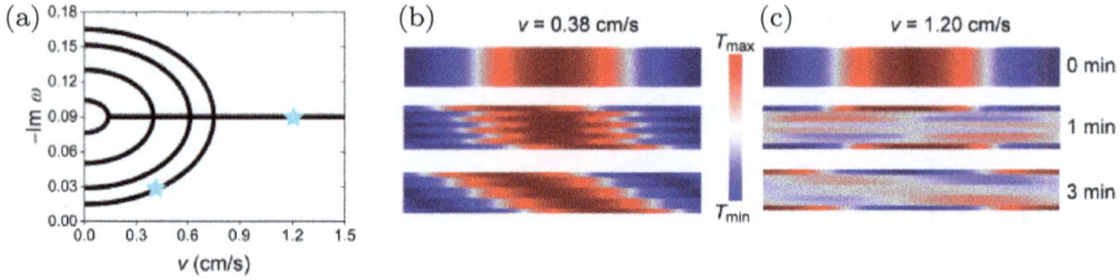

Fig. 7.18. Convection-induced phase-locking effect in 1D coupled thermal channels of the particle-hole symmetry. (a) Decay rate *vs.* the convection velocity. (b) Phase-locking effect for the uniform excitation before the EP at $v = 0.38$ cm/s, as marked by the star in (a). (c) Boundary localization of temperature fields for the uniform excitation after the EP at $v = 1.20\,\text{cm/s}$. Note that the excited branch of the eigenstate is in the lowest decay rate, and the phases are unlocked in the case after the EP. Adapted from Ref. [14].

In practical scenarios, the preferred excited state corresponds to the eigenstate characterized by the minimal decay rate. In Fig. 7.18a, this specific branch, indicated by stars at various phases, is selected for excitation. The findings show that the system maintains phase lock under uniform excitation and displays consistent amplitude distribution in the steady state before the EP. As illustrated in Fig. 7.18b, a specific lag phase exists. Beyond the EP, as depicted in Fig. 7.18c, the system exhibits edge field localization with no consistent phase difference across the channels when uniformly excited. It is important to note that this edge localization, driven by convection, lacks topological protection and can be disrupted by irregularities in the convection flow.

By adjusting the convection velocity and asymmetric coupling, one can effectively modulate both the phase and amplitude of temperature fields in a non-Hermitian lattice. This property is detailed through a Hamiltonian that integrates both onsite convection in each channel and asymmetric couplings between them. The phase diagram, as depicted in Fig. 7.19a, distinctly shows the impact of these parameters on the system's behavior, particularly highlighting an asymmetric coupling scenario. The decay rates, detailed in Fig. 7.19b, illustrate how thermal convection influences the system, splitting a static nth-order EP into four second-order EPs.

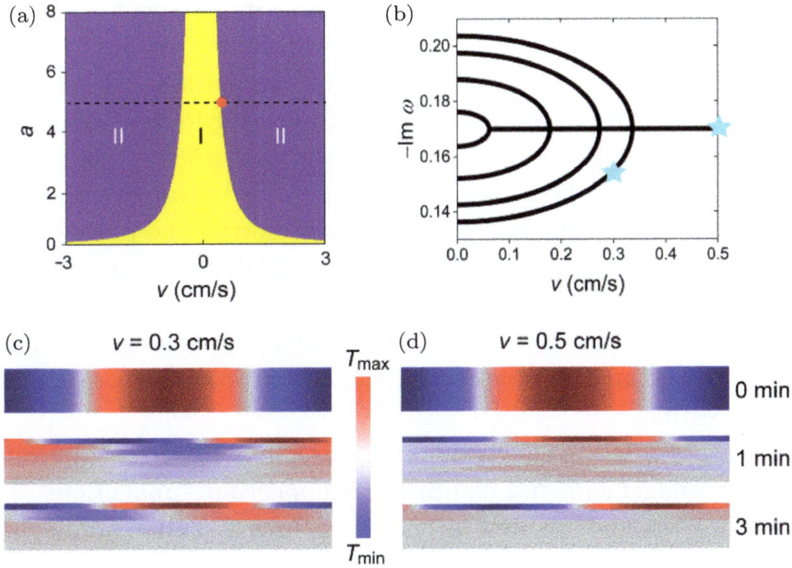

Fig. 7.19. The phase-locking diffusive skin effect. (a) Phase diagram *vs.* the convection velocity and asymmetric coupling. Phase I before the EP (at the branch with the lowest decay rate in (b)) is distinguished from phase II after the EP by different colors. (b) Decay rate at the asymmetric coupling $a = 5$ *vs.* the convection velocity. (c) Phase-locking diffusive skin effect at phase I. (d) Diffusive skin effect at phase II, where the phase difference between the adjacent channels is no longer locked. The cases in (b,d) are located on the dashed line in (a), where the red dot in (a) denotes an EP. Adapted from Ref. [14].

Exciting the eigenmodes at the branch of the lowest decay rate reveals two distinct phases: phase I before the EP shows a phase-locking diffusive skin effect, where temperature fields in adjacent channels exhibit a fixed phase lag and localized amplitude at the edges, as shown in Fig. 7.19c. Conversely, phase II after the EP, depicted in Fig. 7.19d, shows disordered phase relationships across channels due to increased thermal convection, though the underlying diffusive skin effect remains intact despite these changes. This robustness underscores the adaptability of the system under varying thermal dynamics.

This study elucidates the crucial role of asymmetric coupling in achieving non-reciprocal thermal effects akin to phenomena observed in non-Hermitian wave systems. Introducing convection as a bias field significantly enhances the control over thermal field distributions, enabling applications such as thermal cloaking and heat flow modulation. The observed high-order EPs are particularly significant for applications in precise thermal sensing, showcasing how thermal system behavior can be finely tuned through parameters like convection velocity and coupling asymmetry. The phase-locking diffusive skin effect illustrates the potential for controlling the amplitude and phase of temperature fields through non-Hermitian topology, which could be crucial for developing advanced thermal devices exploiting thermoelectric effects for power generation and heat harvesting.

These topological effects highlight the potential for innovative thermal management and equipment engineering applications. This feature has opened up new avenues for micro- and macro-control of heat flux, which is significant for future technological progress in energy efficiency and materials science.

7.4. Higher-Dimensional Topological States

Higher-order topological insulators have been realized in classical wave systems, such as photonics [67, 69], acoustics [69, 70], and electronics [71]. They exhibit essential differences from traditional 1D or 2D topological insulators in multiple aspects, especially in their applications in systems such as thermal lattices. These differences are reflected in physical properties and potential advantages in technology and application. Higher-order topological insulators exhibit more complex boundary state behaviors. For example, second-order topological insulators do not exhibit topologically protected edge states on 2D boundaries but have localized topological states at lower-dimensional corners, namely corner states. Third-order topological insulators do not possess topological states at the edges of 3D structures but appear on 1D lines or 0D angles. Compared with 1D topological insulators, high-order topological insulators exhibit higher design flexibility, stronger localization control, and better thermal stability and isolation performance in thermal topology applications. These characteristics make high-order topological materials important in future technological applications, especially in advanced materials science and nanotechnology.

7.4.1. *Pure Diffusion Model*

The study advances the theoretical framework for topological insulators in non-electronic systems and demonstrates practical applications in thermal management. The ability to localize and control heat dissipation through topological states opens up new avenues for designing advanced thermal materials and devices.

The 1D pure diffusion model mentioned above is transformed into a new model of a high-dimensional topological insulator, showcasing the distinct behaviors of in-bulk corner states compared to traditional in-gap states, with implications for enhanced thermal management.

Figure 7.20a illustrates a sphere-rod structure arranged in a 2D array. The intra-cell and inter-cell rods are distinguished by their

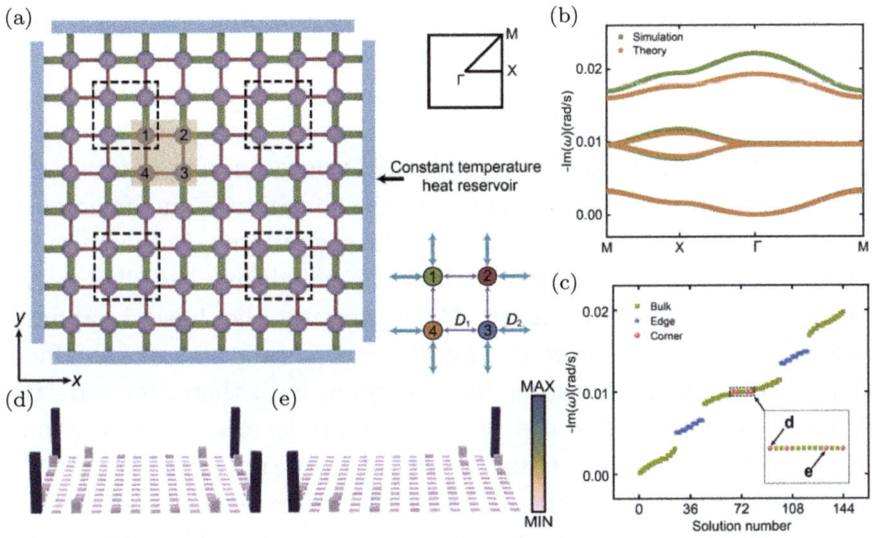

Fig. 7.20. Diffusive SSH model with in-bulk corner states. (a) Schematic diagram of sphere-rod structure with four constant temperature boundaries set at room temperature. (b) Band structure. (c) Fixed boundary condition spectrum. (d,e) The mode distributions of two in-bulk corner states. Adapted from Ref. [71].

radii, $R_{0,1}$ and $R_{0,2}$, respectively, and R denotes the sphere radii. To ensure consistent diffusivity at the boundaries, we implement a fixed boundary condition by connecting the outer rods to constant temperature heat reservoirs. This arrangement results in a discretized thermal lattice system that effectively mirrors the 2D SSH model's dynamics in a diffusive setting. The Bloch Hamiltonian of this system, derived after applying a Fourier transformation, encapsulates the essence of this system,

$$\hat{H}(\boldsymbol{k}) = (-i) \times \begin{pmatrix} 2(D_1+D_2) & -D_1-D_2e^{ik_x} & 0 & -D_1-D_2e^{ik_y} \\ -D_1-D_2e^{-ik_x} & 2(D_1+D_2) & -D_1-D_2e^{ik_y} & 0 \\ 0 & -D_1-D_2e^{-ik_y} & 2(D_1+D_2) & -D_1-D_2e^{-ik_x} \\ -D_1-D_2e^{-ik_y} & 0 & -D_1-D_2e^{ik_x} & 2(D_1+D_2) \end{pmatrix} \tag{7.32}$$

where D_1 and D_2 are thermal diffusivities of intracell and intercell rods.

These diffusivities are determined by aligning the simulated eigenvalues of the sphere-rod structure with those predicted by the tight-binding model. The model's band structure exhibits two band gaps, specifically between the first and second bands and the third and fourth bands, with notable hybridization occurring between the second and third bands. This arrangement facilitates the emergence of higher-order in-bulk corner states. The eigenvalue ω in this thermal diffusion context is purely imaginary, highlighting the dissipative nature of the system. Figure 7.20c presents the fixed boundary condition spectrum, showing four thermal corner states embedded within the bulk states rather than merging with them, termed "in-bulk corner states." These states are notably localized at either four or two corners of the structure, as detailed in Fig. 7.20d,e.

Experiments are conducted to observe the body angle state within the baseball bat structure, as shown in Fig. 7.21a. The experiment focused on a specific corner sphere, exciting it while applying fixed boundary conditions along two intersecting boundaries at this corner. This setup altered the C4 symmetry, embedding a single thermal

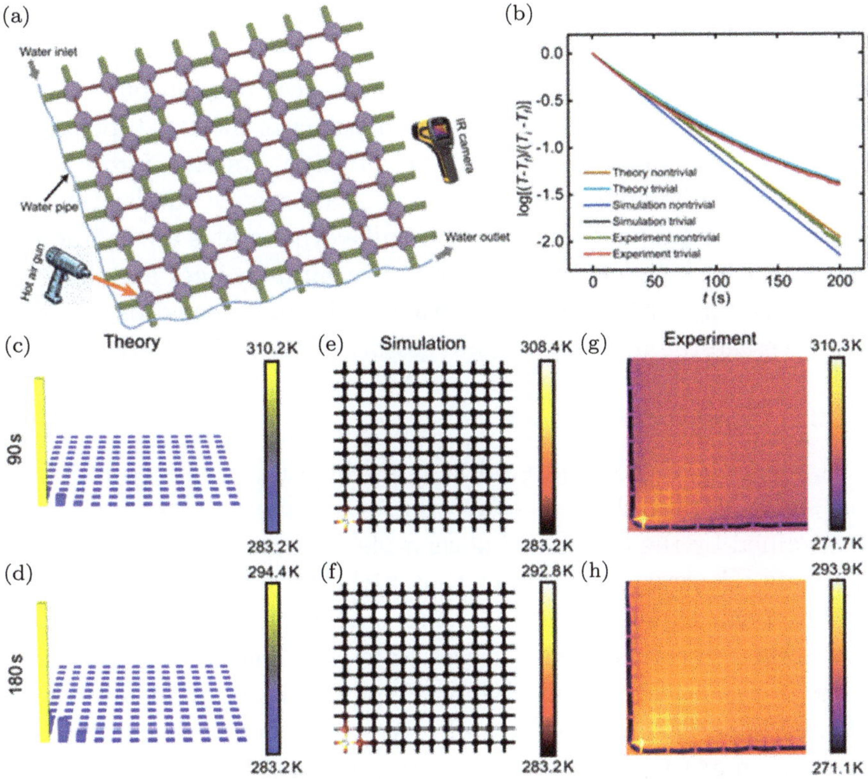

Fig. 7.21. In-bulk corner state in a nontrivial lattice. (a) Experimental setup. (b) Normalized temperature evolution of corner sphere for nontrivial and trivial lattices. (c,d) Theoretical temperature distributions for the nontrivial lattice at (c) 90 s and (d) 180 s. (e,f) Simulated temperature distributions for the nontrivial lattice at (e) 90 s and (f) 180 s. (g,h) Experimental temperature distributions for the nontrivial lattice at (g) 90 s and (h) 180 s. Adapted from Ref. [71]

corner state in the bulk with a decay rate consistent with the C4 symmetry-preserving states. This in-bulk corner state was localized at the stimulated corner, whereas the other three corner states adapted to hybridize with the edge states.

The temperature evolution of this particular corner sphere was monitored to study the thermal behavior of the in-bulk corner state. The results revealed that the corner sphere in the non-trivial phase showed an exponential temperature decay and minimal heating impact on adjacent spheres, as depicted in Fig. 7.21. This localized temperature distribution supports efficient heat dissipation.

This work broadens the theoretical horizons of thermal topological insulators by extending topological diffusion systems into higher dimensions and demonstrates their potential practical applications in thermal management. The observed bulk corner states are characterized by their robustness against environmental changes and defects, paving the way for the design of advanced thermal regulation systems that leverage the inherent advantages of topological materials.

7.4.2. *Convection Model*

This research marks a groundbreaking investigation into topological states within thermal convection systems, an area previously constrained by the absence of observable quantized bulk quadrupole moments. Contrary to the fact that higher-order features can only be observed experimentally in the real value band of classical wave fields, researchers have successfully identified these topological states in real and imaginary bands. Experiments have demonstrated thermal quadrupole topology and revealed pronounced temperature localization at the bulk, edges, and corners of the fabricated samples. This advancement offers fresh insights into creating quantized bulk quadrupole moments in thermal contexts, facilitating the exploration of complex topological phase transitions across various diffusion processes and advancing the study of higher-order topological insulators in thermal conduction diffusion systems.

This study delves into topological states in thermal diffusion systems using a convective fluid model with discrete sites (Fig. 7.22a,b). Unlike classical wave dynamics, this model adheres to conservation laws through continuous conditions necessary for quantization. Effective oscillations are created by applying tunable advections at each site, forming unit structures of four neighboring sites, as shown in Fig. 7.22. These units are arranged into a 2D square lattice with 16 sites for heat transfer.

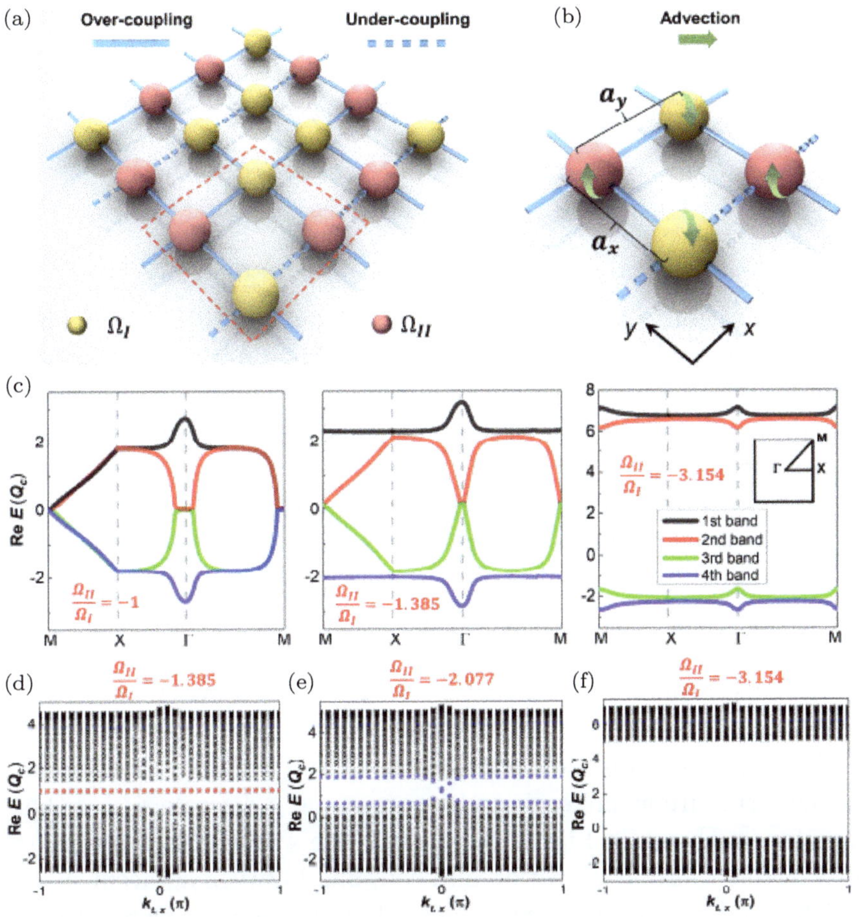

Fig. 7.22. Quadrupole topological phases in thermal transport influenced by Hermitian advection. (a) A square lattice with a four-site unit structure highlighted by a red border. (b) Schematic unit structure. (c) Real-valued phase diagrams for the first Brillouin zones affected by specific advections. (d–f) The dispersion spectra resulting from Hermitian advection highlight the in-gap corner states, gapped edge states, and trivial bulk states. Adapted from Ref. [72]

The general heat energy equation [36] for each site can be expressed as

$$\frac{\partial T_{ij}}{\partial t} = \frac{\kappa}{\rho c}\nabla(\nabla T_{ij}) \pm \Omega_{I/II} R(\theta)\cdot\nabla T_{ij} + \underbrace{\sum \frac{h}{\rho c a_{x/y}}\Delta T}_{\text{intracell}} + \underbrace{\sum \frac{\beta h}{\rho c a_{x/y}}\Delta T}_{\text{intercell}}. \tag{7.33}$$

which incorporates parameters such as density (ρ), specific heat (c), and thermal conductivity (κ) of each site. Additional variables include the angular velocities, radial and azimuth components, and the heat transfer coefficient influencing the thermal couplings between sites.

The system's design allows for the decoupling of x- and y-directional components from the site-specific advection, potentially leading to a diffusive analog of quantized bulk quadrupole moments when adjusting advection and thermal coupling parameters. This setup, governed by skew-Hermitian physics, enables complex band formations through real and imaginary angular frequencies influenced by intrinsic conduction and thermal couplings.

The experimental setup consisted of a meticulously designed 2D lattice structure where each node represented a discrete site for heat transfer, as shown in Fig. 7.23a. The lattice was engineered to allow for variable thermal advection and coupling between adjacent nodes, which could be controlled externally. Thermal sensors were strategically placed at various nodes to measure temperature variations and gradients, which were then analyzed to observe topological states such as edge and corner states. To ensure topological transitions were induced solely by Hermitian advection, uniform structures were maintained across all coupling channels, keeping the intercell and intracell thermal coupling strengths consistent ($\beta = 1$). Specific advections were applied to diagonal sites within a unit structure (Fig. 7.23a), with Ω set based on the dispersion outcomes from Fig. 7.23d, and Ω varied to identify the corresponding real angular

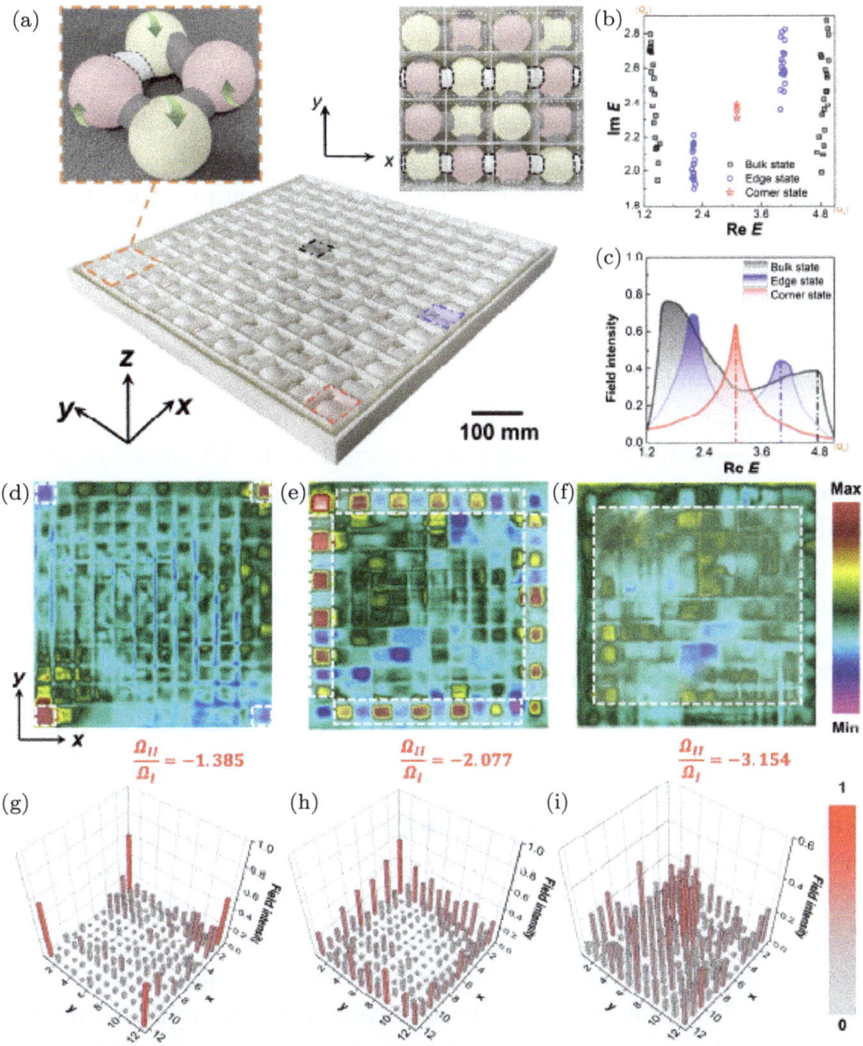

Fig. 7.23. Quadrupole topological phases in heat transport induced by Hermitian advection. (a) Fabricated sample with nine square lattices with the same intercell and intracell coupling channels. (b) Calculated spectrums of the thermal quadrupole topological phases. (c) Measured temperature field intensities at corresponding boundaries. (d–f) Captured temperature profiles at the corner, edge, and bulk spectra peaks. (g–i) Temperature field intensity profiles. Adapted from Ref. [72].

frequency. This setup facilitated a quantized quadrupole moment, leading to distinct hierarchical features within the real-valued band and a singular gapless imaginary-valued band (Fig. 7.23b).

Figure 7.23c offers empirical evidence of temperature localization at the corner and edge of the sample grid, indicating the characteristics of high-order topological insulators. The temperature profiles under different conditions display the predicted topological angle states, highlighting the role of Hermite advection in stabilizing these states.

We adjusted the heat exchange mechanics within intercell and intracell channels to ensure varied thermal coupling strengths, which prevent the closure of these bands and the emergence of trivial bulk states. These adjustments allow β to vary from 0 to 1, showing phase transitions as illustrated in the imaginary phase diagrams of the modified lattice.

This setup reveals both 1D edge and 0D corner states within the imaginary-valued bands as the β value is modulated. The resulting dispersion relations confirm the presence of these states, as presented in Fig. 7.23c–e. Furthermore, the system expanded to include nine modified square lattices, depicted in Fig. 7.24a, where tuning β near zero and one indicates trivial bulk states (Fig. 7.24b) and a mix of two gapped edges and one in-gap corner states (Fig. 7.24c). The experimental data corroborates these findings, demonstrating consistent field intensities across these configurations (Fig. 7.24d–i).

By adjusting thermal advection parameters, researchers can induce and detect the presence of these moments, which manifest as local temperature gradients at specific lattice positions. This observation is crucial for verifying the theoretical model and proving the feasibility of replicating electronic topology phenomena in thermal systems. The transition between trivial and nontrivial topologies can be induced by changing the intensity and direction of thermal advection. This tunability is similar to the gating effect in electronic topological insulators and demonstrates the potential for controlling topological states in thermal materials.

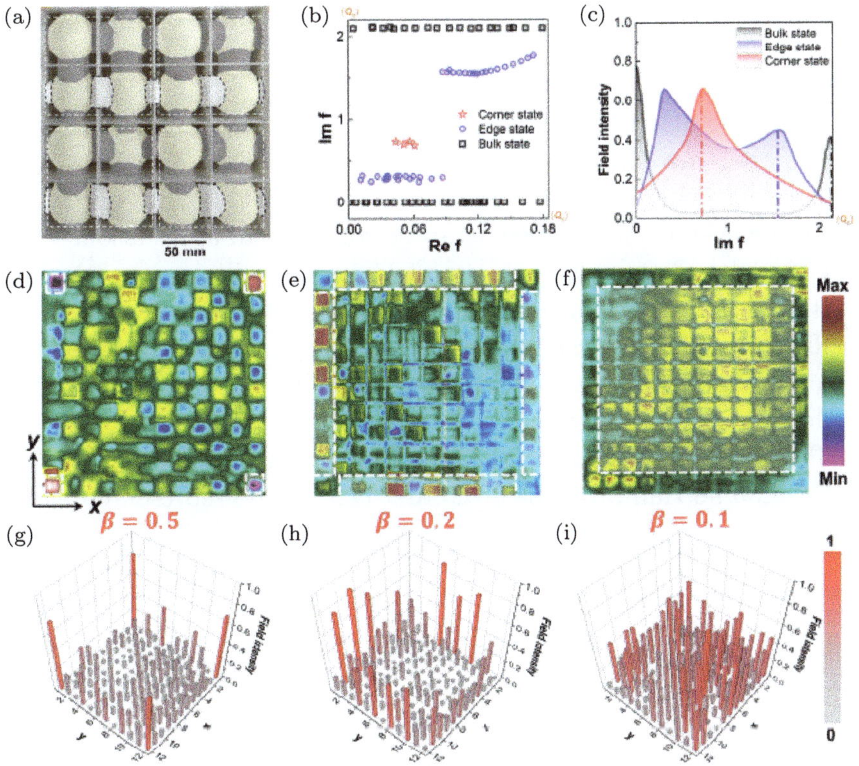

Fig. 7.24. Observation of quadrupole topological phases in heat transport solely induced by non-Hermitian couplings. (a) Fabricated square lattice. (b) Calculated spectra vary based on the thermal coupling strength ratios β. (c) Measured temperature field intensities at marked regions of the sample. (d–f) Temperature distributions at the peak activity points of corner, edge, and bulk states. (g–i) Temperature field intensity profiles. Adapted from Ref. [72].

7.4.3. *Asymmetric Coupling Model*

The concept of thermal regulation using non-Hermitian physics is relatively unexplored. Here, two non-Hermitian SSH models are introduced to control thermal fields in two dimensions. These models harness the NHSE, previously applied in various other physics domains, to achieve controlled thermal localization and propagation in designed structures. This approach is pioneering in its application and promises significant advancements in thermal management technologies.

One aims at creating edge states and the other at corner states. Both models are constructed on the foundation of the non-Hermitian SSH model, which is renowned for its topological characteristics in electronic systems and is now adapted for thermal applications. The first model and the edge state model arrange cell arrays with dual cavities, each interconnected with specific nonreciprocal coupling coefficients. The coefficients are adjusted asymmetrically to yield a directional heat flow bias, effectively channeling thermal energy along predetermined pathways.

The edge state model comprises cell arrays with two cavities in each cell. The Hamiltonian of the first cell can be represented as

$$H_{1-2} = -i \begin{bmatrix} h_{12} & -h_{12} \\ -h_{21} & h_{21} \end{bmatrix} \tag{7.34}$$

which is inherently anti-Hermitian. Non-Hermiticity arises from adjusting material or geometric variables to ensure asymmetric coupling strengths ($h_{12} \neq h_{21}$). Experimentally, this can involve modifying conductivity and mass density by incorporating air gaps in the material, a process achieved through methods like punching holes. The effective parameters for this setup can be calculated using the Maxwell-Garnett formula. Adjusting geometric parameters provides a more direct experimental approach. These modifications render the Hamiltonian non-reciprocal, leading to the NHSE, effectively directing temperature aggregation within the system.

Adjusting the non-reciprocal coupling coefficient within a cell sets the coupling strength to h_0/a in the positive x direction and to ah_0 in the negative x direction while maintaining h_0 between cells. The coefficient a represents the degree of nonreciprocity influenced by material and geometric factors. Bloch boundary conditions applied in the y direction ensure reciprocal coupling, aiding in forming a larger edge state model. This setup allows modulation depth to be controlled by a, and the temperature field adheres to specific equations that reflect these boundary conditions.

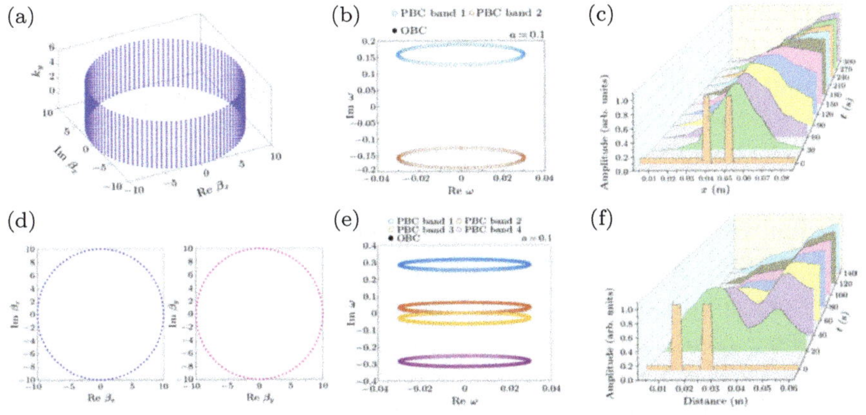

Fig. 7.25. (a) Generalized Brillouin zone of the edge state model. The radius of the generalized Brillouin zone is always $1/a$, irrespective of the Bloch wave number in the y direction (k_y). (b) The complex plane energy band diagram of the edge state thermal skin effect model. (c) The time-evolution diagram of the edge state thermal skin effect model. (d,f) The corner state scenarios depicted in (a–c). Adapted from Ref. [73].

The behavior of the temperature field is analyzed by examining the relationship between the generalized Brillouin zone and the Brillouin zone, depicted in Fig. 7.25a, where the generalized Brillouin zone, at $a = 0.1$, forms a circle with a radius of $1/a$, indicating thermal accumulation in the positive x direction when $|\beta_x| > 1$. The system's non-Hermitian properties, including the emergence of energy loops in the complex plane, signal the presence of the NHSE under open boundary conditions, as shown by the energy distribution in Fig. 7.25b. The presence of these loops and the disparity between the energy gaps under periodic and open conditions show the significant nonreciprocity in the system, with changes in a affecting the visibility and closure of these gaps.

The models were tested under various parameters to observe the localization and spread of thermal fields. The edge state model demonstrated a clear, directed thermal flow along the edge of the model, corroborated by thermal images shown in Fig. 7.25c. These significant results confirmed the theoretical predictions about NHSEs driving thermal behavior.

The corner state model showed even more intriguing results. The corner state model extends this concept by applying nonreciprocal coupling in both the x and y directions, thus localizing the thermal effect at the array's corners. When activated, the heat localized dramatically at the corners of the model, as depicted in Fig. 7.25d, where the generalized Brillouin zones in the x and y directions do not coincide with the Brillouin zones, indicating the occurrence of corner state NHSEs. Figure 7.25e shows point energy gaps under periodic boundary conditions, suggesting the occurrence of NHSE under open boundary conditions. This effect directly results from the unique coupling strengths applied in both directions, demonstrating the model's capability to control thermal behavior in a multi-dimensional setup. Both models exhibited robustness against initial conditions and structural disturbances variations, highlighting their potential for real-world application in thermal management systems. This model is particularly innovative because it leverages the intrinsic geometric properties of the system to manipulate the thermal field extensively, a technique that could be revolutionary in creating highly efficient thermal insulators or conductors, depending on the configuration.

The time-dependent evolution of the corner state model is analyzed through simulations depicted in Fig. 7.25f, which includes a layout of four cavities arranged in both the x and y directions. The simulation applies various thermal conditions (heat, cold, and alternating heat and cold) to these cavities. The thermal field stabilizes into a corner skin mode swiftly, achieving stability at 80 seconds for consistent heat or cold conditions and at 120 seconds for alternating conditions. These findings underscore the model's effective control over thermal regulation within the designated paths.

The ability to localize and direct thermal energy without physical movement or external forces is a significant advancement in material science and could lead to the development of new thermal metamaterials. These materials could be designed to adjust their thermal conductivity automatically in response to environmental conditions, potentially leading to smarter building materials, more efficient heat exchangers, and innovative cooling systems. Furthermore, the study validates the application of non-Hermitian physics in a new realm, suggesting that the principles of NHSEs could be more universally applicable than previously thought. Exploring these models also raises interesting questions about the limits of thermal control at the nanoscale and the possible quantum mechanical effects that could be harnessed. The robustness and efficiency of these models under various conditions suggest a promising future for applying topological concepts in everyday thermal applications, marking a substantial contribution to thermal physics and engineering.

7.5. Conclusion

The thermal lattice links diffusion phenomena with classical band theory, enabling many novel effects from band theory to be realized in heat transfer systems. Both thermal and electronic band theories involve the formation of band structures due to periodicity in a system. In electronic systems, the periodic potential of atoms in a solid creates a band structure for electrons. Similarly, in thermal systems, periodic arrangements of different materials can make a thermal band structure. Both theories explore the effects of boundary conditions on the system's physical states. For example, edge states in electronic systems have their counterpart in thermal systems where localized thermal modes can exist at the edges or interfaces of different thermal materials, often driven by discontinuities in material properties. However, thermal systems are inherently dissipative due to the nature of heat transfer, involving entropy generation and energy dispersion. The thermal bands have unique characteristics, with pure diffusion being anti-Hermitian and purely imaginary. Thermal band structures often involve anti-Hermitian components due to the dissipative nature of the system. This feature starkly contrasts with electronic systems, typically described by Hermitian operators in quantum mechanics. In summary, while thermal and traditional band theories share foundational concepts, the unique challenges and behaviors of heat transfer bring about novel phenomena not observed in electronic materials. These differences highlight the adaptability of band theory across different physical realms and open up innovative approaches to controlling and utilizing energy in various forms.

The study of pure diffusion, thermal convection, and asymmetric coupling reveals topological phenomena such as local thermal diffusion, thermal diodes, and NHSEs. These developments provide a new perspective for the design and development of thermal lattices and indicate their application prospects in thermal management and regulation fields.

Studying purely diffusive topological effects reveals untapped potential in artificial structures determined by purely diffusive topological properties. Although the purely dissipative nature of diffusion and the broad nature of its modes pose challenges, systematic studies fill this gap. Analytical derivation of band structure and geometric phase provides a definite connection between geometric phase and edge states without relying on tight-binding approximations. This connection enables the experimental demonstration of local thermal diffusion as a topological edge state, thereby validating the bulk boundary correspondence in thermal diffusion. This work provides rigorous proof of this correspondence and points toward exploiting topological edge states in innovative thermal management applications.

Furthermore, exploring topological effects in convection models introduces a class of topological transitions in thermal diffusion. By establishing a comprehensive parameter space in a coupled convective system, non-Hermitian topology in thermal diffusion is revealed for the first time. The discovery of diffusion-based topological states, supported by spatiotemporally regulated advection, marks a significant advance in this field. These states help control diffusion phenomena to achieve robust heat and mass transfer, establishing a new framework for topological diffusion and edge/bulk states.

The asymmetric coupling model demonstrates the possibility of realizing asymmetric coupling and skin effect in diffusion systems through direct contact with the thermal metamaterial. The diffusion skin effect shows that all eigenstates in the body band can become local edge states under the open boundary, a fundamental departure from the Hermitian system. This novel temperature field modulation mechanism of coupled diffusion systems highlights the application potential of non-Hermitian topology in thermal diffusion and opens up new avenues for precise thermal sensing. The research results lay a solid foundation for developing thermal lattice technology.

By proving the existence of topological phenomena in thermal systems and demonstrating their practical applications, we pave the way for the design of innovative thermal metamaterials and devices. This property not only helps the topological physics field but is also expected to revolutionize the future of thermal management strategies.

References

1. Heeger, A. J., Kivelson, S., Schrieffer, J. R., and Su, W.-P. (1988). Solitons in conducting polymers. *Rev. Mod. Phys.*, 60, pp. 781–850.
2. Qi, M., Wang, D., Cao, P.-C., Zhu, X.-F., Qiu, C.-W., Chen, H., and Li, Y. (2023). Geometric phase and localized heat diffusion. *Adv. Mater.*, 34, pp. 2202241.
3. Yariv, A. and Yeh, P. (1983). *Optical Waves in Crystals: Propagation and Control of Laser Radiation*. John Wiley & Sons.
4. Zak, J. (1989). Berry's phase for energy bands in solids. *Phys. Rev. Lett.*, 62, pp. 2747.
5. Delplace, P., Ullmo, D., and Montambaux, G. (2011). Zak phase and the existence of edge states in graphene. *Phys. Rev. B*, 84, pp. 195452.
6. Li, Y., Peng, Y.-G., Han, L., Miri, M.-A., Li,W., Xiao, M., Zhu, X.-F., Zhao, J., Alù, A., Fan, S., and Qiu, C.-W. (2019). Anti-parity-time symmetry in diffusive systems. *Science*, 364, pp. 170–173.
7. Cao, P., Li, Y., Peng, Y.-G., Qiu, C., and Zhu, X. (2019). High-order exceptional points in diffusive systems: robust APT symmetry against perturbation and phase oscillation at APT symmetry breaking. *ES Energy Environ.*, 7, pp. 48–55.
8. Kawabata, K., Shiozaki, K., Ueda, M., and Sato, M. (2019). Symmetry and topology in non-Hermitian physics. *Phys. Rev. X*, 9, pp. 041015.
9. Li, Z., Cao, G., Li, C., Dong, S., Deng, Y., Liu, X., Ho, J. S., and Qiu, C.-W. (2021). Non-Hermitian electromagnetic metasurfaces at exceptional points. *Prog. Electromagn. Res.*, 171, pp. 1–20.
10. Regensburger, A., Bersch, C., Miri, M.-A., Onishchukov, G., Christodoulides, D. N., and Peschel, U. (2012). Parity-time synthetic photonic lattices. *Nature*, 488, pp. 167.

11. Chong, Y. D., Ge, L., Cao, H., and Stone, A. D. (2010). Coherent perfect absorbers: time-reversed lasers. *Phys. Rev. Lett.*, 105, pp. 053901.
12. Peng, B., Özdemir, Ş. K., Rotter, S., Yilmaz, H., Liertzer, M., Monifi, F., Bender, C. M., Nori, F., and Yang, L. (2014). Loss-induced suppression and revival of lasing. *Science*, 346, pp. 328–332.
13. Weidemann, S., Kremer, M., Helbig, T., Hofmann, T., Stegmaier, A., Greiter, M., Thomale, R., and Szameit, A. (2020). Topological funneling of light. *Science*, 368, pp. 311–314.
14. Cao, P.-C., Li, Y., Peng, Y.-G., Qi, M., Huang, W.-X., Li, P.-Q., and Zhu, X.-F. (2021). Diffusive skin effect and topological heat funneling. *Commun. Phys.*, 4, pp. 1–7.
15. Cao, P.-C., Peng, Y.-G., Li, Y., and Zhu, X.-F. (2022). Phase-locking diffusive skin effect. *Chin. Phys. Lett.*, 39, pp. 057801.
16. Xu, G., Li, Y., Li, W., Fan, S., and Qiu, C.-W. (2021). Configurable phase transitions in a topological thermal material. *Phys. Rev. Lett.*, 127, pp. 105901.
17. Xu, G., Yang, Y., Zhou, X., Chen, H., Alù, A., and Qiu, C.-W. (2022). Diffusive topological transport in spatiotemporal thermal lattices. *Nat. Phys.*, 18, pp. 450–456.
18. Xu, G., Li, W., Zhou, X., Li, H., Li, Y., Fan, S., Zhang, S., Christodoulides, D. N., and Qiu, C.-W. (2022). Observation of Weyl exceptional rings in thermal diffusion. *Proc. Natl. Acad. Sci. U.S.A.*, 119, pp. e2110018119.
19. Hasan, M. Z., and Kane, C. L. (2010). Colloquium: Topological insulators. *Rev. Mod. Phys.*, 82, pp. 3045–3067.
20. Chen, Z., and Segev, M. (2021). Highlighting photonics: looking into the next decade. *eLight*, 1, pp. 2.
21. Shan, Q., Yu, D., Li, G., Yuan, L., and Chen, X. (2020). One-way topological states along vague boundaries in synthetic frequency dimensions including group velocity dispersion. *PIER*, 169, pp. 33–43.
22. Rhim, J.-W., Behrends, J., and Bardarson, J. H. (2017). Bulk-boundary correspondence from the intercellular Zak phase. *Phys. Rev. B*, 95, pp. 035421.
23. Yang, Y., Gao, Z., Xue, H., Zhang, L., He, M., Yang, Z., Singh, R., Chong, Y., Zhang, B., and Chen, H. (2019). Realization of a three-dimensional photonic topological insulator. *Nature*, 565, pp. 622–626.
24. Lustig, E., Weimann, S., Plotnik, Y., Lumer, Y., Bandres, M. A., Szameit, A., and Segev, M. (2019). Photonic topological insulator in synthetic dimensions. *Nature*, 567, pp. 356–360.
25. Zilberberg, O., Huang, S., Guglielmon, J., Wang, M., Chen, K. P., Kraus, Y. E., and Rechtsman, M. C. (2018). Photonic topological

boundary pumping as a probe of 4D quantum Hall physics. *Nature*, 553, pp. 59–62.
26. Liu, Y., Leung, S., Li, F. F., Lin, Z. K., Tao, X., Poo, Y., and Jiang, J. H. (2021). Bulk–disclination correspondence in topological crystalline insulators. *Nature*, 589, pp. 381–385.
27. Stützer, S., Plotnik, Y., Lumer, Y., Titum, P., Lindner, N. H., Segev, M., Rechtsman, M. C., and Szameit, A. (2018). Photonic topological Anderson insulators. *Nature*, 560, pp. 461–465.
28. Peng, B., Özdemir, S. K., Lei, F., Monifi, F., Gianfreda, M., G. Long, L., Fan, S., Nori, F., Bender, C. M., and Yang, L. (2014). Parity-time-symmetric whispering-gallery microcavities. *Nat. Phys.*, 10, pp. 394–398.
29. Zhao, H., Miao, P., Teimourpour, M. H., Malzard, S., El-Ganainy, R., Schomerus, H., and Feng, L. (2018). Topological hybrid silicon microlasers. *Nat. Commun.*, 9, pp. 1–6.
30. Zhao, H., Qiao, X., Wu, T., Midya, B., Longhi, S., and Feng, L. (2019). Non-Hermitian topological light steering. *Science*, 365, pp. 1163–1166.
31. Zeng, Y., Chattopadhyay, U., Zhu, B., Qiang, B., Li, J., Jin, Y., Li, L., Davies, A. G., Linfield, E. H., Zhang, B., Chong, Y., D., and Wang, Q. J. (2020). Electrically pumped topological laser with valley edge modes. *Nature*, 578, pp. 246–250.
32. Bandres, M. A., Wittek, S., Harari, G., Parto, M., Ren, J., Segev, M., Christodoulides, D. N., and Khajavikhan, M. (2018). Topological insulator laser: Experiments. *Science*, 359, pp. eaar4005.
33. Harari, G., Bandres, M. A., Lumer, Y., Rechtsman, M. C., Chong, Y. D., Khajavikhan, M., Christodoulides, D. N., and Segev, M. (2018). Topological insulator laser: Theory. *Science*, 359, pp. eaar4003.
34. Choi, J. H., Hayenga, W. E., Liu, Y. G. N., Parto, M., Bahari, B., Christodoulides, D. N., and Khajavikhan, M. (2021). Room temperature electrically pumped topological insulator lasers. *Nat. Commun.*, 12, pp. 1–7.
35. Bahari, B., Ndao, A., Vallini, F., Amili, A. E., Fainman, Y., and Kanté, B. (2017). Nonreciprocal lasing in topological cavities of arbitrary geometries. *Science*, 358, pp. 636–640.
36. Liang, G. Q. and Chong, Y. D. (2013). Optical resonator analog of a two-dimensional topological insulator. *Phys. Rev. Lett.*, 110, pp. 203904.
37. Takata, K. and Notomi, M. (2018). Photonic topological insulating phase induced solely by gain and loss. *Phys. Rev. Lett.*, 121, pp. 213902.
38. Fan, C. Z., Gao, Y., and Huang, J. P. (2008). Shaped graded materials with an apparent negative thermal conductivity. *Appl. Phys. Lett.*, 92, pp. 251907.

39. Han, T., Bai, X., Gao, D., Thong, J. T. L., Li, B., and Qiu, C.-W. (2014). Experimental demonstration of a bilayer thermal cloak. *Phys. Rev. Lett.*, 112, pp. 054302.
40. Zhou, X., Xu, G., and Zhang, H. Y. (2021). Binary masses manipulation with composite bilayer metamaterial. *Compos. Struct.*, 267, pp. 113866.
41. Zhou, X. and Xu, G. (2021). Self-adaptive field manipulation with thermal logic material. *Int. J. Heat Mass Transf.*, 172, pp. 121147.
42. Xu, G., Dong, K., Li, Y., Li, H., Liu, K., Li, L., Wu, J., and Qiu, C.-W. (2020). Tunable analog thermal material. *Nat. Commun.*, 11, pp. 6028.
43. Zhu, X., Ramezani, H., Shi, C., Zhu, J., and Zhang, X. (2014). PT-symmetric acoustics. *Phys. Rev. X*, 4, pp. 031042.
44. Feng, L., Wong, Z. J., Ma, R.-M., Wang, Y., and Zhang, X. (2014). Single-mode laser by parity-time symmetry breaking. *Science*, 346, pp. 972–975.
45. Assawaworrarit, S., Yu, X., and Fan, S. (2017). Robust wireless power transfer using a nonlinear parity-time-symmetric circuit. *Nature*, 546, pp. 387–390.
46. Wang, Q., Che, W., Dionigi, M., Mastri, F., Mongiardo, M., and Monti, G. (2019). Gains maximization via impedance matching networks for wireless power transfer. *PIER*, 164, pp. 135–153.
47. Lee, T. E. (2016). Anomalous edge state in a non-Hermitian lattice. *Phys. Rev. Lett.*, 116, pp. 133903.
48. Kunst, F. K., Edvardsson, E., Budich, J. C., and Bergholtz, E. J. (2018). Biorthogonal bulk-boundary correspondence in non-Hermitian systems. *Phys. Rev. Lett.*, 121, pp. 26808.
49. Yao, S. and Wang, Z. (2018). Edge states and topological invariants of non-Hermitian systems. *Phys. Rev. Lett.*, 121, pp. 86803.
50. Yao, S., Song, F., and Wang, Z. (2018). Non-Hermitian Chern bands. *Phys. Rev. Lett.*, 121, pp. 136802.
51. Yokomizo, K. and Murakami, S. (2019). Non-Bloch band theory of non-Hermitian systems. *Phys. Rev. Lett.*, 123, pp. 66404.
52. Okuma, N., Kawabata, K., Shiozaki, K., and Sato, M. (2020). Topological origin of non-Hermitian skin effects. *Phys. Rev. Lett.*, 124, pp. 86801.
53. Li, L., Lee, C., and Gong, J. (2021). Impurity induced scale-free localization. *Commun. Phys.*, 4, pp. 42.
54. Ghatak, A., Brandenbourger. M., Wezel, J. V., and Coulais, C. (2020). Observation of non-Hermitian topology and its bulk-edge correspondence in an active mechanical metamaterial. *Proc. Natl. Acad. Sci. U. S. A.*, 47, pp. 29561.

55. Xiao, L., Deng, T., Wang, K., Zhu, G., Wang, Z., Yi, W., and Xue, P. (2020). Non-Hermitian bulk-boundary correspondence in quantum dynamics. *Nat. Phys.*, 16, pp. 761–766.
56. Helbig, T., Hofmann, T., Imhof, S., Abdelghany, M., Kiessling, T., Molenkamp, L. W., Lee, C. H., Szameit, A., Greiter, M., and Thomale, R. (2020). Generalized bulk-boundary correspondence in non-Hermitian topolectrical circuits. *Nat. Phys.*, 16, pp. 747–750.
57. Budich, J. C. and Bergholtz, E. J. (2020). Non-Hermitian topological sensors. *Phys. Rev. Lett.*, 125, pp. 180403.
58. Xu, L., Dai, G., Wang, G., and Huang, J. (2020). Geometric phase and bilayer cloak in macroscopic particle-diffusion systems. *Phys. Rev. E*, 102, pp. 32140.
59. Bender, C. M. (2007). Making sense of non-Hermitian Hamiltonians. *Rep. Prog. Phys.*, 70, pp. 947.
60. Liu, Y.-K., Cao, P.-C., Qi, M., Huang, Q.-K.-L., Gao, F., Peng, Y.-G., Li, Y., and Zhu, X.-F. (2024). Observation of non-Hermitian skin effect in thermal diffusion. *Sci. Bull.*, 69, pp. 1228–1236.
61. Gu, Z., Gao, H., Cao, P.-C., Liu, T., Zhu, X.-F., and Zhu, J. (2021). Controlling sound in non-Hermitian acoustic systems. *Phys. Rev. Appl.*, 16, pp. 057001.
62. Shi, C., Dubois, M., Chen, Y., Cheng, L., Ramezani, H., Wang, Y., and Zhang, X. (2016). Accessing the exceptional points of parity-time symmetric acoustics. *Nat. Commun.*, 7, pp. 11110.
63. Wang, S., Hou, B., Lu, W., Chen, Y., Zhang, Z. Q., and Chan, C. T. (2019). Arbitrary order exceptional point induced by photonic spin-orbit interaction in coupled resonators. *Nat. Commun.*, 10, pp. 832.
64. Longhi, S., Gatti, D., and Della Valle, G. (2015). Non-Hermitian transparency and one-way transport in low-dimensional lattices by an imaginary gauge field. *Phys. Rev. B*, 92, pp. 094204.
65. Yan, Q., Chen, H., and Yang, Y. (2021). Non-Hermitian skin effect and delocalized edge states in photonic crystals with anomalous parity-time symmetry. *PIER*, 172, pp. 33–40.
66. Mittal, S., Orre, V. V., Zhu, G., Gorlach, M. A., Poddubny, A., and Hafezi, M. (2019). Photonic quadrupole topological phases. *Nat. Photonics*, 13, pp. 692–696.
67. He, L., Addison, Z., Mele, E. J., and Zhen, B. (2020). Quadrupole topological photonic crystals. *Nat. Commun.*, 11, pp. 3119.
68. Qi, Y., Qiu, C., Xiao, M., He, H., Ke, M., and Liu, Z. (2020). Acoustic realization of quadrupole topological insulators. *Phys. Rev. Lett.*, 124, pp. 206601.

69. Du, J., Li, T., Fan, X., Zhang, Q., and Qiu, C.-W. (2022). Acoustic realization of surface-obstructed topological insulators. *Phys. Rev. Lett.*, 128, pp. 224301.
70. Imhof, S., Berger, C., Bayer, F., Brehm, J., Molenkamp, L. W., Kiessling, T., Schindler, F., Lee, C. H., Greiter, M., Neupert, T., *et al.* (2018). Topolectrical-circuit realization of topological corner modes. *Nat. Phys.*, 14, pp. 925–929.
71. Liu, Z., Cao, P.-C., Xu, L., Xu, G., Li, Y., and Huang, J. (2024). Higher-order topological in-bulk corner state in pure diffusion systems. *Phys. Rev. Lett.*, 132, pp. 176302.
72. Xu, G., Zhou, X., Yang, S., Wu, J., and Qiu, C.-W. (2023). Observation of bulk quadrupole in topological heat transport. *Nat. Commun.*, 14, pp. 3252.
73. Huang, Q.-K.-L., Liu, Y.-K., Cao, P.-C., Zhu, X.-F., and Li, Y. (2023). Two-dimensional thermal regulation based on non-Hermitian skin effect. *Chin. Phys. Lett.*, 40, pp. 106601.

Chapter 8

Outlook

Although various new functions for controlling heat transfer have been realized with thermal metamaterials, there remain fundamental questions to be answered, important challenges to be tackled, and interesting opportunities to be considered for future developments. In the following section, we suggest ideas for each part of this book.

8.1. Steady-State Heat Transfer

8.1.1. *Transformation Thermotics*

The transformation theory has been used in many conductive systems. The remaining questions concern its validity and application in combining thermal conduction with other effects or heat transfer processes.

Although the transformation theory has been developed for nonlinear thermal conduction by Li *et al.* [1], the nonlinear effects have not been fully explored. Specifically, the thermal conductivity tensor $\boldsymbol{\kappa}(x, T)$ has a spatial variance,

$$\frac{d}{dx}\boldsymbol{\kappa}(x, T) = \frac{\partial}{\partial x}\boldsymbol{\kappa}(x, T) + \frac{\partial}{\partial T}\boldsymbol{\kappa}(x, T)\frac{dT}{dx}. \tag{8.1}$$

The first term on the right-hand side comes from the spatial distribution of the constituent materials, while the second term comes from the spatial distribution of the temperature field. In previous works, the temperature was mostly used as a constant rather than a field, so the second term was ignored. However, in systems with large temperature gradients, the second term may significantly contribute to the variance of the thermal conductivity. An open question is how a temperature-dependent transformation performs on such systems and influences functionalities such as switchable cloaking.

In addition, a general transformation theory for heat transfer systems with a source/drain is lacking. In the presence of a heat source/drain introduced by Joule heating or Peltier cooling, the steady-state governing equation changes from homogeneous to inhomogeneous. The question is how the source/drain term changes under a coordinate transformation. How to realize the transformed heat source/drain in physical space is also unclear.

Developing transformation theories for thermal convection and thermal radiation has been an important topic in recent years. It is successful for thermal convection in porous media [2], where relatively simpler equations govern the creeping flow. On the other hand, the Navier-Stokes equation is much more complex, so a transformation theory for thermal convection in general fluid fields is lacking. The question remains as to whether it is possible to construct such a theory or a modified version to control heat transfer in general thermal convection. The transformation theory for thermal radiation in solids has been developed by Xu *et al.* [3] under the Rosseland diffusion approximation. It is unclear whether a more general form of the theory exists for the many-body near-field thermal radiation.

Finally, other multiphysical effects require further study. For example, in thermoelectric systems, we should consider not only the changes in heat source/drain under coordinate transformation but also the changes in the thermoelectric material parameters, such as the Seebeck coefficient. Developing such a theory is challenging because of the intrinsic non-homogeneous relations between thermal and electric fields. Another unexplored issue is the nonreciprocal heat transfer in magnetic fields induced by thermomagnetic effects. It would be interesting to have a theory on transforming the asymmetric effective thermal conductivity tensor.

8.1.2. *Effective-medium Method*

In companion with the transformation theories for various heat transfer processes, the effective-medium method should also be developed to realize the corresponding material parameters. For example, Dai *et al.* [4] have calculated the effective thermal conductivity of nonlinear periodic components. Li *et al.* [5] have given general expressions for the effective thermal conductivity of rotating components. Many other effective material parameters do not have systematic schemes for their realization. The effective Seebeck coefficient is one important yet challenging example. The difficulty comes from its dependence on the loading conditions. Sometimes, calculating an effective field is also needed. For example, the velocity field is entered as a parameter in the governing equation of thermal convection. Still, its calculation can become difficult or even impossible when it is strongly coupled with the temperature field, like in natural convection. Similar challenges exist for systems with strong coupling between the thermal and mechanical fields, such as those with large thermal deformation.

The theoretical problems may be resolved with the help of algorithms. Topology optimization has recently become useful for designing thermal metamaterials with complex shapes [6]. The method has been proven effective for both the direct inverse design from the desired functionality [7] and the indirect inverse design of the effective thermal conductivity [8]. To further speed up the optimization process, a deep-learning algorithm has been developed by Wang *et al.* [9]. Numerical simulations on the heat transfer through different unit cells were performed to form a database (Fig. 8.1a). With the database, a neural network was trained to minimize the difference between the temperature field and the ideal one (Fig. 8.1b,c). In 2024, Jin *et al.* [10] showed that a deep-learning algorithm is also useful for designing active thermal metamaterials with incorporated convective fields. The thermal metamaterial could adaptively modify its functionality based on the ambient temperature. In the future, new algorithms should be developed to design more complex 3D structures under various conditions, such as applying mechanical, electrical, or magnetic external fields. It is also

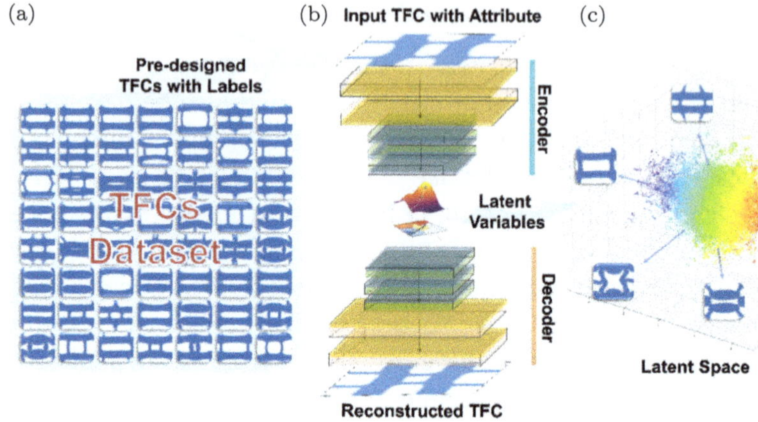

Fig. 8.1. Deep-learning topological functional cells (TFC) for the design of thermal metamaterials [9]. (a) Predesigned TFC dataset. (b) Deep-generative model. (c) Training data in the latent space.

important to uncover the mechanisms behind the algorithms to give physical interpretations of them.

8.2. Time-Harmonic Heat Transfer

8.2.1. *Scattering of Thermal Fields*

The thermal scattering theory was developed mainly to analyze nonreciprocal heat transfer, but it can have much broader applications, especially in heat signal manipulation. For such purposes, the theory must be further developed. For example, the thermal ports that have been studied are assumed to be identical, quasi-1D, and purely conductive. A more generalized theory should consider other cases where the ports are made of different materials, in more complex shapes, and even containing convective fields. The thermal modes are then not limited to simple plane-wave modes. In analog to electromagnetic waveguides, a more general thermal port or channel may support multiple thermal modes with different distributions on the cross-section. In purely conductive channels, the thermal wave decays quickly. If convection is introduced by fluid flowing through the channel to the scatterer, the decay can be significantly suppressed to maintain the heat signal.

Moreover, the scattering coefficients of a scatterer are closely related to its Green function, but the Green functions of many heat transfer systems have not been well studied. To date, only 1D systems, and 2D systems with very simple geometries and material distributions, have been calculated. Some fundamental questions remain to be answered. We list a few that immediately come to mind: (1) How do the anisotropy and asymmetry of the thermal conductivity tensor influence the Green function? (2) How do the channels and the boundary conditions influence the Green function? (3) How do active sources and multi-physical effects influence the Green function? (4) How is the Green function related to the eigenmodes of the system?

Finally, the thermal scattering theory should be able to reveal more physics behind time-harmonic heat transfer, thereby inspiring heat signal processing applications. The symmetries of the scattering matrix can reveal essential properties of the heat transfer system but have not been systematically studied. For example, it is only recently that the symmetry of the scattering matrix for an APT-symmetric system has been obtained by Wang *et al.* [11], which interestingly links the coefficients at positive and negative frequencies. The resonant properties of the scattering coefficients are also important. The resonant frequency is expected to be zero for a purely conductive system. The resonant frequencies can be nonzero for convective or time-modulated systems, and the resonant strengths can be compared. More studies need to be performed on this topic. For example, as shown by Ju *et al.* [12] for a thermal circulator, the resonant frequencies of the nonreciprocity show a nontrivial distribution, but the mechanism remains unclear.

Another prominent challenge in studying thermal scattering is the experimental demonstration, particularly in realizing a stable and accurate heat signal source. At nonzero frequency, the temperature of the source should oscillate, so both heating and cooling sources are needed. The common method of using a pulsed heat source is not good enough to generate an ideal waveform because it relies on natural heat dissipation to the environment, which may not be sufficient to balance the temperature rise in the heating period. An efficient cooling source is thus crucial. A good control system

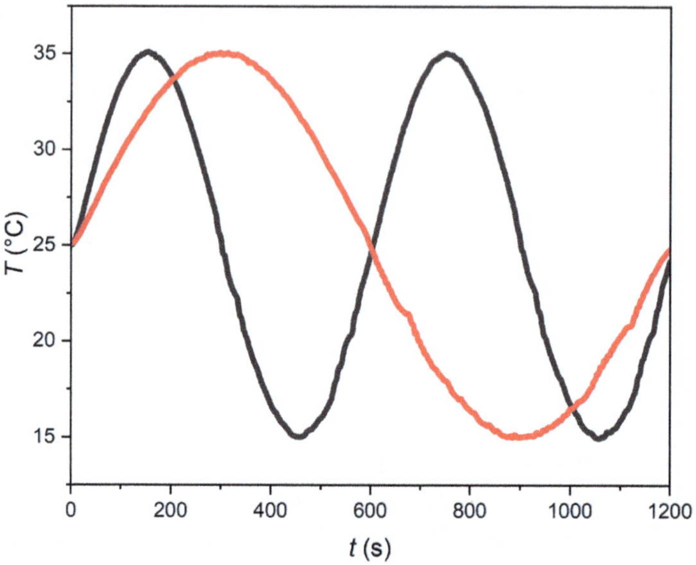

Fig. 8.2. Temperature curves of time-harmonic heat signals with periods of 600 s (grey) and 1200 s (red) generated by thermoelectric sources.

is also necessary to monitor the temperature curve and finetune the source power (Fig. 8.2).

8.2.2. *Temporal Modulation*

Nonreciprocity and resonance are the two major effects of dynamic modulation on a heat transfer system. Both effects require further study on their mechanisms and applications. It is also expected to find new possibilities in time-modulated systems. The readers can refer to the perspective of Liu *et al.* [13], which gives more details.

The nonreciprocal heat transfer in time-modulated systems is both promising and controversial. Many theoretical works have shown that it is an effective method to break thermal reciprocity by simultaneously modulating the boundary temperatures and the thermal conductivity distribution [14, 15]. This effect is a form of heat shuttling. However, it has only been experimentally tested with zero average bias at both ends [16]. The geometric contribution at nonzero bias should be tested. It has been theoretically proved and experimentally shown that the simultaneous modulation of thermal conductivity and density cannot break thermal reciprocity under constant boundary conditions [17]. Therefore, we should find other temporal modulation methods for steady-state heat rectification. Modulating the heat capacity could be one choice, but an effective mechanism is needed for its realization.

Temporal modulation also strongly influences the resonant property of thermal scattering, as mentioned above. For purely conductive systems, the resonant frequency is zero. Then, a modulation at frequency ω will shift the resonant frequency to $n\omega$ (n is an integer). Simple as it is, such an effect has not been well studied, mainly due to the lack of a good oscillating heat source. The modulation is expected to introduce other interesting phenomena for more complex systems with intrinsic frequencies, such as convective systems, considering the recent rich results on time-modulated or Floquet electromagnetic metamaterials [18].

8.3. Transient Heat Transfer

8.3.1. *Effective Hamiltonian and Thermal Modes*

A major limitation of present studies on thermal modes is that they only consider their free decay. The reason is that a single thermal mode can be observed in the process if an appropriate initial condition is used. In other processes, multiple thermal modes can exist. The mode that we are interested in may soon become unobservable due to the existence of other modes that decay more slowly. To study more practical cases when the system is heated or in contact with heat sinks, efficient methods must be developed to single out a particular mode, either during the experiment or after post-processing.

Another challenging issue is constructing good, effective Hamiltonians for various heat transfer systems. Existing works have shown qualitative consistency between the eigenvalues/eigenmodes analytically solved from the effective Hamiltonian and numerically solved from the actual model. However, the model should be further improved to predict the position of the exceptional point and determine the nonlinear response near it. The accuracy of the effective Hamiltonian is important for non-Hermitian sensing [19], mode conversion [20], and other exceptional point-based applications. Therefore, more in-depth studies on estimating the heat transfer coefficient between typical components are needed to obtain the coupling strength in the effective Hamiltonian. For some convective systems, it could be difficult and rely on enormous numerical and experimental results.

Conversely, finding good physical systems that implement the desired effective Hamiltonians is equally important. For example, the coupling coefficient for purely conductive heat exchange is purely imaginary [21]. Cao *et al.* [22] realized a purely real coupling coefficient using a fast-rotating intermediate layer. A question arises about constructing a system with an arbitrary complex coupling coefficient, which would intuitively require a combination of convective and conductive heat exchange. As for the onsite potential, it also can be a complex number whose real part can be either positive or negative [21], but it is unclear and appears to be extremely challenging to realize a negative imaginary part under the $\exp(-i\omega t)$ convention. A negative imaginary part is similar to the optical gain in photonics, which amplifies the onsite field. The amplification of the thermal field cannot be achieved in any passive system because that would violate the *second law* of thermodynamics. Thus, we must design a thermal amplifier based on some active sources with feedback. The gain component is equivalent to a thermal metamaterial with negative effective thermal conductivity [23], which has not been realized. Finally, the effective Hamiltonian should be better tunable for studying its topological properties related to its geometry in the parameter space [24]. More efforts are thus needed to realize tunable thermal properties [25].

8.3.2. *Thermal Lattices*

The band structure is at the center of existing studies on thermal lattices. However, we are still unclear about its physical meaning. The eigenvalues are all purely imaginary for a purely conductive thermal lattice [28], but there is no good interpretation for an imaginary band gap. Recall that a real band gap for a photonic crystal means that the electromagnetic field at any frequency ω inside the band gap does not propagate [29] because the wavenumber k in the band gap is purely imaginary, corresponding to an exponential decay in space. The wavenumber has the same property as a conductive thermal lattice: it is purely imaginary in the band gap and purely real outside. However, we cannot similarly claim that an imaginary band gap is signified by the fact that the temperature field does not propagate. The field still does not propagate outside the band gap because of the imaginary eigenfrequency. That is, the field decays irrespective of the wavenumber. Such difficulties also exist for convective and other types of thermal lattices.

The localization is a controversial issue in another sense. If decaying states at imaginary frequencies are considered, there are both extended and localized states. However, there is no effective method to selectively excite a particular decaying state without using itself as an initial state. Therefore, even after distinguishing between a localized and an extended state, the verification is not easy. The common method used in wave physics is to excite a state with a time-harmonic point source. However, for thermal states at real frequencies, the wavenumber always has a finite imaginary part due to the dissipative nature of heat transfer. Then, all states are localized in space. The contrasting behaviors of states with different time evolution should be considered. The mechanism behind the difference remains unexplored, which may shed light on the recent studies about the disorder-induced localization transitions in 1D [26] and 2D [27] lattices (Fig. 8.3).

Constructing the effective Hamiltonian for a thermal lattice also shares similar difficulties as for a general heat transfer system: the accuracy of some models is insufficient. As Qi *et al.* [28] show, the intuitive tight-binding model may give incorrect predictions mainly because of the nonlocal effects. For example, the thermal modes may not be well localized on each site. This problem is solved using a continuum model and plane wave to expand the thermal modes [28]. However, this method is impractical for 2D or 3D systems. In the

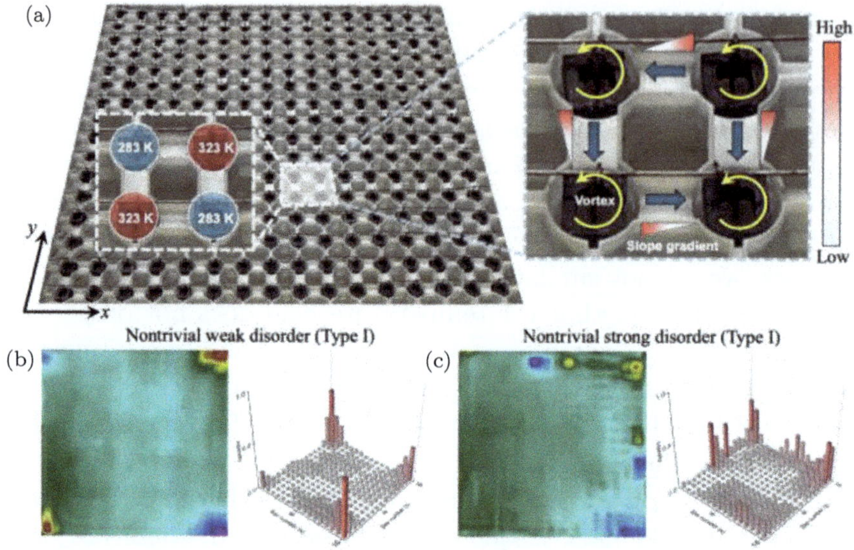

Fig. 8.3. Topological Anderson phases in a thermal lattice [27]. (a) The lattice sample. (b) The unit cell with convective perturbations. (c) Field localized at corners under weak disorder. (d) Field localized at random sites under strong disorder.

future, a balance must be found between the accuracy and complexity of the model. If the coupling coefficient can be easily estimated, introducing next-nearest-neighbor coupling into the tight-binding model might be one solution.

There are also fundamental effective Hamiltonians that need physical implementation. Taking intra-cell coupling as an example, Cao *et al.* [30] have proposed a model with gradient density to realize asymmetric coupling strength effectively. However, the design breaks the spatial periodicity that can only be realized with unit cells of different sizes [31], limiting its applications. Therefore, it remains unclear how to break the symmetry of coupling strength in periodic lattices. Since reciprocity is broken, nonlinearity or some bias must be introduced. Another difficult task is to realize negative coupling. Although Xu *et al.* [32] have fabricated a square lattice with effective negative coupling to realize quadrupole states, the coupling strength was calculated by subtracting a reference value, and the absolute value is still positive. A negative thermal coupling in the absolute sense, which requires active sources, has not been realized.

8.4. Conclusion

In summary, many issues are worth to be studied. Two common themes are (1) establishing theoretical frameworks for more complex and multi-physical processes and (2) finding methods to deal with the spatiotemporal decay of states. Finally, the importance of pushing existing research results towards real-world applications cannot be overstated.

References

1. Li, Y., Shen, X., Wu, Z., Huang, J., Chen, Y., Ni, Y., and Huang, J. (2015). Temperature-dependent transformation thermotics: From switchable thermal cloaks to macroscopic thermal diodes. *Phys. Rev. Lett.*, 115, pp. 195503.
2. Dai, G., Shang, J., and Huang, J. (2018). Theory of transformation thermal convection for creeping flow in porous media: Cloaking, concentrating, and camouflage. *Phys. Rev. E*, 97, pp. 022129.
3. Xu, L., Dai, G., and Huang, J. (2020). Transformation multithermotics: controlling radiation and conduction simultaneously. *Phys. Rev. Appl.*, 13, pp. 024063.
4. Dai, G. and Huang, J. (2020). Nonlinear thermal conductivity of periodic composites. *Int. J. Heat Mass Transf.*, 147, pp. 118917.
5. Li, J., Li, Y., Wang, W., Li, L., and Qiu, C.-W. (2020). Effective medium theory for thermal scattering off rotating structures. *Opt. Express*, 28, pp. 25894–25907.
6. Sha, W., Xiao, M., Wang, Y., Huang, M., Li, Q., and Gao, L. (2024). Topology optimization methods for thermal metamaterials: A review. *Int. J. Heat Mass Transf.*, 227, pp. 125588.
7. Fujii, G., Akimoto, Y., and Takahashi, M. (2018). Exploring optimal topology of thermal cloaks by CMA-ES. *Appl. Phys. Lett.*, 112, pp. 061108.
8. Sha, W., Hu, R., Xiao, M., Chu, S., Zhu, Z., Qiu, C.-W., and Gao, L. (2022). Topology-optimized thermal metamaterials traversing full-parameter anisotropic space. *NPJ Comput. Mater.*, 8, pp. 179.

9. Wang, Y., Sha, W., Xiao, M., Qiu, C.-W., and Gao, L. (2023). Deep-learning-enabled intelligent design of thermal metamaterials. *Adv. Mater.*, 35, pp. 2302387.
10. Jin, P., Xu, L., Xu, G., Li, J., Qiu, C.-W., and Huang, J. (2024). Deep learning-assisted active metamaterials with heat-enhanced thermal transport. *Adv. Mater.*, 36, pp. 2305791.
11. Wang, D., Chen, H., Qiu, C., and Li, Y. (2023). Non-Hermitian scattering symmetry revealed by diffusive channels. arXiv.2303.13104.
12. Ju, R., Cao, P.-C., Wang, D., Qi, M., Xu, L., Yang, S., Qiu, C.-W., Chen, H., and Li, Y. (2024). Nonreciprocal heat circulation metadevices. *Adv. Mater.*, 36, pp. 2309835.
13. Liu, J., Xu, L., and Huang, J. (2024). Spatiotemporal diffusion metamaterials: Theories and applications. *Appl. Phys. Lett.*, 124, pp. 210502.
14. Xu, L., Xu, G., Huang, J., and Qiu, C.-W. (2022). Diffusive Fizeau drag in spatiotemporal thermal metamaterials. *Phys. Rev. Lett.*, 128, pp. 145901.
15. Xu, L., Xu, G., Li, J., Li, Y., Huang, J., and Qiu, C.-W. (2022). Thermal Willis coupling in spatiotemporal diffusive metamaterials. *Phys. Rev. Lett.*, 129, pp. 155901.
16. Wang, Z., Chen, J., and Ren, J. (2022). Geometric heat pump and no-go restrictions of nonreciprocity in modulated thermal diffusion. *Phys. Rev. E*, 106, pp. L032102.
17. Li, J., Li, Y., Cao, P.-C., Qi, M., Zheng, X., Peng, Y.-G., Li, B., Zhu, X.-F., Alù, A., Chen, H., and Qiu, C.-W. (2022). Reciprocity of thermal diffusion in time-modulated systems. *Nat. Commun.*, 13, pp. 167.
18. Yin, S., Galiffi, E., and Alù, A. (2022). Floquet metamaterials. *eLight*, 2, pp. 8.
19. Budich, J. C. and Bergholtz, E. J. (2020). Non-Hermitian topological sensors. *Phys. Rev. Lett.*, 125, pp. 180403.
20. Wang, D., Huang, W.-X., Zhou, B., Yu, W., Cao, P.-C., Peng, Y.-G., Zhou, Z.-Y., Chen, H., Zhu, X.-F., and Li, Y. (2024). Efficient evolution framework for chirality control in non-Hermitian systems with adiabaticity engineering. *Phys. Rev. B*, 110, pp. 064308.
21. Li, Y., Peng, Y.-G., Han, L., Miri, M.-A., Li, W., Xiao, M., Zhu, X.-F., Zhao, J., Alù, A., Fan, S., and Qiu, C.-W. (2019). Anti-parity-time symmetry in diffusive systems. *Science*, 364, pp. 170–173.
22. Cao, P.-C., Ju, R., Wang, D., Qi, M., Liu, Y.-K., Peng, Y.-G., Chen, H., Zhu, X.-F., and Li, Y. (2024). Observation of parity-time symmetry in diffusive systems. *Sci. Adv.*, 10, pp. eadn1746.

23. Liu, Y., Chao, K., Sun, F., Chen, S., Dai, H., and Chen, H. (2023). Active thermal metasurfaces for remote heating/cooling by mimicking negative thermal conductivity. *Adv. Mater.*, 35, pp. 2210981.
24. Xu, G., Li, Y., Li, W., Fan, S., and Qiu, C.-W. (2021). Configurable phase transitions in a topological thermal material. *Phys. Rev. Lett.*, 127, pp. 105901.
25. Liu, C., Wu, C., Zhao, Y., Chen, Z., Ren, T.-L., Chen, Y., and Zhang, G. (2024). Actively and reversibly controlling thermal conductivity in solid materials. *Phys. Rep.*, 1058, pp. 1–32.
26. Liu, Z., Cao, P.-C., Li, Y., and Huang, J. (2024). Extended-localized transition in diffusive quasicrystals. *Phys. Rev. Appl.*, 21, pp. 064035.
27. Gao, H., Xu, G., Zhou, X., Yang, S., Su, Z., and Qiu, C.-W. (2024). Topological Anderson phases in heat transport. *Rep. Prog. Phys.*, 87, pp. 090501.
28. Qi, M., Wang, D., Cao, P.-C., Zhu, X.-F., Qiu, C.-W., Chen, H., and Li, Y. (2022). Geometric phase and localized heat diffusion. *Adv. Mater.*, 34, pp. 2202241.
29. Joannopoulos, J. D., Johnson, S. G., Winn, J. N., and Meade, R. D. (2008). *Photonic Crystals: Molding the Flow of Light*, 2nd Ed. Princeton University Press.
30. Cao, P.-C., Li, Y., Peng, Y.-G., Qi, M., Huang, W.-X., Li, P.-Q., and Zhu, X.-F. (2021). Diffusive skin effect and topological heat funneling. *Commun. Phys.*, 4, pp. 230.
31. Liu, Y.-K., Cao, P.-C., Qi, M., Huang, Q.-K.-L., Gao, F., Peng, Y.-G., Li, Y., and Zhu, X.-F. (2024). Observation of non-Hermitian skin effect in thermal diffusion. *Sci. Bull.*, 69, pp. 1228–1236.
32. Xu, G., Zhou, X., Yang, S., Wu, J., and Qiu, C.-W. (2023). Observation of bulk quadrupole in topological heat transport. *Nat. Commun.*, 14, pp. 3252.

Index

www.ingramcontent.com/pod-product-compliance
Ingram Content Group UK Ltd.
Pitfield, Milton Keynes, MK11 3LW, UK
UKHW021106240325
456617UK00014B/2

9 789811 294402